Industrial Design of Experiments

Sammy Shina

Industrial Design of Experiments

A Case Study Approach for Design and Process Optimization

 Springer

Sammy Shina
Department of Mechanical Engineering
University of Massachusetts Lowell
Framingham, MA, USA

ISBN 978-3-030-86269-5   ISBN 978-3-030-86267-1   (eBook)
https://doi.org/10.1007/978-3-030-86267-1

This Springer imprint is published by the registered company Springer Nature Switzerland AG
The registered company address is: Gewerbestrasse 11, 6330 Cham, Switzerland

In memory of a dear friend Masood Baig
and
A former student Val Carvalho

Other Books by the Author

Engineering Project Management for the Global High-Technology Industry
Green Electronic Design and Manufacturing: Implementing Lead-Free and RoHS
    Compliant Global Products
Six Sigma for Electronics Design and Manufacturing
Manufacturing Costs for Electronic Products, volume three of the *Encyclopedia of*
    Materials
Successful Implementation of Concurrent Engineering Products and Processes
Concurrent Engineering and Design for Manufacture of Electronic Products

Preface

Design of experiments (DoE) principles and techniques are becoming more important as technology companies compete in a worldwide market for customers desiring high-quality and low-cost products. Professional engineers struggling daily with different methodologies use DoE to design better products, improve processes, and resolve quality problems. In the early adoption of DoE in US companies, management realized the need for expertise in solving complex interdisciplinary problems using statistical techniques. They encouraged the recruitment of highly trained statistical experts or DoE gurus, provided in-depth training of DoE for their employees, and encouraged classifying that knowledge through the Six Sigma Belt programs.

This book was written as a guide for enhancing professional engineers' knowledge of DoE and its proper use for product design and process optimization. It is intended to be used as a reference for all phases of DoE project lifecycle, including problem definition, resource allocation, outcome measurement, scope, timeline, data analysis, and verification. It augments the basic principles of DoE and its many common and sometimes conflicting techniques by using case studies to help engineers navigate through the maze of design space, factor and level selection, interactions and confounding, error handling, mean and variability analysis, pooling and significance, single or sequential experiments, and balancing conflicting goals of satisfying multiple outcomes.

This book focuses on the use of professional experience to reduce DoE projects' efforts and making sound decisions by providing a balance between performing full factorial experiments or using alternative techniques for sequential experiments to achieve desired outcomes with acceptable risks. The adverse consequences of taking shortcuts and assumptions in executing DoE projects are highlighted by confounding and interactions issues and alternatives are demonstrated for making DoE projects more robust. In addition, it merges several terminologies, nomenclatures, and methodologies used by DoE practitioners as well as DoE software providers into one universal set of elements in the formulation and analysis of DoE projects. It reconstructs different DoE tools from basic statistical elements and shows their equivalency. The aim is to allow the reader to recognize that different DoE approaches are similarly based on statistical principles discussed earlier in the book.

Successful DoE projects can be completed efficiently and successfully when careful considerations are made in selecting DoE teams, project goals, methodologies used, data analysis, and verification using the alternatives outlined in each chapter.

About the Book's Organization

This book is intended to introduce and familiarize newly minted as well as experienced professional engineers in industrial companies to many of the issues regarding the use of DoE tools and techniques, and how to effectively apply them for new product creation, process enhancement, as well as solving quality problems. It is based on the author's experience in researching, practicing, consulting, and teaching DoE for the last 35 years. The book chapters deal with the step-by-step DoE successful implementations with tools and case studies for using them effectively. The book chapters deal with the different aspects of DoE as follows:

1. Chapter 1 begins with quality definitions and basic statistical tools and what are the proper conditions of using these tools, why they were created, and how they became widely adopted. The relationship of these tools to each other are explored to allow DoE teams to use them effectivity and in the right order.
2. Chapters 2 and 3 proceed with using basic statistical tests for comparing two levels of one factor to analyzing multiple levels of one factor in treatments to full factorial comparing multiple factors with multiple levels. Tools and techniques discussed in these two chapters can be used as initial investigations prior to launching DoE projects.
3. Chapters 4 and 5 investigate special conditions of using two- or three-level factors in orthogonal arrays (OA) for DoE projects simplification. Detailed model coefficients and ANOVA are used to illustrate the use of popular OA such as L8, L9, L16, L27, and L32. Each OA use is demonstrated with case studies discussed with DoE teams decision-making reasons, results obtained, and critiques of each.
4. Chapters 6 and 7 round off two- and three-level OA use by discussing the overall DoE project lifecycle, including goal setting and error handling. Experiment reduction through use of interactions and confounding are discussed, outlining their relationships visually and mathematically. Techniques for using professional experience for managing and understanding acceptable risk in allowing confounding are discussed. Previous case studies in earlier chapters are re-analyzed for their handling of these issues.
5. Chapters 8 and 9 address alternative approaches to DoE reduction and simplification by discussing sequential DoE projects rather than single large full factorial experiments. L12, L18, and L36 non-interacting OA are introduced with case

studies each as alternatives to confounding two-and three-level arrays. Multiple-level OA made by modifying standard two- or three-level arrays are shown as survey DoE to evaluate new technologies, materials, or methodologies. Variability analysis is introduced as an adjunct to mean analysis for more comprehensive DoE decisions.

6. Chapter 10 discusses the handling of multiple outcomes in a single DoE as well as providing a summary of 22 case studies used in the book. Different DoE techniques are listed separately and cross-referenced with these case studies

This book is recommended for both neophytes and the experienced DoE practitioners in industrial companies, in particular, the small to medium size companies that do not have the support staff and the resources necessary to have a well-organized DoE process. It is beneficial to try out some of the principles and DoE tools and techniques outlined in this book and meld them into the company culture. The experiences documented here should be helpful to encourage many companies to venture out and develop new world-class products that can make them grow and prosper in the future.

Framingham, MA, USA   Sammy Shina
December 2021

Acknowledgments

The principles of design of experiments discussed in this book were learned, collected, and practiced through my almost 55 years in industry and academia. After graduation from MIT, I worked in the high-technology industry for 22 years, followed by another 33 years in the faculty of the University of Massachusetts Lowell. At the university, I worked as a teacher; researcher, and consultant to different companies, increasing my personal knowledge and experience in the fields of design of experiments, product development and design, manufacturing, and quality.

I am indebted to several organizations for supporting and encouraging me during the lengthy time to collect my materials, write the chapters, and edit the book. I would like to thank the University of Massachusetts Lowell for its continuing support of my courses, programs, and certificates, especially Provost Joseph Hartman, Dean of Engineering James Sherwood, and ME Department faculty. They supported me in my research and work for developing the book's material, approved my plans for academic programs and certificates, and encouraged me to organize, write, and edit this book.

In addition, I want to convey my thanks to Michael McCabe, Senior Editor, Applied Sciences, at Springer Nature.

Mike was my editor for this and my previous book on engineering project management. Mike's humor, encouragement, and good spirit guided me through this book, and for that I am very grateful. I also wish to extend my gratitude to Ms. Srividya Subramanian, Books Project Coordinator for Springer Nature, and her team for their prompt and efficient editing and production of this book. In addition, I want to thank Gedy Budreckis, North America Six Sigma lead, Takeda Corporation, and Osei Serebour, ASQ-CQE, CBA, CSSBB, Quality Engineer/Auditor at BK Medical, who both lectured to my DoE classes with clarity and focus.

I also want to thank the many family members who hosted me through the very long period of writing the book, including daughters Gail Brown of Amesbury Massachusetts and Nancy Shina of Ogden Utah.

Finally, many thanks to my family for emotional support during the writing, editing, and production of the book, including my wife Jackie, who edited the book with her superb English, and our children, Mike, Gail, Nancy, and Jon, as well as my grandchildren who brought me great joy between the many days of writing and editing. I also wish to thank the many students who attended my classes in DoE and quality engineering over many years and peppered me with questions and challenges to explain the many topics, which cleared and refocused my mind. I wish them best success in implementing DoE principles and methods in their companies.

Contents

1   **Data Presentations, Statistical Distributions, Quality Tools,**
    and Relationship to DoE 1
    1.1   Graphical Presentation of Data 2
    1.2   Probability Distributions and Their Use in Modern Quality
           Systems 13
           1.2.1   The Binomial Distribution 13
           1.2.2   The Poisson Distribution 15
           1.2.3   Continuous Distributions and Reliability 17
           1.2.4   The Use of the Standard Normal Distribution
                    in Quality Methods 20
    1.3   Six Sigma and Its Relationship with DoE 31
           1.3.1   Converting Defect Rate to Cp and Sigma Design
                    with No Mean Shift from Nominal 34
           1.3.2   The Implied Mean Shift to Nominal
                    of Cp and Six Sigma 36
           1.3.3   Process Capability Studies for Quality
                    Enhancement 38
           1.3.4   Process Capability for Prototype and Early
                    Production Parts 39
           1.3.5   Corrective Action for Process Capability
                    Problems 40
           1.3.6   DoE Effects on Six Sigma 41
    1.4   Control Charts and Their Relationship with DoE 41
           1.4.1   Selection of Control Charts 42
           1.4.2   Variable Control Charts 43
           1.4.3   Relationship of Control Limits to Specification
                    Limits 46
           1.4.4   Variable Control Chart Usage Guidelines 49
           1.4.5   Control Charting and Process Capability
                    for Low-Volume Production 53

        1.4.6   Attribute Control Charts   55

        1.4.7   Use of Control Charts in Factories with High

                    Process Capability   57

        1.4.8   DoE Effects on Control Charts   58

  1.5   Conclusions   59

  References   59

        Additional Reading Material   59

2   Samples and Populations: Statistical Tests for Significance

**   of Mean and Variability**   61

  2.1   Sample and Population Statistics   62

        2.1.1   Standard Deviation Estimation Methodologies

                    and Data Collection   64

        2.1.2   Measurement System Error (GR&R)

                    and Its Impact on Statistical Measurements   67

  2.2   Tests for Sample and Population   69

  2.3   Tests for Means   71

        2.3.1 z-Test for Population Means   72

        2.3.2   The Wilcoxon Test for Non-normal Population

                    Mean Test   75

        2.3.3 t-Tests for Sample Means: Single Sample   76

        2.3.4 t-Tests for Comparing Two Sample Means

                    with Unknown Variance   79

        2.3.5 d- or Paired t-Test   81

        2.3.6   Confidence Interval (CI) of the Mean   83

        2.3.7   Determination of Sample Size Based on Error   87

  2.4   Tests for Variability   88

        2.4.1 X^2 (Chi-Square) Significance Test   89

        2.4.2 X^2 Goodness of Fit Test and Checking

                    for Normality   91

        2.4.3   F-Test   94

  2.5   Conclusions   96

  References   97

        Additional Reading Material   97

3   Regression, Treatments, DoE Design,

**   and Modelling Tools**   99

  3.1   Regression Analysis   101

        3.1.1   Least Squares Regression   102

        3.1.2   Linear Regression Analysis Using Model

                    Coefficients Estimates   103

        3.1.3   Linear Regression Analysis Using ANOVA   108

        3.1.4 R^2 and Accuracy of Model Estimate   110

        3.1.5   Using Linear Regression for Normality

                    Checking   111

	3.2	Treatment Design and Analysis	118
		3.2.1 Treatment Design and Analysis Example	119
		3.2.2 Significance Determination Techniques and p% Contribution	121
	3.3	Full Factorial DoE Design and Analysis	123
		3.3.1 Limiting DoE Scope with Design Space and Process Map	123
		3.3.2 Full Factorial DoE Design Analysis Using Interactions	124
	3.4	Full Factorial DoE Design and Analysis Case Study: Green Electronics Manufacturing	128
		3.4.1 Summary of Phase I Green Electronics DoE Case Study	129
		3.4.2 Phase II of Green Electronics DoE Case Study	129
		3.4.3 Analysis of Phase II DoE	132
	3.5	Conclusions	137
		Additional Reading Material	138

4 Two-Level Factorial Design and Analysis Techniques 139
	4.1	DoE Definitions, Expectations, and Processes	141
		4.1.1 DoE Lifecycle Process	143
		4.1.2 DoE Project Timing and Error Source	148
	4.2	Two-Level Factorial DoE Design	150
		4.2.1 Commonly Used Two-Level Orthogonal Arrays	153
		4.2.2 Types of Uses for Two-Level OA	155
		4.2.3 Interaction, Confounding, and Interconnecting Graphs	160
	4.3	Two-Level OA Analysis and Model Reductions	161
	4.4	Two-Level DoE Case Studies	163
		4.4.1 Full Factorial L8 Case Study, Hipot DoE: Selecting Best Alternative Among Equally Performing Designs	163
		4.4.2 Partial Factorial L8 Case Study, Underfill Voids DoE: Selecting Process Parameters for Zero Defects	170
		4.4.3 Partial Factorial L16 Example Case Study, Rivet Design DoE: Selecting Part Dimension Design for Best Product Performance	174
		4.4.4 Partial Factorial L32 Case Study, APOS for Robotics DoE: Selecting Process Parameters for Multiple Adjustment Production Machines	179
	4.5	Conclusions	185
		Additional Reading Material	186

5   Three-Level Factorial Design and Analysis Techniques   187

5.1   Three-Level Factorial Design   188

     5.1.1   Commonly Used Three-Level Orthogonal Arrays   191

     5.1.2   Use of Three- Versus Two-Level DoE   193

5.2   Three-Level DoE Analysis and Model Reductions   194

5.3   Three-Level DoE Case Studies   195

     5.3.1   Screening Design L9 Case Study: Bonding I DoE   195

     5.3.2   Screening Design L9 Case Study: Zero-Defect

             Mixed Soldering DoE   201

     5.3.3   Partial Factorial DoE L27 Case Study: Green

             Electronics Phase I DoE   205

     5.3.4   Screening Design Software L27 Case Study:

             Minimizing Half-Adder Chip Delay Time DoE   212

5.4   Conclusions   215

Additional Reading Material   215

6   DoE Error Handling, Significance, and Goal Setting   217

6.1   DoE Error Handling Techniques for Significance Testing   218

     6.1.1   Regression Equation and Predicted Outcome with

             Interaction   221

     6.1.2   DoE Error Handling Types   222

     6.1.3   Error Handling and Significance Technique L8 Case

             Study: Plastics Injection Molding DoE   223

     6.1.4   Error Handling and Significance for Single Repetition

             DoE Analysis   224

     6.1.5   Error Handling and Significance for Multiple

             Replication DoE Analysis   230

     6.1.6   Error Handling and Significance for Multiple

             CenterPoint Replications   233

     6.1.7   Error Handling and Significance for Some

             Experiment Outcome Replications   236

6.2   Project Goal Setting and Design Space   238

     6.2.1   Types of DoE Project Goals   239

     6.2.2   Design Space and Level Selection   241

6.3   Experiment Blocking (Dividing)   243

6.4   Conclusions   246

Additional Reading Material   246

7   DoE Reduction Using Confounding and Professional

Experience   247

7.1   Design Resolution and Confounding   248

     7.1.1   Techniques for Managing Confounding   251

	7.2	Interactions and Confounding for L8 for Reduced Experiments	252
		7.2.1 L8 Half Fraction Interaction and Confounding	252
		7.2.2 L8 Partial Factorial Design and Confounding	254
		7.2.3 L8 Screening Design Confounding	257
		7.2.4 L8 Factor Conversion Tables for Labeling Numeric and Alphabetic Factors	258
	7.3	Interactions and Confounding for L16 for Reduced Experiments	260
		7.3.1 L16 Half Fraction Interaction and Confounding	261
		7.3.2 Interactions and Confounding in L16 Rivet DoE Case Study	261
		7.3.3 L16 Partial Factorial Design and Confounding	263
		7.3.4 L16 Partial Factorial Design and Confounding with DoE Interaction Matrix Method	265
		7.3.5 Interaction Matrix L32 Case Study: Solder Wave DoE Design	267
		7.3.6 Resolving Confounded Interactions	269
		7.3.7 L16 Partial Factorial Design and Confounding for Eight or More Factors	271
	7.4	Interaction and Confounding for Large OA	274
	7.5	Conclusions	279
		Additional Reading Material	279
8	**Multiple-Level Factorial Design and DoE Sequencing Techniques**		**281**
	8.1	Multi-level OA Arrangements	282
		8.1.1 Multi-level Arrangement DoE L8 Case Study: Machining I Pin Fin Heat Sinks	284
		8.1.2 Multi-level Arrangement DoE L16 Case Study: Machining II Stencil Forming	287
	8.2	DoE Sequencing Techniques	291
		8.2.1 Foldover Sequencing Techniques: Folding on One Factor	292
		8.2.2 Foldover Sequencing Techniques: Folding on All Factors	296
		8.2.3 DoE Sequencing Technique Case Study: Printer Design DoE	296
	8.3	Non-interacting Orthogonal Array Use in DoE	302
		8.3.1 Non-interacting OA Case Study I: L18 Painting DoE	303
		8.3.2 Non-interacting OA Case Study II: L36 Air Knife DoE	309
	8.4	Conclusions	312
		Additional Reading Material	312

9   Variability Reduction Techniques and Combining with Mean
      Analysis..   313
      9.1   Controlled and Noise Factors in DoE.....................   313
      9.2   Variability Reduction Definitions and Analysis............   314
      9.3   Balancing Mean and Variability Outcomes.................   314
      9.4   Conclusions..   314
             9.4.1   Controlled and Noise Factors in DoE.............   314
             9.4.2   Variability Reduction Definitions and Analysis......   316
             9.4.3   Balancing Mean and Variability Outcomes.........   329
             9.4.4   Conclusions.................................   331
      Additional Reading Material....................................   331

10   Strategies for Multiple Outcome Analysis and Summary
      of DoE Case Studies and Techniques........................   333
      10.1   Summary of Previous Chapters..........................   333
      10.2   Combining Multiple Desired Outcomes with Mean
             and Variability Analysis................................   335
             10.2.1   Interaction Matrix L32 Case Study: Solder
                      Wave DoE Analysis...........................   336
      10.3   Summary of DoE Case Studies and Techniques............   344
             10.3.1   DoE Case Studies List by OA Size for Two
                      and Three Levels.............................   345
             10.3.2   DoE Techniques Used and Demonstrated
                      in Chapters and Case Studies..................   349
      10.4   Conclusions..   352

Index..   353

List of Figures

Fig. 1.1   Bar chart and Pareto diagram of fabrication failures 3
Fig. 1.2   Pie chart and Pareto diagram of fabrication failures 3
Fig. 1.3   Ranked bar chart and Pareto diagram of fabrication failures 3
Fig. 1.4   Spider diagram of fabrication failures 4
Fig. 1.5   Stem-and-leaf presentation of data 5
Fig. 1.6   Dot diagram of hole location deviation for two samples 5
Fig. 1.7   Two-year progression of solder process defect rate 6
Fig. 1.8   Control chart for the mean (Xbar) of solder paste thickness 6
Fig. 1.9   Cause-and-effect diagram for solder height DoE 7
Fig. 1.10   Box and whisker plots for three data sets (experimental lots) 7
Fig. 1.11   Changing one factor at a time (OFAT) yield 8
Fig. 1.12   Multivariate charts for two variables 9
Fig. 1.13   Multifactor plots for variance analysis of solder paste height 9
Fig. 1.14   Half adder IC delay DoE outcomes 10
Fig. 1.15   Process quality production before and after DoE 11
Fig. 1.16   Weibull probability plots using two different materials 19
Fig. 1.17   Z transformation of normal data to standard normal distribution
            (SND) 29
Fig. 1.18   One- or two-sided distribution of quality defects 30
Fig. 1.19   Acceptance rate for multiple sigma specifications 31
Fig. 1.20   Typical Six Sigma design or process capability $Cp = 2.0$ 33
Fig. 1.21   Four sigma design with mean shift or process capability
            $Cp = 1.33$ 34
Fig. 1.22   Example of DoE outcomes with four factors 40
Fig. 1.23   Types of control charts 44
Fig. 1.24   Relationship of control limits to specification limits 47
Fig. 1.25   R control chart example of solder paste thickness 50
Fig. 1.26   Surface cleanliness measurement example 51

Fig. 2.1   Student's t distribution 77
Fig. 2.2 t distribution with various DOFs (ν) compared with standard
normal distribution (SND) 78
Fig. 2.3   Sample confidence interval (CI) versus population mean (μ) 85
Fig. 2.4 X^2 distribution 89

Fig. 3.1   Data variation types versus model prediction 102
Fig. 3.2   Visual illustration of least squares method 103
Fig. 3.3   Scatter diagram of linear regression sample data with regression
equation and R^2 111
Fig. 3.4   Linear regression plot and histogram of observed versus ranked
data example 114
Fig. 3.5   Normal score regression plot and histogram for example of
observed data 115
Fig. 3.6   Normal score regression plot and histogram of transformed
data 117
Fig. 3.7   Visual presentation of 30 experiments for 3 factors
at 5, 3, and 2 levels 125
Fig. 3.8   Visual presentation of interaction for factors B and C 126
Fig. 3.9   Phase II DoE test vehicle 130
Fig. 3.10   Phase II DoE plot of total defects versus factor coefficients 134
Fig. 3.11   Phase II DoE interaction plots 136

Fig. 4.1   Evolving quality definitions 141
Fig. 4.2   Arrangement of DoE elements 142
Fig. 4.3   Modelling seven factors (A–G) in screening versus full
factorial mode 159
Fig. 4.4   Model coefficient plots for Hipot DoE 166
Fig. 4.5   Hipot DoE two-way Contact*Cable interaction plot 169
Fig. 4.6   Cause-and-effect diagram for underfill voids DoE 171
Fig. 4.7   Coefficient plots for underfill voids DoE 172
Fig. 4.8   Model coefficient plots for Rivet DoE 178
Fig. 4.9   SMART® robotic line concept 180
Fig. 4.10   Spring and walking beam APOS parts 182
Fig. 4.11   APOS DoE coefficient plots for two parts 183

Fig. 5.1   Bonding I DoE coefficient plots 199
Fig. 5.2   Mixed soldering DoE coefficient plots 203
Fig. 5.3   Green Electronics Phase I DoE test vehicle 206
Fig. 5.4   Green Electronics Phase I DoE coefficient plots 209
Fig. 5.5   Green Electronics Phase I DoE interaction A*B 211
Fig. 5.6   Half-adder chip design 213

Fig. 6.1   Plastics DoE model coefficient plots 219
Fig. 6.2   Ranked Pareto diagram of coefficient t-values for Plastics
Single DoE after pooling 220
Fig. 6.3   Velocity*Pressure interaction plot for Plastics Single L8 DoE ... 228

Fig. 6.4   Design space for L8 full factorial Hipot experiment 242
Fig. 6.5   Selection of factor levels 242
Fig. 6.6   Blocking L8 into groups of four 244
Fig. 6.7   Blocking L8 into groups of two 244

Fig. 7.1   L8 half fraction interconnection diagrams 253
Fig. 7.2   L16 half fraction (pentagon) interconnection diagram
            and its use in rivet project 261
Fig. 7.3   L16 six-factor assignments and interactions 266
Fig. 7.4   L16 seven-factor assignments and interactions (hourglass) 267
Fig. 7.5   L16 eight-factor assignments and interactions (windmill) 273
Fig. 7.6   L16 eight-factor assignments and interactions (star) 273
Fig. 7.7   L27 factor assignments and interactions 275

Fig. 8.1   Pin fin Machining I outcomes 285
Fig. 8.2   DoE Machining I bent pins mean analysis 286
Fig. 8.3   Printer DoE test pattern 297
Fig. 8.4   Painting DoE L18 mean coefficient analysis 307

Fig. 9.1   Controlled and noise factors for variability measurements 315
Fig. 9.2   Bonding II DoE mean and variability coefficient plots 322
Fig. 9.3   Painting DoE L18 variability coefficient plots 326
Fig. 9.4   Stencil DoE L9 mean and variability coefficient plots 328

Fig. 10.1   L32 Solder Wave DoE lead and touch defect coefficient
            plots 341
Fig. 10.2   L32 Solder Wave DoE Speed*Width and Speed*Preheat
            interaction plots for lead defects 342
Fig. 10.3   L32 Solder Wave DoE quality performance after DoE 343

List of Tables

Table 1.1	Commonly used graphical presentations of data	2
Table 1.2	TQM tools and techniques	12
Table 1.3	Use of TQM tools and techniques	13
Table 1.4	Probability distributions	14
Table 1.5	Probabilities of obtaining X number of heads in five (n) tosses	15
Table 1.6	Probabilities for the Poisson distribution	16
Table 1.7	Standard normal distribution (SND)	21
Table 1.8	Control chart factors	45
Table 1.9	Control chart calculation example	48
Table 1.10	Equivalent probabilities for out-of-control conditions in X_{bar} control charts	52
Table 2.1	Statistical terms used in sample mean and standard deviation common to DoE	64
Table 2.2	Summary of methods 1–3 for calculating mean and standard deviation of 32 data points	65
Table 2.3	Moving range method standard deviation calculations for 32 data points	66
Table 2.4	GR&R example	67
Table 2.5	Summary of statistical tests for mean and variability	71
Table 2.6	Selected values of $t_{\alpha,\nu}$ of Student's t distribution	78
Table 2.7	d-test example 2 and comparison with t-test	84
Table 2.8	Lipid-lowering effects of Zocor® (simvastatin) in adolescent patients for 24 weeks	86
Table 2.9	Error of $t_{\alpha,30}$ compared with standard normal distribution (SND)	87
Table 2.10	Selected values of X^2 distribution	90
Table 2.11	Selected F table values for 95% confidence or $\alpha = 0.05$ significance	94

Table 3.1   Linear regression example 105
Table 3.2   Model fitted versus residual data for linear regression
             example 106
Table 3.3   Model coefficient estimates of linear regression example 107
Table 3.4   Regression analysis terms 108
Table 3.5   ANOVA testing in linear regression 109
Table 3.6   ANOVA for linear regression example 110
Table 3.7   Examples of data transformation for improved statistical
             analysis 112
Table 3.8   Example of observed data and checking for normality 113
Table 3.9   Transformed data for example and checking for normality 116
Table 3.10   ANOVA testing of treatments 118
Table 3.11   Treatment decomposition into groups, deviations, and
             residuals 119
Table 3.12   Treatment example of ANOVA 120
Table 3.13   Treatment example of pairwise t-tests 121
Table 3.14   Full factorial DoE distribution of degrees of freedom (DOF) ... 127
Table 3.15   Visual defect-free results of phase I Green Electronics DoE 129
Table 3.16   Green Electronics Phase II DoE design matrix 131
Table 3.17   Green Electronics Phase II DoE ANOVA 132
Table 3.18   Green electronics statistically significant outcome effects'
             summary 133
Table 3.19   Green electronics phase II DoE factor multiple range test
             for total defects 135

Table 4.1   DoE project process steps 144
Table 4.2   One-factor-at-a-time (OFAT) DoE 150
Table 4.3   Successive two-level orthogonal array arrangements 151
Table 4.4   Two-way interactions for two-level assignments 152
Table 4.5   Orthogonal array L4 with two main factors (1/A and 2/B)
             and their interaction 3/C = 1*2/A*B 154
Table 4.6   Orthogonal array L8 with three main factors (A/1, B/2,
             and C/3) and their interactions 154
Table 4.7   Orthogonal array L16 with four main factors (A/1, B/2, C/4,
             D/8) and their 11 interactions 155
Table 4.8   Orthogonal array L32 with 5 main factors (A/1, B/2, C/4, D/8,
             and E/16) and their 26 interactions 156
Table 4.9   Two-level OA used in full, half, and partial factorial and
             screening modes 157
Table 4.10   L8 Full factorial Hipot DoE design and results 164
Table 4.11   Hipot DoE ANOVA with no pooling 164
Table 4.12   Hipot DoE ANOVA after first pooling 165
Table 4.13   Hipot DoE ANOVA with second pooling 165
Table 4.14   Hipot DoE model coefficient estimates 167
Table 4.15   Hipot DoE predicted results based on two models 167

Table 4.16   L8 partial factorial underfill voids DoE factor and level
              assignments 172
Table 4.17   Underfill voids DoE confounding 172
Table 4.18   Voids DoE statistical analysis 173
Table 4.19   Rivet DoE factor and level assignments 175
Table 4.20   Rivet DoE factor confounding 175
Table 4.21   Rivet DoE design and results 176
Table 4.22   Rivet DoE coefficient analysis 177
Table 4.23   Rivet DoE ANOVA after pooling 177
Table 4.24   APOS DoE factors and interactions 180
Table 4.25   APOS DoE factor assignment and experiment results 181
Table 4.26   APOS DoE ANOVA of two parts 182
Table 4.27   APOS DoE outcome validation for two parts 184

Table 5.1   L9 with two main factors (1/A and 2/B) and interactions
              (3/C/AB and 4/D/AB$_2$) 189
Table 5.2   L27 with three main factors (1/A, 2/B and 5/C) and ten
              interactions 190
Table 5.3   Column assignments for three-leve cl OA 190
Table 5.4   Three-level OA use in full and partial factorial and
              screening modes 191
Table 5.5   Bonding I DoE design and outcomes 197
Table 5.6   Bonding I DoE ANOVA before pooling 197
Table 5.7   Bonding I DoE ANOVA after pooling 198
Table 5.8   Bonding I DoE coefficient estimates 198
Table 5.9   Mixed soldering DoE and results 202
Table 5.10   Mixed soldering DoE ANOVA after pooling 203
Table 5.11   Mixed soldering DoE coefficient estimates 204
Table 5.12   Green Electronics Phase I DoE design and results 207
Table 5.13   Green Electronics Phase I DoE first ANOVA 208
Table 5.14   Green Electronics Phase I DoE final pooled ANOVA 209
Table 5.15   Green Electronics Phase I DoE coefficient analysis 210
Table 5.16   L27 half-adder chip design DoE results 214

Table 6.1   Plastics DoE error handling case study 224
Table 6.2   Plastics Single DoE analysis before pooling 225
Table 6.3   Plastics Single DoE analysis after pooling 225
Table 6.4   Plastics three-repetition DoE analysis before pooling 231
Table 6.5   Plastics three-repetition DoE analysis after pooling 233
Table 6.6   Plastics DoE CenterPoint replication analysis before
              pooling 234
Table 6.7   Plastics DoE with replication of some experiment
              outcomes 236
Table 6.8   Plastics DoE with replication of some experiment
              outcome analysis before pooling 237
Table 6.9   Blocking L16 into groups of four 245

Table 7.1   L8 half fraction interactions and confounding 250
Table 7.2   L16 half fraction interactions and confounding 251
Table 7.3   Resolution of two-level OA with added factors 255
Table 7.4   L8 OA and interaction table 259
Table 7.5   DoE two-level factor conversions 260
Table 7.6   L16 interaction table 263
Table 7.7   L16 additional factor assignments 264
Table 7.8   L16 interaction matrix solder wave DoE 268
Table 7.9   L16 confounding for solder wave DoE resolution III 269
Table 7.10   L16 confounding for solder wave DoE resolution IV 270
Table 7.11   Additional runs to resolve confounding in L16 seven-factor
             resolution IV design 271
Table 7.12   L16 eight-factor star resolution IV design 274
Table 7.13   Additional runs to resolve confounding in column 2 for
             eight-factor star resolution IV design 274
Table 7.14   L16 eight-factor star resolution III design 275
Table 7.15   L32 half fraction division into two L16s 277
Table 7.16   L16 half fraction division into two L8s 278
Table 7.17   L8 half fraction division into two L4 278

Table 8.1   Multi-level arrangements for L8 with four levels 283
Table 8.2   DoE Machining I Pins factor-level assignments
             and outcomes 285
Table 8.3   DoE Machining I Pins ANOVA 287
Table 8.4   DoE Machining II Stencil factor and interaction
             assignments 289
Table 8.5   DoE Machining II Stencil ANOVA 290
Table 8.6   Foldover on column (4) and confounding for two L8
             screening designs 293
Table 8.7   Foldover column (4) and confounding for combined L16
             screening design 293
Table 8.8   Chemical efficiency folding screening DoE example and
             outcomes 294
Table 8.9   Chemical efficiency example of folded factor coefficients 295
Table 8.10   Printer DoE defect definitions 297
Table 8.11   Printer DoE screening L8 DoE 298
Table 8.12   Printer DoE screening L8 outcomes versus defect types 298
Table 8.13   Printer DoE first screening L8 ANOVA summary 299
Table 8.14   Print□er DoE second screening L9 DoE 300
Table 8.15   Printer DoE second screening L9 outcomes 301
Table 8.16   Printer DoE second screening L9 ANOVA summary 301
Table 8.17   Non-interacting L12 (2^{11}) OA 303
Table 8.18   Non-interacting L18 OA ($2^1 \times 3^7$) 304
Table 8.19   Non-interacting L36 OA ($2^{11} \times 3^{12}$) 305
Table 8.20   Painting L18 DoE factors and levels 306

Table 8.21   Painting L18 DoE and outcomes 306
Table 8.22   Painting L18 DoE mean ANOVA 307
Table 8.23   Painting L18 DoE mean coefficient ranking 308
Table 8.24   Air Knife DoE defect outcome level analysis summary 311
Table 8.25   Air Knife DoE defect prediction and validation 311

Table 9.1   Bonding II DoE L8 factors, levels and outcomes 320
Table 9.2   Bonding II DoE mean coefficient analysis 321
Table 9.3   Bonding II DoE mean ANOVA 322
Table 9.4   Bonding II DoE variability coefficient analysis 323
Table 9.5   Bonding II DoE variability ANOVA 323
Table 9.6   Painting L18 DoE variability ANOVA 325
Table 9.7   Painting L18 DoE variability coefficient ranking 325
Table 9.8   Stencil II DoE L9 factors, levels, and outcomes 327
Table 9.9   Stencil II DoE L9 mean and variability coefficients 329
Table 9.10   Stencil II mean DoE L9 ANOVA before and after pooling 329

Table 10.1   L32 Solder Wave DoE project DoE matrix and results 339
Table 10.2   L32 Solder Wave DoE lead defect coefficient and ANOVA 340
Table 10.3   L32 Solder Wave DoE touch defect coefficient and
              ANOVA 340
Table 10.4   L32 Solder Wave DoE factor level assignments 343

About the Author

Sammy Shina Ph.D., P.E., is Professor of Mechanical Engineering at the University of Massachusetts Lowell and has previously lectured at the University of Pennsylvania's EXMSE Program and at the University of California Irvine. He is the coordinator of the Engineering Management Program, Design and Manufacturing and Quality Engineering Certificates. Previously, he coordinated ME senior Capstone Projects and COOP education for the College of Engineering at UML. He is a past chairman of the Society of Manufacturing Engineers (SME) Robotics/FMS and a founding member of the Massachusetts Quality Award. He was the founder of the New England Lead Free Consortium, with of over 30 contributing companies working on electronics products and supply chain. The consortium was engaged in researching, testing, and evaluating materials and processes for lead-free RoHS compliance and converting to nanotechnology. It was funded by TURI, EPA, NSF, and member companies. The consortium has published over 40 papers; some were translated into Asian languages and won a regional EPA Environmental Merit Award for business category in May 2006. He is the author of several best-selling books on concurrent engineering, Six Sigma, green design, and engineering project management. He contributed 2 chapters and over 170 technical publications in his fields of research.

Dr. Shina is an international consultant, trainer, and seminar provider on project management, quality methods in design and manufacturing, Six Sigma, and DoE, as well as technology supply chains, product design and development, and electronics manufacturing, testing, and automation. He worked for 22 years in high-technology companies developing new products and state-of-the-art manufacturing technologies. He was the speaker for the Hewlett Packard Executive Seminars on Concurrent Product/Process Design, Mechanical CAD Design and Test, and the Motorola Six Sigma Institute. He received B.S. degrees in electrical engineering and industrial management from MIT, M.S. degree in computer science from WPI, and Ph.D. degree in mechanical engineering from Tufts. He resides in Framingham, Massachusetts.

Chapter 1
Data Presentations, Statistical Distributions, Quality Tools, and Relationship to DoE

Many of the misconceptions of statistics in general and Design of Experiments (DoEs) stem from the confusion surrounding the use of data to prove a particular hypothesis and multiple methods that data can be manipulated to attain a desired outcome. This chapter will review the basic building blocks of statistics and DoE and explore how these blocks are used in the quality tools as we know them today. The chapter will also review the basic assumptions that are required to make these tools perform as intended.

The principles of basic statistics will be explored, in terms of several techniques of presenting data for visual and qualitative analysis, as well as an introduction to common quality tools such as control charts in manufacturing and inspection. In addition, process capability to measure manufacturing performance and Six Sigma used in product design quality will be explored. A discussion of how these tools are interrelated through the use of the standard normal distribution (SND) for significance testing is presented. The z distribution, a universal standard normal distribution (SND) version for transposing individual data sets for analysis, will be demonstrated with case studies and testing examples.

The purpose of the chapter is to allow students, scientists, and engineers to interpret and perform data analysis and presentations and arrive at correct decisions and conclusions readily without the burden of losing their intended audience and teammates with heavy statistical jargon.

Topics to be discussed in this chapter are as follows:

1.1. Graphical presentation of data. Several data presentations and graph types are discussed, especially those that are used for DoE and will be presented from actual industrial case studies. The discussion will include the best use of each data presentation technique for the interpretation of a particular set of data and intent of the data collector. This section focuses on the qualitative analysis of data to improve quality by using team tools and techniques.

1.2. Probability distributions and their use in quality systems. A quick introduction is given to the several types of distributions (discrete and continuous) used in basic statistical analysis and significance testing that form the foundation for DoE. The use of these distributions in quality tools such as control charts and Six Sigma will be demonstrated, as well as the positioning of DoE in the proper perspective and sequencing of using these tools.

1.3. Six Sigma and its relationship to DoE. The origins and concepts of Six Sigma will be explained, with emphasis on new product design and relationship to manufacturing quality performance. Process capability as input to Six Sigma will be discussed, including two common versions (Cp and Cpk). Different uses of each version will be explored for their applications in design and manufacturing. Examples of using Six Sigma to estimate new product quality performance and ways to improve it will be shown, with emphasis on quantitative quality defect analysis using common manufacturing operations.

1.4. Control charts and their relationship to DoE. Two different types of control charts will be discussed, variable and attribute data, as well as their use with constant or variable sample sizes. The development of control charts and their proper use for maintaining quality will be highlighted. In addition, a hierarchical approach of using quality tools including TQM, DoE, and control charts will be shown to achieve world-class quality. The proper sequence of using these tools is dependent on whether optimizing current or new processes and designs.

1.5. Conclusions

1.1   Graphical Presentation of Data

The following graphical presentations of data are common in industry. Examples are used as for different types of data presentations as shown in Table 1.1.

1. Pareto diagrams are common in quality data presentations. They are used to rank the variables that are being plotted, either singly or in groups. They can be used as bar charts or pie charts. In the examples shown in Figs. 1.1 and 1.2, quality data of failures created through the different process types in fabrications are plotted in various Pareto diagram forms. Figure 1.3 shows a ranked set of diminishing effect bars and cumulative bars for each process in fabrication step. Figure 1.4 shows the same data in spider or radar chart form, with equidistant axis points to

Table 1.1 Commonly used graphical presentations of data	
1. Pareto, bar, pie, and spider diagrams
2. Histograms and stem-and-leaf and dot diagrams
3. Line and control charts
4. Cause-and-effect diagrams
5. Box and whisker plots
6. Multivariate plots (response surface)

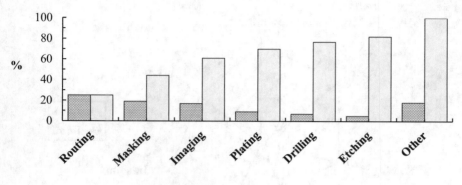

Fabrication Defect Cause

Fig. 1.1   Bar chart and Pareto diagram of fabrication failures

Fig. 1.2   Pie chart and Pareto diagram of fabrication failures

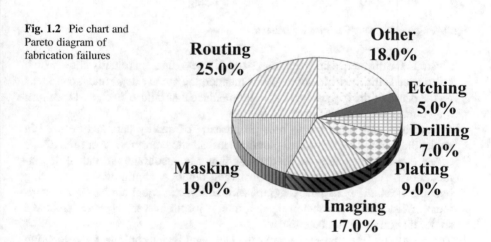

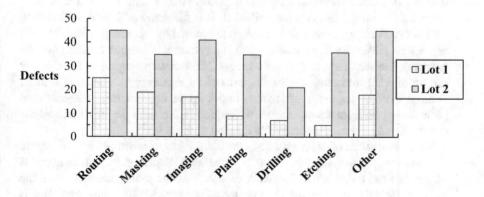

Fabrication Defect Cause

Fig. 1.3   Ranked bar chart and Pareto diagram of fabrication failures

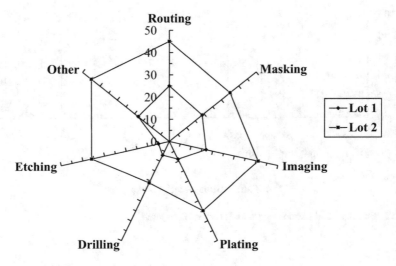

Fig. 1.4 Spider diagram of fabrication failures

illustrate the difference between Lots 1 and 2. These diagrams clearly show which failure needs the most attention. In this manner, the quality department can focus on the cause of the highest level of failure, reduce that failure level, and then turn their attention to the next cause of failures.

An interesting approach to time progression of quality data can be used in spider diagrams. This method of plotting data can show progress over time of one defect cause against another. Spider diagrams can encourage individual departments or managers to compete against each other. A continuously improving operation can show a spider diagram with concentric equal circles. If a department or function falls down in quality, then it would be very apparent since the spider diagram will not be circular.

2. The next set of data presentation is the stem-and-leaf technique for presenting data in an ordered fashion, shown in Fig. 1.5 as a process instruction to employees working in a factory. In this method, tabulation of data is measured as the number of data sets (curved lines) being collected. It is the same as Pareto approach to data but converting the data from random to ordered by frequency, similar to FM rather than AM radio transmission. This is a common technique to examine data in successive sets, bins, ranges, or segments, known as histograms. The data set can be divided into range segments and ranked accordingly. Data can also be sectioned off in segments of tens. A two-level or ordered stem-and-leaf data presentation of segments of ten can be divided further into the first five and the second five as a second-order leaf belonging to the stem.

Another method of ordering data is the dot diagram, shown in Fig. 1.6, where the hole location deviation from true positioning is displayed for two samples. If data points can be plotted individually on a scale in the x-axis, then the plots can indicate visually whether the data is normal or not. Making sure that data is normal is very important in statistical analysis. When performing statistical tests

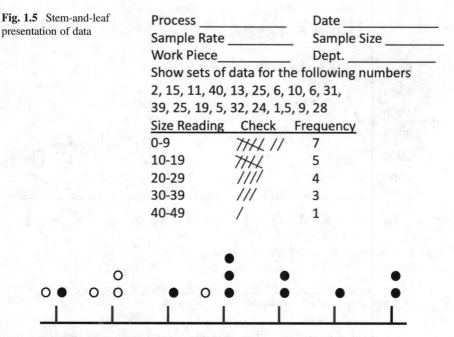

Fig. 1.5 Stem-and-leaf presentation of data

Process _____ Date _____

Sample Rate _____ Sample Size _____

Work Piece_____ Dept. _____

Show sets of data for the following numbers

2, 15, 11, 40, 13, 25, 6, 10, 6, 31,

39, 25, 19, 5, 32, 24, 1,5, 9, 28

Size Reading	Check	Frequency
0-9	𝟕𝑯𝑳 //	7
10-19	𝟕𝑯𝑳	5
20-29	////	4
30-39	///	3
40-49	/	1

Fig. 1.6 Dot diagram of hole location deviation for two samples

or DoE analysis, the parent distribution of the data must be normal. If not, other techniques must be used to successfully apply these tests.

3. The next set of data presentations is the use of annotation to mark line diagrams. In the example in Fig. 1.7, the solder process defect rate during a 2-year progression was continuously brought down using different quality techniques. The total quality management (TQM) qualitative phase was first used to reverse the increasing trend of poor quality. Statistical process control (SPC) techniques such as control chart phase were used next to maintain quality at a particular level. Design of Experiments (DoEs) was used in the last phase in order to bring quality to world-class levels using advanced quantitative analysis techniques. This is the recommended progression of using quality tools in improving quality for current products' design or manufacture. For new products, the order is reversed: TQM quality tools are implied throughout the design or process functions. DoE is performed to initially determine the optimum design or process parameters, and then SPC is used to monitor and maintain the quality levels. This solder case study will be described in detail later in the book.

Figure 1.8 shows a typical multi-level variable control chart for the mean (X_{bar}) or $\overline{X}$ in the case of monitoring the solder paste thickness in a typical electronic manufacturing plant for a month. In this case, three levels are shown: upper control limit (UCL), process mean or average ($X_{double\ bar}$) or $\overline{\overline{X}}$, and lower control limits (LCL). Many equal sample sizes of four readings were taken each day, with only the mean of the daily readings being plotted for each day. The control charts are essentially line plots of sample averages. Whenever a plotted

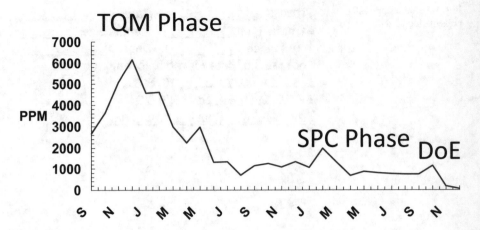

Fig. 1.7 Two-year progression of solder process defect rate

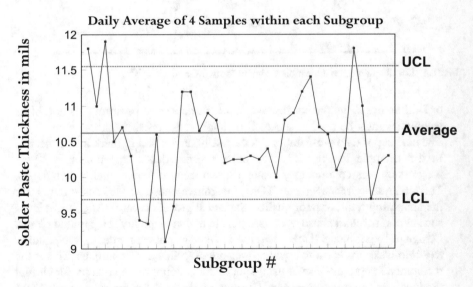

Fig. 1.8 Control chart for the mean (Xbar) of solder paste thickness

sample mean exceeds UCL or its value is lower than LCL, then a quality corrective action is required.

The decision-making process for quality problem detection and resolution is thus simplified. An operator is required to contact a manufacturing, process, or quality engineer to resolve the problem whenever a plotted sample mean is outside the control limits. It is obvious that the paste thickness process is out of control having seven points above or below the control limits. Three sample means are above UCL, and four sample means are below LCL. An eighth sample

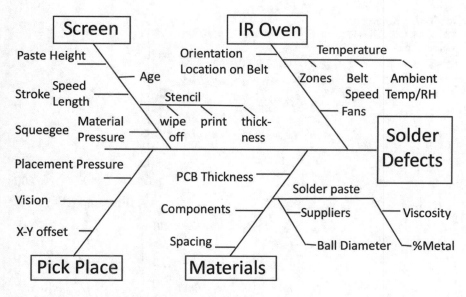

Fig. 1.9 Cause-and-effect diagram for solder height DoE

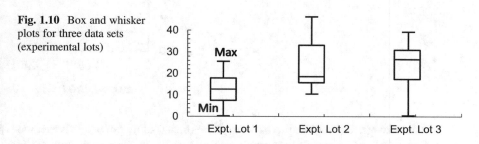

Fig. 1.10 Box and whisker plots for three data sets (experimental lots)

mean is right at the LCL value. This is a typical out-of-control process that prompts the initiation of a DoE project to set the paste thickness to the recommended level (10 mils) and reduce variability. This case study will be discussed later in the book.

4. Figure 1.9 is a cause-and-effect diagram used for the solder paste thickness DoE. It has many other names, such as Ishikawa, 4M, or fishbone diagram. The 4M stands for Methods, Machines, "Men," and Materials. Cause-and-effect diagrams organize all the possible reasons for defects, in contrast to the Pareto diagrams which prioritize the reasons for the defects. It is a team problem-solving approach where the team knowledge is pooled to collect all possible causes of quality defects. Pareto is relational, while cause-and-effect is hierarchical.

5. The box and whisker plots in Fig. 1.10 give more information about the distribution of data around the mean. For each data set (lot number), the minimum and maximum points are shown. The square in the middle represents the 25% and

Fig. 1.11 Changing one factor at a time (OFAT) yield

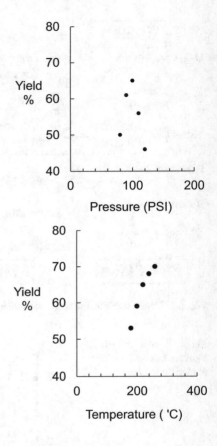

75% data levels. The middle line is the average (mean) of all points. This method of data presentation provides a quick estimate as to whether each data set is normally distributed or not. For experimental Lot 1, the data looks very normally distributed. For experimental Lot 2, the data looks skewed toward the minimum, while for experimental Lot 3, the data is skewed to the maximum.

6. A DoE technique is to change one factor at a time (OFAT) and record the results and then change another factor again. The combined results can be plotted to show both factors' effect on the desired result.

In Fig. 1.11, the yield results of changing pressure and temperature independently are shown. Pressure versus yield is measured individually, while temperature is kept constant. Temperature versus yield is then measured, while pressure is kept constant. In performing DoEs, both factors should be changed simultaneously, so that the interaction between them and its effect on the yield results should be considered, rather than changing one factor at a time.

DoE is a methodology to show the effect of changing all factors at the same time, as shown in Fig. 1.12, called multivariate charts for two factors. In this case, the pressure and temperature factor variation versus yield is shown in one multivariate chart. The data sets used in Fig. 1.11 are applied again in Fig. 1.12

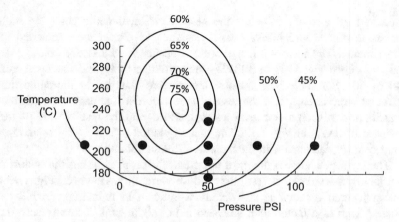

Fig. 1.12 Multivariate charts for two variables

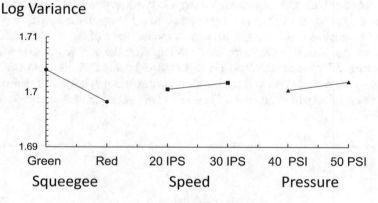

Fig. 1.13 Multifactor plots for variance analysis of solder paste height

versus the yield percentage. Individual data points are shown as dark filled small circles, and the yield percentages are shown as circles and large unfilled curves. An area of a particular percentage of yield, 60%, for example, is shown versus different values of temperature and pressure. Multivariate charts are very similar to topographical maps with one line for the same elevation on the map. It can be seen from Fig. 1.12 that as the yield increases, the range of temperature and pressure narrows and reaches an optimum value. In this manner, DoE can focus on selecting the appropriate levels of temperature and pressure to maximize yield. It is preferable to account for the interaction of the two variables on each other and not assume that the two are independent, as used in OFAT experiments.

Figure 1.13 represents multifactor plots for variance analysis of paste thickness. The data set for Fig. 1.13 was generated after a DoE was performed to resolve the out-of-control condition shown in Fig. 1.8. This method is commonly used in DoE analysis to select the best factor level while changing multiple factors to reduce variability. Two levels for each factor were included in the experiments. The factors selected were squeegee hardness (by color, green or red), squeegee speed (20 or 30 IPS—inches per second), and squeegee pressure (40 or 50 PSI). The three factors' levels were changed at the same time in different experiments, and the results were recorded for each experiment. The contribution of each factor level was calculated, and the results were plotted by each factor levels in Fig. 1.13. The data is plotted in Excel® by separating the chart data for each factor in the table in different columns.

The squeegee color is the most important factor, since it has the widest span, and its lower-level choice (red) may reduce variability. However, Fig. 1.13 does not indicate if a factor is statistically significant in reducing variability. If a smaller span of a factor such as speed is not significant, then any selection of speed levels (20 or 30 IPS) does not influence the desired outcome of low variability. In future chapters, statistical data analysis tools will be used for significance testing.

Another DoE technique is called steps in the direction of steepest ascent. This is a suggested set of DoE to be performed to try to reach the optimum points of performance as demonstrated in the multivariate charts of Fig. 1.13. In this case of the design of a half adder chip, shown in Fig. 1.14, the chip fabrication was being optimized for the lowest delay (input to output) using DoE. Several DoE experiments were conducted to reach the absolute minimum delay performance. As the number of experiments noted in the x-axis value increases, no substantial

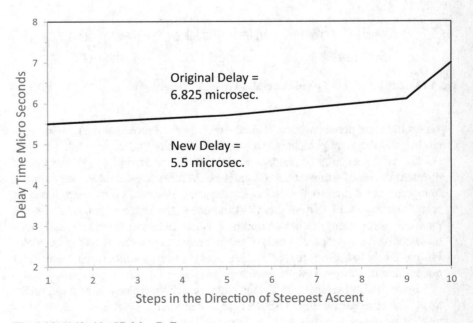

Fig. 1.14 Half adder IC delay DoE outcomes

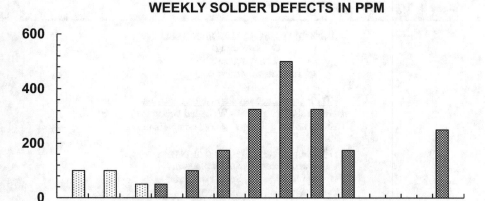

Fig. 1.15 Process quality production before and after DoE

improvement is occurring in the optimal delay operating point of 5.5 micro-seconds. This is an indication that an optimum point has been reached early in the first experiment and cannot be further enhanced by additional experiments or analysis.

Figure 1.15, process quality production before and after DoE, is a bar graph showing the impact resulting from the solder DoE experiment performed at the end of 2-year progression of reducing of quality defects shown in Fig. 1.7. In this case, the plot is a histogram of defect data distribution using 6 months of quality defect data before and after the DoE. This plot indicates that the quality improvements were maintained long after the DoE results were implemented. DoE has reduced both the mean defect and the variability of the defect distribution. It is interesting to note that the defects before the experiments were lognormally distributed. After the experiment, only half of the normal curve distribution is shown since there are no negative defects. An alternative to the histogram presentation of the before and after DoE data is to use normal plots of the defects' distribution based on the mean and standard deviation of the before and after DoE.

The use of these common graphical presentations of data can help in documenting the progression of quality problem detection, analysis, and resolution, especially when using DoE techniques. They can effectively construct a detailed progression of achieving world-class quality goals. They are classified into two categories:

- Qualitative tools such as total quality management (TQM), DMIAC, or Lean Six Sigma
- Quantitative tools as such statistical process control, DoE, and defect-based Six Sigma

Table 1.2 TQM tools and techniques

Tools for generating ideas/information
Brainstorming
Check sheets
Interviewing/surveying

Tools for decision-making and consensus	
Balance sheet	Weighted voting
Criteria rating	Paired comparison

Tools for data analysis and display	
Cause and effect	Spider diagrams
Histograms	Scatter diagrams
Pareto analysis	Sampling
Pie/time charts	Cost-benefit analysis

The qualitative tools will be briefly shown next, and the quantitative tools and their relationships to DoE will be discussed in the next section.

Table 1.2 is a summary of total quality management (TQM) tools and techniques. It organizes the tools of TQM into specific elements for generating ideas, decision-making consensus, and data analysis and display. In a DoE project, it is important that the first step should be identifying the causes of poor performance of design or process and selecting the proper factor and levels. While these tools were discussed individually in the graphical presentation of data, placing the TQM tools in the proper perspective is shown in Table 1.2. Why and when to use a particular TQM tool is relevant to the success of a quality improvement project, as well as the timing and proper tool to use as shown in Table 1.3.

TQM has evolved into continuous process improvement with five elements (DMIAC): Define the project based on its impact on the profits of the company, Measure the performance of the targeted area to be Improved by collecting data, Analyze the data, and then take Corrective action to improve quality. DoE is placed in the data analysis portion of TQM/DMIAC continuum. Lean Six Sigma uses many of the tenets of TQM/DMIAC qualitative tools to focus on quality improvements. Defect-based Six Sigma is the classical origin of Six Sigma which relates quality defects to the relationship between technical specifications in product design as measured against statistical performance of product manufacturing process.

Table 1.5 Probabilities of obtaining X number of heads in five (n) tosses

x	$C_{5,x}$	$p^x * q^{5-x}$	$pB(x)$
0	1	0.03125	0.03125
1	5	0.03125	0.15625
2	10	0.03125	0.31250
3	10	0.03125	0.31250
4	5	0.03125	0.15625
5	1	0.03125	0.03125
			Total 1.0000

Examples of Using the Binomial Distribution

1. If the probability of failure of one part is 25%, what is the probability that the next two parts out of 4 are also failures?

$$pB\left(2; 4, \tfrac{1}{4}\right) = \tfrac{4!}{2!2!}\left(\tfrac{1}{4}\right)^2\left(\tfrac{3}{4}\right)^2 = 6 * \tfrac{1}{16} * \tfrac{9}{16} = \tfrac{54}{256} = 0.21$$

2. The probability of a failed part is 5%. If 20 parts are made from the same machine, what is the mean (expected value E) and standard deviation of failure? What is the cumulative (total) probability that 5 parts will fail out of 20?

$E(x) = 20 * 0.05 = 1$, Standard deviation $= \sqrt{(20 * 0.05 * 0.95)} = 0.97$

Probability of five parts failing $= \Sigma$ (Probability (P) of part 0 fail + P part 1 fail + P part 2 fail + P part 3 fail + P part 4 fail).

$pB(x = 1, 2, 3, 4) = \Sigma P(1, 2, 3, 4; 20, 0.05)$

x	$C_{20,x}$	$p^x * q^{20-x}$	$pB(x)$
0	1	0.35848592	0.3585
1	20	0.01886768	0.3774
2	190	0.00099304	0.1887
3	1140	0.00005227	0.0596
4	4845	0.00000275	0.0133
$= 0.997426$ or 99.74%			

1.2.2   The Poisson Distribution

The Poisson distribution is more appropriate when each event has an equal probability of failure, producing a "defect." It is especially useful in complex production operations, where the possibilities or opportunities of defects increase very rapidly, and the probability of getting a single defect at a specific place or time is small. The Poisson distribution-based control charts (C or U charts) should be used when the area of opportunity or boundary of finding defects is kept constant. Examples are:

- Defects in a one shift operation
- Solder defects in one electronic product
- Defects in 20,000 units of production
- Total number of defects in a production line per day/week/month

The Poisson distribution implies occurrences of events or defects within a boundary of time, space, or region. It has no "memory," that is, the outcome or defect during one interval is in proportion to its length and independent of other intervals. The average rate of defects is constant, and two defects or more cannot occur at the same time.

The Poisson distribution is a discrete distribution based on the binomial distribution as n approaches infinity ($n = \infty$). It is defined solely in one term, (λ), sometimes called (η), which is the mean of the distribution. The Poisson distribution approximates the binomial distribution when the number of trials (n) is large and the probability of each trial (p) is small. In this case, the variable λ, sometimes called the outcome parameter of the distribution, is equal to np. The mean of Poisson distribution $= \lambda$, and the standard deviation (σ) of Poisson distribution $= \sqrt{\lambda}$. The formula for the Poisson distribution is as follows:

$$p(x, \lambda) = e^{-\lambda}\left(\frac{\lambda^x}{x!}\right) \tag{1.2}$$

where x is the outcome during a specific time or region and λ is the average number of outcomes in the time interval or region.

The probability of a single outcome based on the Poisson distribution is shown in Table 1.6, with initial terms for 0–2. It is used to determine the probability of time spent in line when waiting for service at a bank or for the next bus. In this regard, the Poisson distribution looks very much like the exponential distribution.

The binomial and Poisson distributions are discrete distributions, indicating the probability of a certain event or sequence of events. For example, a binomial distribution would return the probability of producing the mean of faulty parts in a production operation when control charts are used to monitor the quality level in factory. A Poisson distribution will return a probability of when (number of minutes) the next bus will arrive at the bus stop or the probability of finding a blemish on a car door for quality control in a paint line.

Table 1.6 Probabilities for the Poisson distribution

$P(x) = e^{-\lambda} * (\lambda^x / x!)$
$P(0) = e^{-\lambda}$
$P(1) = e^{-\lambda} * \lambda$
$P(2) = e^{-\lambda} * \lambda^2/2$
$P(>1) = 1 - P(0) = 1 - e^{-\lambda} = 0.9$

Examples of Using the Poisson Distribution

1. Assuming the number of defects in a car door has a $\lambda = 5$, what is the expected number of defects in a car door? What is the probability of two defects? Up to two defects?

   Expected number of defects $= 5$

   Probability of 2 defects $= p(2,5) = \frac{e^{-5}\,5^2}{2!} = 0.0842$

   Probability   of   up to 2 defects $= \sum p(012) = e^{-5}\left(1 + \lambda + \frac{\lambda^2}{2}\right) =$
   $e^{-5}\left(1 + 5 + \frac{25}{2}\right) = 0.12465$

2. Assuming that the number of defects in a production line during a single hour is $\lambda = 4$, what is the probability that five defects will occur in that hour?

   Probability of 5 defects/hour $= p(5,4) = \frac{e^{-4}\,4^5}{5!} = 0.1563$

3. Assuming the probability of obtaining a defective part is 0.01, what is the probability of obtaining at least 3 defective parts out of a lot of 100, using binomial and Poisson distributions?

   For binomial:

   $p(0,1,2,3) = C_{100,0}\,(0.01)^0\,(0.99)^{100} + C_{100,1}\,(0.01)^1\,(0.99)^{99} + C_{100,2}\,(0.01)^2$
   $(0.99)^{98} + C_{100,3}\,(0.01)^3\,(0.99)^{97}$
   $= 0.366032 + 0.36973 + 0.18486 + 0.060999 = 0.9816$

   For Poisson:

   $\lambda = np = 1$
   $P(0,1,2,3) = e^{-1}(1^0/0!) + e^{-1}(1^1/1!) + e^{-1}(1^2/2!) + e^{-1}(1^3/3!)$
   $= e^{-1}(1 + 1 + \frac{1}{2} + 1/6) = 0.367879 + 0.367879 + 0.18394 + 0.061313 = 0.9810$

   The result of the Poisson distribution is in good agreement with the value of the binomial distribution for small p and large n but computed much easier.

1.2.3   Continuous Distributions and Reliability

These distributions determine the probability of events that are greater or less than a certain set of value or within a range of values. A normal distribution will return the probability of a percentage of students getting a score of more than 90 in an exam or less than 50 or even the number who got between 50 and 90, but not the exact number of students who scored 75. This can be determined by subtracting the number of students that scored 74 in the exam from the number of students that scored 76.

The lognormal distribution is a transformation of plotting y as a function of log x in a normal distribution. This causes a positive skew to the right in the lognormal distribution which tails off longer than the normal distribution. This is best captured by a variant of the lognormal distribution called the Weibull probability distribution, which is used to determine product or system reliability.

Reliability is commonly defined as "quality as a function of time." It is the ability of a product or system to continue to perform as intended as after being subjected to the extremes of operational conditions. These include customer usage, aging, transportation, and environmental exposure. It is usually quantified by accelerated life testing of the system to simulate these extreme conditions. Two parameters define a particular Weibull distribution: the "shape parameter" indicated by β and the "scale parameter" indicated by η.

A Weibull probability distribution can predict the time to fail for a particular design. The data for the Weibull is based on stressing parts to failure over time using cycling temperature and humidity or other conditions such as vibrations and power fluctuations. Several important reliability targets can then be calculated, called time to x percent failure or N_x:

- N_{63} is the most common, which is a good indicator of lifetime and a standard reliability target. It is the time or cycle level to determine when 63% of the parts will fail. Otherwise, a large degree of uncertainty will exist in the lifetime predictions.
- N_{50}, called the median lifetime, is a level when 50% of the parts will fail and can be used to determine stock levels for spare parts.
- $N_{0.01}$ is a level used on high reliability products, when only 1% of the parts will fail.

Another measure of reliability is mean time between failures (MTBF). This metric is most used to describe the performance of complex multi-part or function systems, where repair and maintenance activities are commonly performed. In such applications, the expectation is that some failures will occur over the lifetime of systems. These failures should not occur with great frequency. They can be mitigated with better (more reliable) materials or parts. Another common method is to duplicate critical functions. An example would be to have excess power-generating capacity in power plants to ensure uninterrupted availability of power to consumers.

Weibull Distribution Example

The probability plot shown in Fig. 1.16 is described in the author's edited book *Green Electronics Design and Manufacturing* [1] and calculated by David Pinsky using a commercially available software package (ReliaSoft Weibull++). A green solder material was contrasted to legacy solder to show the probability of failure over time.

Life testing was performed on a population set (no. 1) of 12 IC packages that were soldered to a printed circuit board (PCB). Life testing using extreme temperature and humidity cycling was continued until all 12 failed, and the time to failure was recorded for each. The concept of "time" in this example is equally applicable to the "number of thermal cycles," duration of exposure to vibration, or length of time exposed to any particular test environment. The hypothetical lifetimes are 473, 826, 947, 1252, 1389, 1746, 2012, 2046, 2243, 2684, 2903, and 3500. They are denoted

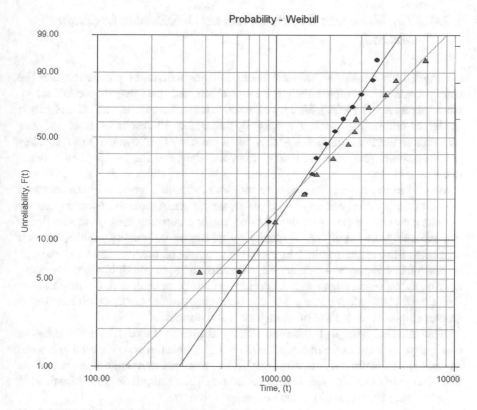

Fig. 1.16   Weibull probability plots using two different materials

by the oval data points. The Weibull function, when plotted on this particular scale, defines a straight line with a slope determined by β and the intercept determined by η.

A second set of data using different solder material was then tested to failure for another population of the same 12 packages. The lifetimes for data set (no. 2) are 372, 986, 1437, 1686, 2092, 2543, 2761, 2802, 3312, 4118, 4692, and 6841. They are denoted by the triangular points in Fig. 1.16.

The example in Fig. 1.16 shows two separate straight line approximation results of evaluating reliability for the two alternate materials. The two lines crossed at approximately 1500th hours of life testing, indicating one material use is better in the short term, while the other material use is better in the long term.

The crossover occurred because the two data sets exhibit different values of β, the shape parameter or slope. This creates a dilemma when attempting to determine which material provides "better" reliability. If the principal concern is how long it takes for half of the parts to fail, then the relative values of N_{50} should be compared. Material set no. 2 is "better" by this definition. However, if the principal concern is reducing early failures, by comparison of the relative values of $N_{0.01}$, material set no. 1 is preferred.

1.2.4   The Use of the Standard Normal Distribution in Quality Methods

Quality efforts began as statistical methodologies to control production and the supply chain through tools such as control charts and incoming inspection acceptance quality levels (AQL). Modern companies initially used the vertical integration model in controlling product quality by managing all resources from the raw materials to the final product assembly, as in the early Ford company model. They felt they could guarantee the quality from the source to the end in their product realization chain.

As companies moved into the core competency model, where they concentrated on their primary competitive advantage, the supply chain took on major importance to reduce cost and enhance quality. At first, major companies took on the challenge of developing the supply chain quality by managing the quality training of their suppliers. This effort resulted in the developments of quality standards through professional societies and industry-wide organizations. In addition, global competition in quality which began with Japanese products competing with US manufacturing in the late twentieth century was also becoming a major factor which led to even greater emphasis on the use of quality tools and standards.

The standard normal distribution (SND), shown in Table 1.7, has a mean or average $(\mu) = 0$ and standard deviation $(\sigma) = 1$. A z transformation is used to convert any normally distributed data with a particular mean and standard deviation into the SND (sometimes called the z distribution). To calculate all possible combinations of μ and σ, the z transformations can be used as follows:

$$z = \frac{X - \mu}{\sigma} \tag{1.3}$$

Figure 1.17 shows the z transformation of normal data to standard normal distribution (SND), where data points X_1 and X_2 are transformed into z_1 and z_2. The probability of data points above or below z_1 or z_2 can thus be determined from the z distribution. The z value is searched in Table 1.7, and the corresponding probability $f(z)$ is extracted.

One advantage of the z distribution table is that when given $-z$, the corresponding $f(-z)$ is the area of the z curve (probability) from $-\infty$ to $-z$. Since the z distribution is symmetrical, it is recommended to use the negative value of z directly rather than use the positive value $+z$ which requires subtraction from $(1 - f(z))$ to extract the probability from $+z$ to $+\infty$.

Examples of Using SND for Determining Defect Rates

1. Given the mean (μ) and standard deviation (σ) of a quiz, several probabilities can be obtained. The probability of how many students would score greater than 90 in

Table 1.7 Standard normal distribution (SND)

Standard normal distribution

Z	f(Z)	Z	f(Z)	Z	f(Z)	Z	f(Z)	Z	f(Z)	Z	f(Z)
Z<0											
0	0.5	**1**	0.84134	**2**	0.97725	**3**	0.99865	**4**	0.99996833	**5**	0.99999971334
0.01	0.50399	1.01	0.84375	2.01	0.97778	3.01	0.99869	4.01	0.99996963	5.01	0.99999972742
0.02	0.50798	1.02	0.84614	2.02	0.97831	3.02	0.99874	4.02	0.99997089	5.02	0.99999974123
0.03	0.51197	1.03	0.84849	2.03	0.97882	3.03	0.99878	4.03	0.99997210	5.03	0.99999975436
0.04	0.51595	1.04	0.85083	2.04	0.97932	3.04	0.99882	4.04	0.99997326	5.04	0.99999976685
0.05	0.51994	1.05	0.85314	2.05	0.97982	3.05	0.99886	4.05	0.99997438	5.05	0.99999977873
0.06	0.52392	1.06	0.85543	2.06	0.98030	3.06	0.99889	4.06	0.99997545	5.06	0.99999979002
0.07	0.52790	1.07	0.85769	2.07	0.98077	3.07	0.99893	4.07	0.99997648	5.07	0.99999980076
0.08	0.53188	1.08	0.85993	2.08	0.98124	3.08	0.99896	4.08	0.99997747	5.08	0.99999981096
0.09	0.53586	1.09	0.86214	2.09	0.98169	3.09	0.99900	4.09	0.99997842	5.09	0.99999982066
0.1	0.53983	**1.1**	0.86433	**2.1**	0.98214	**3.1**	0.99903	**4.1**	0.99997933	**5.1**	0.99999982988
0.11	0.54380	1.11	0.86650	2.11	0.98257	3.11	0.99906	4.11	0.99998021	5.11	0.99999983864
0.12	0.54776	1.12	0.86864	2.12	0.98300	3.12	0.99910	4.12	0.99998105	5.12	0.99999984696
0.13	0.55172	1.13	0.87076	2.13	0.98341	3.13	0.99913	4.13	0.99998185	5.13	0.99999985487
0.14	0.55567	1.14	0.87286	2.14	0.98382	3.14	0.99916	4.14	0.99998262	5.14	0.99999986238
0.15	0.55962	1.15	0.87493	2.15	0.98422	3.15	0.99918	4.15	0.99998337	5.15	0.99999986952
0.16	0.56356	1.16	0.87698	2.16	0.98461	3.16	0.99921	4.16	0.99998408	5.16	0.99999987630
0.17	0.56749	1.17	0.87900	2.17	0.98500	3.17	0.99924	4.17	0.99998476	5.17	0.99999988274
0.18	0.57142	1.18	0.88100	2.18	0.98537	3.18	0.99926	4.18	0.99998542	5.18	0.99999988885
0.19	0.57535	1.19	0.88298	2.19	0.98574	3.19	0.99929	4.19	0.99998604	5.19	0.99999989465
0.2	0.57926	**1.2**	0.88493	**2.2**	0.98610	**3.2**	0.99931	**4.2**	0.99998665	**5.2**	0.99999990017
0.21	0.58317	1.21	0.88686	2.21	0.98645	3.21	0.99934	4.21	0.99998722	5.21	0.99999990540
0.22	0.58706	1.22	0.88877	2.22	0.98679	3.22	0.99936	4.22	0.99998778	5.22	0.99999991036

(continued)

Table 1.7 (continued)

Standard normal distribution

Z	f(Z)	Z	f(Z)	Z	f(Z)	Z	f(Z)	Z	f(Z)	Z	f(Z)
0.23	0.59095	1.23	0.89065	2.23	0.98713	3.23	0.99938	4.23	0.99998831	5.23	0.9999999991508
0.24	0.59483	1.24	0.89251	2.24	0.98745	3.24	0.99940	4.24	0.99998882	5.24	0.9999999991955
0.25	0.59871	1.25	0.89435	2.25	0.98778	3.25	0.99942	4.25	0.99998930	5.25	0.9999999992380
0.26	0.60257	1.26	0.89617	2.26	0.98809	3.26	0.99944	4.26	0.99998977	5.26	0.9999999992783
0.27	0.60642	1.27	0.89796	2.27	0.98840	3.27	0.99946	4.27	0.99999022	5.27	0.9999999993165
0.28	0.61026	1.28	0.89973	2.28	0.98870	3.28	0.99948	4.28	0.99999065	5.28	0.9999999993528
0.29	0.61409	1.29	0.90147	2.29	0.98899	3.29	0.99950	4.29	0.99999106	5.29	0.9999999993872
0.3	0.61791	**1.3**	0.90320	**2.3**	0.98928	**3.3**	0.99952	**4.3**	0.99999145	**5.3**	0.9999999994198
0.31	0.62172	1.31	0.90490	2.31	0.98956	3.31	0.99953	4.31	0.99999183	5.31	0.9999999994507
0.32	0.62552	1.32	0.90658	2.32	0.98983	3.32	0.99955	4.32	0.99999219	5.32	0.9999999994801
0.33	0.62930	1.33	0.90824	2.33	0.99010	3.33	0.99957	4.33	0.99999254	5.33	0.9999999995079
0.34	0.63307	1.34	0.90988	2.34	0.99036	3.34	0.99958	4.34	0.99999287	5.34	0.9999999995343
0.35	0.63683	1.35	0.91149	2.35	0.99061	3.35	0.99960	4.35	0.99999319	5.35	0.9999999995593
0.36	0.64058	1.36	0.91308	2.36	0.99086	3.36	0.99961	4.36	0.99999349	5.36	0.9999999995830
0.37	0.64431	1.37	0.91466	2.37	0.99111	3.37	0.99962	4.37	0.99999378	5.37	0.9999999996054
0.38	0.64803	1.38	0.91621	2.38	0.99134	3.38	0.99964	4.38	0.99999406	5.38	0.9999999996267
0.39	0.65173	1.39	0.91774	2.39	0.99158	3.39	0.99965	4.39	0.99999433	5.39	0.9999999996469
0.4	0.65542	**1.4**	0.91924	**2.4**	0.99180	**3.4**	0.99966	**4.4**	0.99999458	**5.4**	0.9999999996660
0.41	0.65910	1.41	0.92073	2.41	0.99202	3.41	0.99968	4.41	0.99999483	5.41	0.9999999996842
0.42	0.66276	1.42	0.92220	2.42	0.99224	3.42	0.99969	4.42	0.99999506	5.42	0.9999999997013
0.43	0.66640	1.43	0.92364	2.43	0.99245	3.43	0.99970	4.43	0.99999528	5.43	0.9999999997176
0.44	0.67003	1.44	0.92507	2.44	0.99266	3.44	0.99971	4.44	0.99999550	5.44	0.9999999997330
0.45	0.67364	1.45	0.92647	2.45	0.99286	3.45	0.99972	4.45	0.99999570	5.45	0.9999999997476
0.46	0.67724	1.46	0.92785	2.46	0.99305	3.46	0.99973	4.46	0.99999590	5.46	0.9999999997614

x	Φ	x	Φ	x	Φ	x	Φ	x	Φ	x	Φ
0.47	0.68082	1.47	0.92922	2.47	0.99324	3.47	0.99974	4.47	0.99999609	5.47	0.99999997744
0.48	0.68439	1.48	0.93056	2.48	0.99343	3.48	0.99975	4.48	0.99999626	5.48	0.99999997868
0.49	0.68793	1.49	0.93189	2.49	0.99361	3.49	0.99976	4.49	0.99999644	5.49	0.99999997985
0.5	0.69146	**1.5**	0.93319	**2.5**	0.99379	**3.5**	0.99977	**4.5**	0.99999660	**5.5**	0.99999998096
0.51	0.69497	1.51	0.93448	2.51	0.99396	3.51	0.999776	4.51	0.99999676	5.51	0.99999998201
0.52	0.69847	1.52	0.93574	2.52	0.99413	3.52	0.999784	4.52	0.99999691	5.52	0.99999998301
0.53	0.70194	1.53	0.93699	2.53	0.99430	3.53	0.999792	4.53	0.99999705	5.53	0.99999998395
0.54	0.70540	1.54	0.93822	2.54	0.99446	3.54	0.999800	4.54	0.99999718	5.54	0.99999998484
0.55	0.70884	1.55	0.93943	2.55	0.99461	3.55	0.999807	4.55	0.99999732	5.55	0.99999998568
0.56	0.71226	1.56	0.94062	2.56	0.99477	3.56	0.999815	4.56	0.99999744	5.56	0.99999998648
0.57	0.71566	1.57	0.94179	2.57	0.99492	3.57	0.999821	4.57	0.99999756	5.57	0.99999998723
0.58	0.71904	1.58	0.94295	2.58	0.99506	3.58	0.999828	4.58	0.99999767	5.58	0.99999998794
0.59	0.72240	1.59	0.94408	2.59	0.99520	3.59	0.999835	4.59	0.99999778	5.59	0.99999998862
0.6	0.72575	**1.6**	0.94520	**2.6**	0.99534	**3.6**	0.999841	**4.6**	0.99999789	**5.6**	0.99999998925
0.61	0.72907	1.61	0.94630	2.61	0.99547	3.61	0.999847	4.61	0.99999798	5.61	0.99999998986
0.62	0.73237	1.62	0.94738	2.62	0.99560	3.62	0.999853	4.62	0.99999808	5.62	0.99999999043
0.63	0.73565	1.63	0.94845	2.63	0.99573	3.63	0.999858	4.63	0.99999817	5.63	0.99999999096
0.64	0.73891	1.64	0.94950	2.64	0.99585	3.64	0.999864	4.64	0.99999826	5.64	0.99999999147
0.65	0.74215	1.65	0.95053	2.65	0.99598	3.65	0.999869	4.65	0.99999834	5.65	0.99999999196
0.66	0.74537	1.66	0.95154	2.66	0.99609	3.66	0.999874	4.66	0.99999842	5.66	0.99999999241
0.67	0.74857	1.67	0.95254	2.67	0.99621	3.67	0.999879	4.67	0.99999849	5.67	0.99999999284
0.68	0.75175	1.68	0.95352	2.68	0.99632	3.68	0.999883	4.68	0.99999856	5.68	0.99999999325
0.69	0.75490	1.69	0.95449	2.69	0.99643	3.69	0.999888	4.69	0.99999863	5.69	0.99999999363
0.7	0.75804	**1.7**	0.95543	**2.7**	0.99653	**3.7**	0.999892	**4.7**	0.99999870	**5.7**	0.99999999399
0.71	0.76115	1.71	0.95637	2.71	0.99664	3.71	0.999896	4.71	0.99999876	5.71	0.99999999433
0.72	0.76424	1.72	0.95728	2.72	0.99674	3.72	0.999900	4.72	0.99999882	5.72	0.99999999466
0.73	0.76730	1.73	0.95818	2.73	0.99683	3.73	0.999904	4.73	0.99999888	5.73	0.99999999496

(continued)

Table 1.7 (continued)

Standard normal distribution

Z	f(Z)	Z	f(Z)	Z	f(Z)	Z	f(Z)	Z	f(Z)	Z	f(Z)
0.74	0.77035	1.74	0.95907	2.74	0.99693	3.74	0.999908	4.74	0.99999893	5.74	0.99999999525
0.75	0.77337	1.75	0.95994	2.75	0.99702	3.75	0.999912	4.75	0.99999898	5.75	0.99999999552
0.76	0.77637	1.76	0.96080	2.76	0.99711	3.76	0.999915	4.76	0.99999903	5.76	0.99999999578
0.77	0.77935	1.77	0.96164	2.77	0.99720	3.77	0.999918	4.77	0.99999908	5.77	0.99999999602
0.78	0.78230	1.78	0.96246	2.78	0.99728	3.78	0.999922	4.78	0.99999912	5.78	0.99999999625
0.79	0.78524	1.79	0.96327	2.79	0.99736	3.79	0.999925	4.79	0.99999917	5.79	0.99999999647
0.8	0.78814	**1.8**	0.96407	**2.8**	0.99744	**3.8**	0.999928	**4.8**	0.99999921	**5.8**	0.99999999667
0.81	0.79103	1.81	0.96485	2.81	0.99752	3.81	0.999930	4.81	0.99999924	5.81	0.99999999687
0.82	0.79389	1.82	0.96562	2.82	0.99760	3.82	0.999933	4.82	0.99999928	5.82	0.99999999705
0.83	0.79673	1.83	0.96638	2.83	0.99767	3.83	0.999936	4.83	0.99999932	5.83	0.99999999722
0.84	0.79955	1.84	0.96712	2.84	0.99774	3.84	0.999938	4.84	0.99999935	5.84	0.99999999738
0.85	0.80234	1.85	0.96784	2.85	0.99781	3.85	0.999941	4.85	0.99999938	5.85	0.99999999753
0.86	0.80511	1.86	0.96856	2.86	0.99788	3.86	0.999943	4.86	0.99999941	5.86	0.99999999768
0.87	0.80785	1.87	0.96926	2.87	0.99795	3.87	0.999946	4.87	0.99999944	5.87	0.99999999781
0.88	0.81057	1.88	0.96995	2.88	0.99801	3.88	0.999948	4.88	0.99999947	5.88	0.99999999794
0.89	0.81327	1.89	0.97062	2.89	0.99807	3.89	0.999950	4.89	0.99999950	5.89	0.99999999806
0.9	0.81594	**1.9**	0.97128	**2.9**	0.99813	**3.9**	0.999952	**4.9**	0.99999952	**5.9**	0.99999999818
0.91	0.81859	1.91	0.97193	2.91	0.99819	3.91	0.999954	4.91	0.99999954	5.91	0.99999999828
0.92	0.82121	1.92	0.97257	2.92	0.99825	3.92	0.999956	4.92	0.99999957	5.92	0.99999999838
0.93	0.82381	1.93	0.97320	2.93	0.99831	3.93	0.999958	4.93	0.99999959	5.93	0.99999999848
0.94	0.82639	1.94	0.97381	2.94	0.99836	3.94	0.999959	4.94	0.99999961	5.94	0.99999999857
0.95	0.82894	1.95	0.97441	2.95	0.99841	3.95	0.999961	4.95	0.99999963	5.95	0.99999999865
0.96	0.83147	1.96	0.97500	2.96	0.99846	3.96	0.999963	4.96	0.99999965	5.96	0.99999999873
0.97	0.83398	1.97	0.97558	2.97	0.99851	3.97	0.999964	4.97	0.99999966	5.97	0.99999999881

0.98	0.83646	1.98	0.97615	2.98	0.99856	3.98	0.999966	4.98	0.99999968	5.98	0.9999999888
0.99	0.83891	1.99	0.97670	2.99	0.99861	3.99	0.999967	4.99	0.99999970	5.99	0.9999999895
1	0.84134	**2**	0.97725	**3**	0.99865	**4**	0.999968	**5**	0.99999971	**6**	0.9999999901
Z<0											
0	0.5										
−0.01	0.49601	−1.01	0.15625	−2.01	0.02222	−3.01	0.00131	−4.01	0.00003037	−5.01	0.00000027258
−0.02	0.49202	−1.02	0.15386	−2.02	0.02169	−3.02	0.00126	−4.02	0.00002911	−5.02	0.00000025877
−0.03	0.48803	−1.03	0.15151	−2.03	0.02118	−3.03	0.00122	−4.03	0.00002790	−5.03	0.00000024564
−0.04	0.48405	−1.04	0.14917	−2.04	0.02068	−3.04	0.00118	−4.04	0.00002674	−5.04	0.00000023315
−0.05	0.48006	−1.05	0.14686	−2.05	0.02018	−3.05	0.00114	−4.05	0.00002562	−5.05	0.00000022127
−0.06	0.47608	−1.06	0.14457	−2.06	0.01970	−3.06	0.00111	−4.06	0.00002455	−5.06	0.00000020998
−0.07	0.47210	−1.07	0.14231	−2.07	0.01923	−3.07	0.00107	−4.07	0.00002352	−5.07	0.00000019924
−0.08	0.46812	−1.08	0.14007	−2.08	0.01876	−3.08	0.00104	−4.08	0.00002253	−5.08	0.00000018904
−0.09	0.46414	−1.09	0.13786	−2.09	0.01831	−3.09	0.00100	−4.09	0.00002158	−5.09	0.00000017934
−0.1	0.46017	**−1.1**	0.13567	**−2.1**	0.01786	**−3.1**	0.00097	**−4.1**	0.00002067	**−5.1**	0.00000017012
−0.11	0.45620	−1.11	0.13350	−2.11	0.01743	−3.11	0.00094	−4.11	0.00001979	−5.11	0.00000016136
−0.12	0.45224	−1.12	0.13136	−2.12	0.01700	−3.12	0.00090	−4.12	0.00001895	−5.12	0.00000015304
−0.13	0.44828	−1.13	0.12924	−2.13	0.01659	−3.13	0.00087	−4.13	0.00001815	−5.13	0.00000014513
−0.14	0.44433	−1.14	0.12714	−2.14	0.01618	−3.14	0.00084	−4.14	0.00001738	−5.14	0.00000013762
−0.15	0.44038	−1.15	0.12507	−2.15	0.01578	−3.15	0.00082	−4.15	0.00001663	−5.15	0.00000013048
−0.16	0.43644	−1.16	0.12302	−2.16	0.01539	−3.16	0.00079	−4.16	0.00001592	−5.16	0.00000012370
−0.17	0.43251	−1.17	0.12100	−2.17	0.01500	−3.17	0.00076	−4.17	0.00001524	−5.17	0.00000011726
−0.18	0.42858	−1.18	0.11900	−2.18	0.01463	−3.18	0.00074	−4.18	0.00001458	−5.18	0.00000011115
−0.19	0.42465	−1.19	0.11702	−2.19	0.01426	−3.19	0.00071	−4.19	0.00001396	−5.19	0.00000010535
−0.2	0.42074	**−1.2**	0.11507	**−2.2**	0.01390	**−3.2**	0.00069	**−4.2**	0.00001335	**−5.2**	0.00000009983
−0.21	0.41683	−1.21	0.11314	−2.21	0.01355	−3.21	0.00066	−4.21	0.00001278	−5.21	0.00000009460
−0.22	0.41294	−1.22	0.11123	−2.22	0.01321	−3.22	0.00064	−4.22	0.00001222	−5.22	0.00000008964

(continued)

Table 1.7 (continued)

Standard normal distribution

Z	f(Z)	Z	f(Z)	Z	f(Z)	Z	f(Z)	Z	f(Z)	Z	f(Z)
−0.23	0.40905	−1.23	0.10935	−2.23	0.01287	−3.23	0.00062	−4.23	0.00001169	−5.23	0.0000008492
−0.24	0.40517	−1.24	0.10749	−2.24	0.01255	−3.24	0.00060	−4.24	0.00001118	−5.24	0.0000008045
−0.25	0.40129	−1.25	0.10565	−2.25	0.01222	−3.25	0.00058	−4.25	0.00001070	−5.25	0.0000007620
−0.26	0.39743	−1.26	0.10383	−2.26	0.01191	−3.26	0.00056	−4.26	0.00001023	−5.26	0.0000007217
−0.27	0.39358	−1.27	0.10204	−2.27	0.01160	−3.27	0.00054	−4.27	0.00000978	−5.27	0.0000006835
−0.28	0.38974	−1.28	0.10027	−2.28	0.01130	−3.28	0.00052	−4.28	0.00000935	−5.28	0.0000006472
−0.29	0.38591	−1.29	0.09853	−2.29	0.01101	−3.29	0.00050	−4.29	0.00000894	−5.29	0.0000006128
−0.3	0.38209	**−1.3**	0.09680	**−2.3**	0.01072	**−3.3**	0.00048	**−4.3**	0.00000855	**−5.3**	0.00000005802
−0.31	0.37828	−1.31	0.09510	−2.31	0.01044	−3.31	0.00047	−4.31	0.00000817	−5.31	0.00000005493
−0.32	0.37448	−1.32	0.09342	−2.32	0.01017	−3.32	0.00045	−4.32	0.00000781	−5.32	0.00000005199
−0.33	0.37070	−1.33	0.09176	−2.33	0.00990	−3.33	0.00043	−4.33	0.00000746	−5.33	0.00000004921
−0.34	0.36693	−1.34	0.09012	−2.34	0.00964	−3.34	0.00042	−4.34	0.00000713	−5.34	0.00000004657
−0.35	0.36317	−1.35	0.08851	−2.35	0.00939	−3.35	0.00040	−4.35	0.00000681	−5.35	0.00000004407
−0.36	0.35942	−1.36	0.08692	−2.36	0.00914	−3.36	0.00039	−4.36	0.00000651	−5.36	0.00000004170
−0.37	0.35569	−1.37	0.08534	−2.37	0.00889	−3.37	0.00038	−4.37	0.00000622	−5.37	0.00000003946
−0.38	0.35197	−1.38	0.08379	−2.38	0.00866	−3.38	0.00036	−4.38	0.00000594	−5.38	0.00000003733
−0.39	0.34827	−1.39	0.08226	−2.39	0.00842	−3.39	0.00035	−4.39	0.00000567	−5.39	0.00000003531
−0.4	0.34458	**−1.4**	0.08076	**−2.4**	0.00820	**−3.4**	0.00034	**−4.4**	0.00000542	**−5.4**	0.00000003340
−0.41	0.34090	−1.41	0.07927	−2.41	0.00798	−3.41	0.00032	−4.41	0.00000517	−5.41	0.00000003158
−0.42	0.33724	−1.42	0.07780	−2.42	0.00776	−3.42	0.00031	−4.42	0.00000494	−5.42	0.00000002987
−0.43	0.33360	−1.43	0.07636	−2.43	0.00755	−3.43	0.00030	−4.43	0.00000472	−5.43	0.00000002824
−0.44	0.32997	−1.44	0.07493	−2.44	0.00734	−3.44	0.00029	−4.44	0.00000450	−5.44	0.00000002670
−0.45	0.32636	−1.45	0.07353	−2.45	0.00714	−3.45	0.00028	−4.45	0.00000430	−5.45	0.00000002524
−0.46	0.32276	−1.46	0.07215	−2.46	0.00695	−3.46	0.00027	−4.46	0.00000410	−5.46	0.00000002386

−0.47	0.31918	−1.47	0.07078	−2.47	0.00676	−3.47	0.00026	−4.47	0.00000391	−5.47	0.00000000002256
−0.48	0.31561	−1.48	0.06944	−2.48	0.00657	−3.48	0.00025	−4.48	0.00000374	−5.48	0.00000000002132
−0.49	0.31207	−1.49	0.06811	−2.49	0.00639	−3.49	0.00024	−4.49	0.00000356	−5.49	0.00000000002015
−0.5	0.30854	**−1.5**	0.06681	**−2.5**	0.00621	**−3.5**	0.00023	**−4.5**	0.00000340	**−5.5**	0.00000000001904
−0.51	0.30503	−1.51	0.06552	−2.51	0.00604	−3.51	0.000224	−4.51	0.00000324	−5.51	0.00000000001799
−0.52	0.30153	−1.52	0.06426	−2.52	0.00587	−3.52	0.000216	−4.52	0.00000309	−5.52	0.00000000001699
−0.53	0.29806	−1.53	0.06301	−2.53	0.00570	−3.53	0.000208	−4.53	0.00000295	−5.53	0.00000000001605
−0.54	0.29460	−1.54	0.06178	−2.54	0.00554	−3.54	0.000200	−4.54	0.00000282	−5.54	0.00000000001516
−0.55	0.29116	−1.55	0.06057	−2.55	0.00539	−3.55	0.000193	−4.55	0.00000268	−5.55	0.00000000001432
−0.56	0.28774	−1.56	0.05938	−2.56	0.00523	−3.56	0.000185	−4.56	0.00000256	−5.56	0.00000000001352
−0.57	0.28434	−1.57	0.05821	−2.57	0.00508	−3.57	0.000179	−4.57	0.00000244	−5.57	0.00000000001277
−0.58	0.28096	−1.58	0.05705	−2.58	0.00494	−3.58	0.000172	−4.58	0.00000233	−5.58	0.00000000001206
−0.59	0.27760	−1.59	0.05592	−2.59	0.00480	−3.59	0.000165	−4.59	0.00000222	−5.59	0.00000000001138
−0.6	0.27425	**−1.6**	0.05480	**−2.6**	0.00466	**−3.6**	0.000159	**−4.6**	0.00000211	**−5.6**	0.00000000001075
−0.61	0.27093	−1.61	0.05370	−2.61	0.00453	−3.61	0.000153	−4.61	0.00000202	−5.61	0.00000000001014
−0.62	0.26763	−1.62	0.05262	−2.62	0.00440	−3.62	0.000147	−4.62	0.00000192	−5.62	0.00000000000957
−0.63	0.26435	−1.63	0.05155	−2.63	0.00427	−3.63	0.000142	−4.63	0.00000183	−5.63	0.00000000000904
−0.64	0.26109	−1.64	0.05050	−2.64	0.00415	−3.64	0.000136	−4.64	0.00000174	−5.64	0.00000000000853
−0.65	0.25785	−1.65	0.04947	−2.65	0.00402	−3.65	0.000131	−4.65	0.00000166	−5.65	0.00000000000804
−0.66	0.25463	−1.66	0.04846	−2.66	0.00391	−3.66	0.000126	−4.66	0.00000158	−5.66	0.00000000000759
−0.67	0.25143	−1.67	0.04746	−2.67	0.00379	−3.67	0.000121	−4.67	0.00000158	−5.67	0.00000000000716
−0.68	0.24825	−1.68	0.04648	−2.68	0.00368	−3.68	0.000117	−4.68	0.00000144	−5.68	0.00000000000675
−0.69	0.24510	−1.69	0.04551	−2.69	0.00357	−3.69	0.000112	−4.69	0.00000137	−5.69	0.00000000000637
−0.7	0.24196	**−1.7**	0.04457	**−2.7**	0.00347	**−3.7**	0.000108	**−4.7**	0.00000130	**−5.7**	0.00000000000601
−0.71	0.23885	−1.71	0.04363	−2.71	0.00336	−3.71	0.000104	−4.71	0.00000124	−5.71	0.00000000000567
−0.72	0.23576	−1.72	0.04272	−2.72	0.00326	−3.72	0.000100	−4.72	0.00000118	−5.72	0.00000000000534
−0.73	0.23270	−1.73	0.04182	−2.73	0.00317	−3.73	0.000096	−4.73	0.00000112	−5.73	0.00000000000504
−0.74	0.22965	−1.74	0.04093	−2.74	0.00307	−3.74	0.000092	−4.74	0.00000107	−5.74	0.00000000000475

(continued)

Table 1.7 (continued)

Standard normal distribution

Z	f(Z)	Z	f(Z)	Z	f(Z)	Z	f(Z)	Z	f(Z)	Z	f(Z)
−0.75	0.22663	−1.75	0.04006	−2.75	0.00298	−3.75	0.000088	−4.75	0.00000102	−5.75	0.0000000000448
−0.76	0.22363	−1.76	0.03920	−2.76	0.00289	−3.76	0.000085	−4.76	0.00000097	−5.76	0.0000000000422
−0.77	0.22065	−1.77	0.03836	−2.77	0.00280	−3.77	0.000082	−4.77	0.00000092	−5.77	0.0000000000398
−0.78	0.21770	−1.78	0.03754	−2.78	0.00272	−3.78	0.000078	−4.78	0.00000088	−5.78	0.0000000000375
−0.79	0.21476	−1.79	0.03673	−2.79	0.00264	−3.79	0.000075	−4.79	0.00000083	−5.79	0.0000000000353
−0.8	0.21186	**−1.8**	0.03593	**−2.8**	0.00256	**−3.8**	0.000072	**−4.8**	0.00000079	**−5.8**	0.0000000000333
−0.81	0.20897	−1.81	0.03515	−2.81	0.00248	−3.81	0.000070	−4.81	0.00000076	−5.81	0.0000000000313
−0.82	0.20611	−1.82	0.03438	−2.82	0.00240	−3.82	0.000067	−4.82	0.00000072	−5.82	0.0000000000295
−0.83	0.20327	−1.83	0.03362	−2.83	0.00233	−3.83	0.000064	−4.83	0.00000068	−5.83	0.0000000000278
−0.84	0.20045	−1.84	0.03288	−2.84	0.00226	−3.84	0.000062	−4.84	0.00000065	−5.84	0.0000000000262
−0.85	0.19766	−1.85	0.03216	−2.85	0.00219	−3.85	0.000059	−4.85	0.00000062	−5.85	0.0000000000247
−0.86	0.19489	−1.86	0.03144	−2.86	0.00212	−3.86	0.000057	−4.86	0.00000059	−5.86	0.0000000000232
−0.87	0.19215	−1.87	0.03074	−2.87	0.00205	−3.87	0.000054	−4.87	0.00000056	−5.87	0.0000000000219
−0.88	0.18943	−1.88	0.03005	−2.88	0.00199	−3.88	0.000052	−4.88	0.00000053	−5.88	0.0000000000206
−0.89	0.18673	−1.89	0.02938	−2.89	0.00193	−3.89	0.000050	−4.89	0.00000050	−5.89	0.0000000000194
−0.9	0.18406	**−1.9**	0.02872	**−2.9**	0.00187	**−3.9**	0.000048	**−4.9**	0.00000048	**−5.9**	0.0000000000182
−0.91	0.18141	−1.91	0.02807	−2.91	0.00181	−3.91	0.000046	−4.91	0.00000046	−5.91	0.0000000000172
−0.92	0.17879	−1.92	0.02743	−2.92	0.00175	−3.92	0.000044	−4.92	0.00000043	−5.92	0.0000000000162
−0.93	0.17619	−1.93	0.02680	−2.93	0.00169	−3.93	0.000042	−4.93	0.00000041	−5.93	0.0000000000152
−0.94	0.17361	−1.94	0.02619	−2.94	0.00164	−3.94	0.000041	−4.94	0.00000039	−5.94	0.0000000000143
−0.95	0.17106	−1.95	0.02559	−2.95	0.00159	−3.95	0.000039	−4.95	0.00000037	−5.95	0.0000000000135
−0.96	0.16853	−1.96	0.02500	−2.96	0.00154	−3.96	0.000037	−4.96	0.00000035	−5.96	0.0000000000127
−0.97	0.16602	−1.97	0.02442	−2.97	0.00149	−3.97	0.000036	−4.97	0.00000034	−5.97	0.0000000000119
−0.98	0.16354	−1.98	0.02385	−2.98	0.00144	−3.98	0.000034	−4.98	0.00000032	−5.98	0.0000000000112
−0.99	0.16109	−1.99	0.02330	−2.99	0.00139	−3.99	0.000033	−4.99	0.00000030	−5.99	0.0000000000105
−1	0.15866	**−2**	0.02275	**−3**	0.00135	**−4**	0.000032	**−5**	0.00000029	**−6**	0.0000000000099

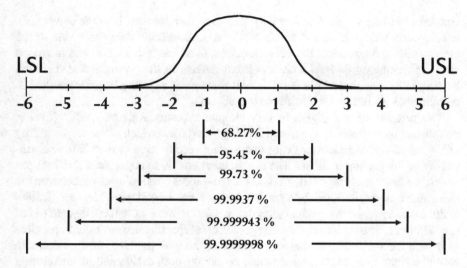

Fig. 1.19 Acceptance rate for multiple sigma specifications

1.3 Six Sigma and Its Relationship with DoE

Six Sigma (6σ) is one of the major quality standards that arose in the late 1980s in US companies to counteract the global quality competition. Its major purpose was to move the quality emphasis from strictly a manufacturing concern into a product design responsibility. The product design emphasis was augmented from focusing on managing the engineering design function and producing a single working prototype into new product realization responsibilities. Design managers are now tasked with the additional responsibilities of focusing on reducing manufacturing costs and improving manufacturing quality, as well as getting involved in negotiating new product specifications with the supply chain, marketing, and customers.

Quality and cost were historically considered to be the responsibility of the manufacturing part of high-technology companies. Technology was the driving factor for these companies, resulting in high product price. The manufacturing cost concerns were minimal, and in many cases, it was about 1–3% of the selling price of technology products. This led to little emphasis being placed on optimizing quality or cost of high-technology products.

As global competition increased for technology products, the need to focus on product quality and manufacturing costs at the design stage increased significantly. While tools such as Design for Manufacture (DFM) focused on reducing cost and manufacturing DoE focused on mean (μ) shift from design nominal (N) and variability reductions, they did not address the need of reducing cost through quality means. Having a higher design quality means a reduction in manufacturing rejects and rework and lower field recalls and returns. In addition, higher design quality resulted in much lower technical resources needed in manufacturing by reducing the amount of testing required to weed out defective components. In some companies, testing costs were equivalent to 1/3 of the manufacturing labor costs, due to the higher skills and salaries of the test technicians. Customer satisfaction was increased with higher-quality designs with less latent defects in the products and therefore better-quality performance in the field. All these design quality benefits resulted in real tangible overall lower cost for new products.

The premise of Six Sigma is very simple: a multiple sigma quality level is established by contrasting the manufacturing variability (which is the responsibility of the manufacturing function) with the product specification (which is the responsibility of the design function). The design team goal is to negotiate a quality level which is either a multiple of sigma (i.e., 4.5σ) or process capability (Cp), between the two functions. Manufacturing's goal is to have the largest possible specification width to easily make the product with the smallest number of defects that are out of specifications. The design engineering goal is to set specifications as tight as possible to reduce the variability in the performance of the new product. At the minimum case, the techniques of Six Sigma would encourage both design and manufacturing engineers to communicate about quality and cost early in the design phase. Some of the formulas for Six Sigma are as follows:

$$Cp = \frac{USL - LSL}{6\,\sigma} = \frac{\pm SL}{\pm 3\sigma} \tag{1.4}$$

$$Cpk = \text{min of } \frac{USL - \mu}{3\sigma} \text{ or } \frac{\mu - LSL}{3\sigma} \tag{1.5}$$

where:

Cp = process capability, negotiated by the design project manager (PM). Process mean (μ) shift from nominal (N) is not considered in this formula.

Cpk = process capability as influenced by the mean of the manufacturing process deviation from specification nominal ($\mu - N$) $\neq 0$.

SL (USL, LSL) = specification limits (upper, lower), determined by the design team and tempered by manufacturing input, using Cp or Cpk.

μ = manufacturing process mean (as contrasted by the specification nominal (N) set by the design engineers).

σ = standard deviation of the manufacturing process.

If the specification limits were set at $\pm 6\sigma$, then $Cp = 2$. If the specification limits were set at 4.5σ, then $Cp = 1.5$, which is alternately called 4.5σ design. The Cpk is a more exact definition of design quality, which is changed when the manufacturing process mean is not equal to the specification nominal or N. Cpk makes the process capability (or the Six Sigma) concept sensitive to the fact that products can be shipped where some of their performance is not equal to the specification nominal, yet they fall within the acceptable range of specification interval ($\pm$ SL). This is the case addressed by the quality loss function (QLF) term, balancing the process mean deviation or reducing the variability of manufactured products. This QLF was

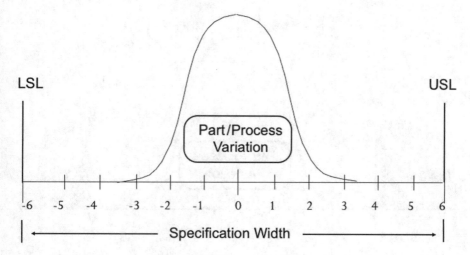

Fig. 1.20   Typical Six Sigma design or process capability $Cp = 2.0$

advocated by Genichi Taguchi, a founder of the Taguchi method for DoE, and discussed in Chap. 9.

Specifying a Six Sigma design for a new product implies that the specification of the product and its parts are much wider than the variability due to manufacturing. Therefore, even if there is a substantial shift in the manufacturing process due to poor quality, whether by changing the process mean or increasing the variability, this shift does not significantly impact the shipping performance of the product. The quality of the product is maintained, with less defects and customer complaints. A typical Six Sigma design or process capability $Cp = 2$ is shown in Fig. 1.20.

If a product design goal was lower than Six Sigma, for example, 4σ design or $Cp = 1.33$, then a small shift in either manufacturing process mean or variability will result in a much larger defect rate in manufacturing, as shown in Fig. 1.21. In this figure, it can be illustrated that a shift in the manufacturing process mean (either to the left/smaller value or right/larger value) will quickly result in a much higher defect rate, hence higher manufacturing costs. This is because the normal distribution is shaped like a bell curve. A shift of the process mean to the right significantly increases defects in the product due to process mean (μ) being > specification nominal (N) and closer to the USL. At the other end, there are slightly less defects due to process mean μ being further away from LSL. To express this condition mathematically, Cpk is measured at both ends of the specification limits, with the right side Cpk < left side Cpk if the mean shifts to the right. The resulting Cpk is the minimum of the two Cpk values. A lower Cpk means lower design quality.

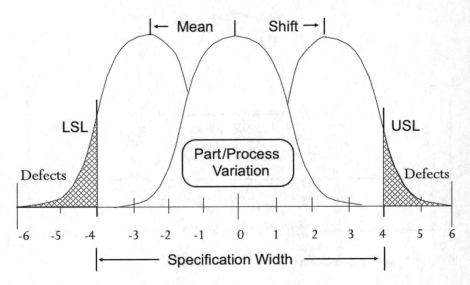

Fig. 1.21 Four sigma design with mean shift or process capability $Cp = 1.33$

1.3.1   Converting Defect Rate to Cp and Sigma Design with No Mean Shift from Nominal

The special case of zero manufacturing process mean (μ) shift from specification nominal (N) results in simplifying the calculations for process capability and defect rates:

$$z = 3 * Cp = 3 * Cpk \text{ when } \mu = N \qquad (1.6)$$

The two terms for process capability Cp and Cpk are thus equal when $\mu = N$.

Examples using SND for determining defect rates in the previous section showed how to determine the defect rate given Cp, process capability (μ and σ), and design specifications (N, USL, and LSL). The defects from both sides of the specification limits should be added if both USL and LSL are specified. In the case of no average shift ($\mu - n = 0$), defects $= 2 * f(-z_1)$.

Conversely, the defect rate can indirectly be used to determine $Cp = Cpk$ from the z distribution. If defects are due to parts being measured outside of one-sided (tailed) specifications, such as defect that results in yielding of test specimen due to external forces, the defect rate can be set $= f(-z)$. If the defect rate is based on two-sided specifications, $f(-z)$ can be set to (defect rate)/2. The corresponding value of z can be extracted from the z distribution in Table 1.7. Cp and Cpk can thus be calculated from $z/3$ based on Eq. 1.6. Several issues must be considered when converting defect rate into Cpk:

1. The defect rate is an attribute of the quality of the parts. There is no direct information given for specification limits (USL, N, and LSL) or the process capability (μ and σ) in Eq. 1.6. It is assumed that the defective parts are outside of some specified but unknown limits.
2. If the unknown specification limits are assumed to be two-sided, then the defect rate will have to be divided by two to make it equal to $f(-z)$ for one side of the z distribution.
3. This mathematical process assumes that measurement distribution of the part is normal and that there is no process mean (μ) shift versus the nominal (N) of the part design specifications or $\mu = N$. There are no actual calculations of the variability σ of the part due to manufacturing process variations.

Many of the measurement of defects are not based on the variability of individual parts, but on counting the defects found in a particular number of parts, assemblies, or operations in manufacturing. These are called defect "attributes" which do not meet a prescribed quality level. Many projects count the number of defects in manufacturing the product, based on opportunities for defects, and then convert the defect rate back to Cpk. This implies that the defects are occurring against a certain complex set of specifications, and the Cpk calculations are derived indirectly assuming that there is no mean shift versus specification nominal N.

An example of conversion methods between Cp and defect rates can be formulated as follows. The process capability is converted into defect rates, and then a defect rate is converted back to Cp/Cpk.

1. Given a Cp of 1.33, what is the expected defect rate?
   Answer: $Cp = 1.33 \rightarrow z = 3* Cp = 4$ or 4 sigma design.
   For $z = 4 \rightarrow f(z = -4) = 0.00032$ or 32 ppm one-sided or 64 ppm two-sided specifications.
   For $Cp = 1$, then the corresponding one- and two-sided defects are 1350 and 2700 ppm.
2. Given a defect rate of 5 in 1000 operations, what is the Cp/Cpk and design sigma, assuming that there is no average shift?
   Answer: 5 in 1000 operations $= 5/1000$ (5000 ppm) $= 0.005$ one-sided and 0.025 two-sided.
   If one-sided, $f(z) = 0.005 \rightarrow z = 2.575$, $Cp = Cpk = z/3 = 0.858$.
   If two-sided, $f(z) = 0.0025 \rightarrow z = 2.81$, $Cp = Cpk = z/3 = 0.937$.

An assembly might have many different opportunities for defects. The amount of defect opportunities is sometimes increased deliberately, by double counting some operations, to reduce the defect rate. For example, in the electronics industry, assemblies such as printed circuit boards (PCBs) have many operations and components embedded in the assembly, as well as joining operations.

There are five major parts of the PCB assembly: (1) the raw (unloaded) PCB, (2) the components that are loaded on the PCB, (3) the placement of solder dots on the PCB surface to prepare the components for soldering onto the PCB, (4) the placement operations to locate the components on the PCB, and, lastly, (5) the joining operation of the components using a conveyorized oven to reflow the solder and attach the components. Any of these operations are opportunities for defects, and the defect number per PCB can be interpreted differently. In the case of PCB defect rates, the relevant industry association for connecting electronics industries (IPC) has published a standard to resolve this issue, specifically counting the opportunities of defect generation in PCB for product and in-process calculations. They are IPC 7912A/9261A End Item and In-Process DPMO (defects per million opportunities). The goal is to level out the different claims of quality in terms of σ level capability for different PCB manufacturing and supplier facilities.

The design team must be careful when tasked with a project designated as Six Sigma. They should keep in mind that Six Sigma is a management tool to encourage communications between design and manufacturing, with the goal of designing and producing high-quality products, with maximum performance in the field and minimum manufacturing cost. Understanding these issues and agreeing beforehand how to handle them will make for a successful Six Sigma project.

1.3.2   The Implied Mean Shift to Nominal of Cp and Six Sigma

There are many variations of Six Sigma quality, the primary one represented by Motorola Incorporated, which was the founding company of Six Sigma. The concepts were introduced by Bill Smith in 1986 under the direction of then Motorola CEO Robert Galvin. It gained wide attention when it was established as a corporate goal for GE through their CEO Jack Welch in 1995 and his successor Jeffrey Immelt. The author had the pleasure of meeting Smith, Galvin, and Immelt and discussed Six Sigma with them. Motorola popularized the Six Sigma concepts through their Motorola University training center in the 1990s where the author taught concurrent engineering. The Cp parameters of mean (μ) and standard deviation (σ) can be obtained from variable control charts, the traditional method of monitoring quality through manufacturing. Control charts were developed by the American Telephone and Telegraph (AT&T) company prior to World War II. They used sampling theory techniques to allow for accepting parts from the supply chain or through manufacturing, if they were within a confidence interval of $\pm 3s$:

$$\overline{\overline{X}} + 3s \tag{1.7}$$

where $\overline{\overline{X}}$ (X$_{doublebar}$) is the grand average of $\overline{X}$ (X$_{bar}$) sample means in variable control charts and s is the sample mean standard deviation. This made the Cp and Six Sigma calculations much easier if data from the manufacturing control charts were used. Control charts use sample data ($\overline{X}$, s), and Six Sigma uses population data (μ, σ). The two are related according to the central limit theorem as follows:

$$\overline{\overline{X}} = \mu \text{ and } s = \frac{\sigma}{\sqrt{n}} \tag{1.8}$$

where n is the sample size. More on these relationship in the next chap. 2 where the error generated by Eq. 1.8 is quantified from the sample size n.

The result of connecting control charts with Cp and Six Sigma posed a dilemma: control charts allowed the manufacturing process mean sample value to be acceptable within $\overline{\overline{X}} + 3s$ interval range. The Motorola approach was to assume that the process mean (μ) can shift as much as $\pm 1.5\sigma$ to account for the acceptability of sample mean variation in the variable control charts. The Cp term does not contain the mean (μ) in Eq. 1.4. The Cpk term uses the actual process mean shift in the calculations and does not assume an implicit process mean shift.

The dilemma of whether to use Cp or Cpk led to confusion when calculating the defect rate for a new product. For a Six Sigma design ($Cp = 2$), the defect rate will be 3.4 parts per million (ppm), given the assumption that the process mean μ inherently can shift by $\pm 1.5\sigma$, based on the half allowable $\overline{\overline{X}}$ shift of $\pm 3s$. The shift causes the z value in Table 1.7 to be reduced from $z = -6$ to $z = -4.5$, making $f(-z) = 0.0000034$ or 3.4 ppm, while the other side of shifted z distribution at $z = 7.5$, which results in near-zero defects. The total implied two-sided defect rate for Six Sigma is thus reduced from 0.002 ppm to 3.4 ppm.

When this mean shift is considered in the Cpk formula, Six Sigma could be more accurately described as $Cpk = 1.5$ or a 4.5 sigma design. A true Six Sigma design ($Cpk = 2$) with no mean shift should have a defect rate of 0.002 ppm or two parts per billion as opposed to 3.4 ppm in the Cp definition.

Some industries have adopted a design quality target of 4σ, which roughly translates to 64 ppm or approximately 1 defect in 10,000 defect opportunities. This is based on using Table 1.7 for $z = -4$, resulting in $f(z) = 0.000032$. Adding up the two sides of the specification limit defects results in 0.000064 or 64 ppm.

There are many interpretation issues with the Six Sigma design, which makes it difficult to use this technique in complex new high-technology products. Some of the issues of implementing a Six Sigma project are:

1. If the goal is to develop and design a Six Sigma product, then does the entire product have to meet Six Sigma, or do individual components of the product have to also meet Six Sigma, or both?
2. If the product is composed of hundreds of parts, each part or subassembly meets Six Sigma individually, but adding up all of the defects from all of the parts makes it difficult for the total product of parts and assemblies to meet the 6σ requirements.
3. One resolution of this issue is to calculate a product defect rate as the sum of all defects of individual parts, added up to the product level. It can then be converted back to Cpk, as shown earlier.
4. Another approach is to have the product design (σ) level assume the level of the lowest quality part in the product or system. In this approach, the lowest quality part can be identified and then targeted for improvement.

1.3.3 Process Capability Studies for Quality Enhancement

Discussions in previous sections outlined a system for investigating and maintaining process capability for the purpose of quality planning. A good indicator of current quality is the use of process capability to monitor and enhance quality using indicators introduced earlier in the chapter such as:

1. Defects per unit (dpu)
2. Defective parts per millions (ppm)
3. Defects per million opportunities (DPMO)
4. Yield (% acceptance rate in first time test)
5. Number of sigma quality achieved (including Six Sigma)

A company can use one or multiple of these indicators, knowing that they are related to each other as discussed earlier. This is especially useful when the company management or major customers contractually require a certain level of quality.

An example would be an enterprise choice of Cpk as the process capability indicator. This requires that all of the fabrication and assembly operations are to report on their process capabilities, as well as major part suppliers and outside contractors. For the supply chain, a supplier management team and contractual processes with quality as well as cost and delivery commitments are required to indicate the need for process capability. The purpose of these activities is to inform new product design teams of current quality status of different operations in manufacturing and the supply chain. If the design team finds the process capability inadequate, manufacturing must perform DoE analysis to improve their processes, purchase better-quality equipment and parts, or select new suppliers that can meet desired goals. The process capability data must be updated regularly to keep the design team abreast of quality and capability enhancements. The frequency of updates should be short enough to comfortably fit inside new product design cycles, as well as meet yearly management goals. A typical frequency of updating process capability is every quarter.

For assembled parts, process capability determination must be compatible with industry standards, as well as the calculations of defect opportunities such as DPMO. For fabricated parts, especially those made in multipurpose machine shops, the process capability determination is more difficult. The machine shop can produce desired parts' geometry through many possible machines in the shop, some with high capability while others with lower quality. The dilemma is whether a particular process should be machine dependent since machine selection is usually not included in part or assembly documentation. If a 1/2 inch hole needs to be drilled, there are many alternative machines in the shop that can perform this operation, with varying process capabilities. What will the design team assume for drilling 1/2″ hole defect rate?

One solution to this fabrication dilemma is to allow for an additional attribute in the Six Sigma methodology. This attribute would be a quality or complexity indicator. The fabrication shop could be divided into several (maximum of three) levels of complexity. As each new part is being designed, the design engineer can select from any of the three process capabilities available, depending on the level of complexity of the part.

For each process, a baseline process capability could be determined. Every quarter, all process capabilities are checked and recalculated if they show a statistically significant shift in mean or variability.

Significance can be determined using statistical comparison tests. The z distribution can be used to compare a large (>30) sample with the baseline population means, and the X^2 test can be used to compare sample to population variability. For smaller size samples, the sample mean shift from the population mean can be tested with the t distribution. These statistical tests will be shown in the next chapter. Some of the process capability data can be obtained from control charts, while others can be calculated directly by taking samples from production lines.

1.3.4 Process Capability for Prototype and Early Production Parts

When prototype parts are acquired, whether through purchase or made in the company's internal factories, the following methodology is recommended for process capability calculations:

1. New parts that are very similar to current parts, or made in the same production line or process, can assume current parts' process capability. Examples would be fabricated, assembled, and machined parts. Process capability can be derived from existing manufacturing statistical control data.
2. The required process capability for new parts and suppliers should be higher than the capability for current parts to ensure future compliance. This strategy is to mitigate possible reduced focus on quality when new products go into full production. This should be applied to parts purchased from the supply chain or internally manufactured.
3. For low-volume manufacturing, smaller sample sizes should be used to measure quality, including the moving range method discussed later in this chapter. Statistical techniques as well as sample size determination can be used to determine ranges of population mean (μ) and standard deviation (σ). Confidence limits can be used to determine the worst-case process capability.
4. To determine the specification limits, especially for Six Sigma designs, product specifications should be related to the customer satisfaction. Many tools exist to quantify these relationships such as Voice of the Customer (VOC) for specification-driven designs and Quality Function Deployment (QFD) for market-driven new products.
5. Six Sigma or Cpk quality level target can be altered for the short versus the long term for new products. In the short term, including prototype and early

production, close attention is given to the parts and manufacturing process by the design team and manufacturing engineers. As production ramps up, more parts are made with a larger workforce and less skilled operators, resulting in less quality, even if a good control system is in place. In the long term, parts' quality levels will increase with good use of corrective action processes as well as increased operators' skills resulting from the learning curve. It is advisable to set a higher quality level in the early production phase to counteract the problems when production ramps up. An example would be to set quality for early production to $Cpk = 1.67$ (5σ design) and then back off to $Cpk = 1.33$ (4σ design) in the long term for mature products.

1.3.5   *Corrective Action for Process Capability Problems*

The previous section described a methodology for calculating process capability for new parts. For existing parts with unacceptable quality, the following suggestions might be followed to bring the process capabilities in compliance with Six Sigma or Cpk targets:

1. Using DoE to decrease process mean shift (μ) from specification nominal (N) and reduce variability.
2. Consider expanding product or part specifications while maintaining system requirements.
3. Increase operator training, process documentation quality, and incoming inspection of parts and contractual agreements with the supply chain to enhance quality and increase process capability.
4. Use of quality tools and techniques discussed in this chapter to foster a world-class quality culture in the company.
5. If current processes remain not capable, consider purchasing new equipment or engaging outside suppliers to provide higher-quality parts and assemblies.

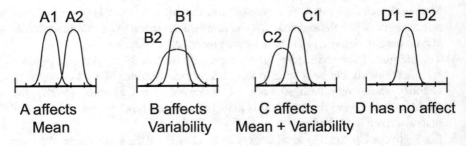

Fig. 1.22 Example of DoE outcomes with four factors

1.3.6   DoE Effects on Six Sigma

The relationship of DoE to Six Sigma is very important. It affects both the numerator and the denominator of Cp and Cpk in Eqs. 1.4 and 1.5. DoE can be used to experiment with design specification (USL, LSL), to determine the optimum performance through proper selection of design parameters. DoE can also be used to shift manufacturing process mean (μ) toward specification nominal (N) and reduce process variability (σ). Another advantage of using DoE is to determine which factors in design or manufacturing have the greatest effect or even no effect on process mean or variability.

These relationships can be shown in Fig. 1.22. It shows an example of DoE with four factors to shift the outcome of a design or reduce variability in manufacturing. Several contributions of each factor are shown in Fig. 1.22 to illustrate different outcomes: factor A affects mean, while factor B affects variability, factor C affects both mean and variability, and factor D has no affect at all. This type of information gained from DoE can help guide the product and process design teams to select the best factor level to meet their requirements. The objective of a typical DoE for use in achieving Six Sigma or high Cpk goals could be:

1. Collect maximum information using minimum resources.
2. Determine which factors affect the mean, variability, and both mean and variability or have no effect on system quality responses.
3. Select factor levels that optimize the quality outcomes.
4. Validate (confirm) DoE suggested factor levels to optimize quality responses.

DoE has been constantly used to meld new materials and technology with current companies' practices, skills, equipment, and processes. It builds on traditional qualitative tools and further enhances quality. It is a quick and efficient methodology to reduce time and effort to reach Six Sigma and world-class quality levels.

1.4   Control Charts and Their Relationship with DoE

Control charts have been traditionally used as the method of determining the performance of manufacturing processes over time. They represent the statistical characterization of a measured parameter which is dependent on the process. They have been used effectively to determine if the manufacturing process is in statistical control, which implies continued adherence to specifications. Control exists when the occurrence of events (failures or defects) follows statistical behavior of the distribution from which the sample was taken.

Control charts are run charts with a centerline drawn at the manufacturing process mean and control limit lines drawn at the three sample standard deviations ($\pm3s$) of the production samples taken. If the manufacturing process is under statistical control, the ($\pm3s$) level or 99.73% of all observations are within the total span of the process. Control charts do not improve quality. They merely indicate that quality is in statistical "synchronization" with the level at the time when the charts were created.

There are two major types of control charts: variable charts, which plot continuous data from observed parameters, and attribute charts, which are discrete and plot reject or percent defect data.

1. Variable control charts are known as $\overline{X}$ and R charts for high volume and moving range (MR) charts for low volume. They measure parts directly and can be compared to their original performance when the control charts were created. A control chart has no relationship to part specifications, and a part could be in statistical control (within control limits) yet fail to meet their design specifications. Process capability and Six Sigma relate manufacturing part control charts directly to part specifications. Control charts are based on sample calculations, while Cpk and Six Sigma are based on population statistics. Every point on a variable control chart is a sample mean $\overline{X}$, and not an individual measurement.

   Control chart centerline is the grand average $X_{\text{doublebar}}$ or $\overline{\overline{X}}$ of sample means $\overline{X}$ s, and its control limits are at $\pm 3s$. The grand average $\overline{\overline{X}}$ and sample standard deviation (s) are related to the population standard deviation (σ) through the central limit theorem shown in Eq. 1.8. The control limits (UCL and LCL) are not related to the specification limits (USL and LSL).

2. Attribute charts are based on actual discrete defects or percent defective. There are four types of attribute charts, P and nP charts and C and U charts. The P and nP charts are based on the binomial distribution, while the C and U charts are based on the Poisson distribution. Each pair differs by the use of constant or variable samples.

   Attribute charts are indirect measures of quality, since they are determined by the number of parts not meeting specifications. Attribute sample sizes are selected based on defect rates, either as percentage or bound by time or space. Unlike the variable control charts, there is no direct mathematical relationship to population statistics such as μ and σ. Attribute chart means, presented by $\overline{p}$ and $\overline{c}$, can be made equal to defect rates and related to Six Sigma and Cpk as mentioned earlier. Converting attribute defect rates to Cpk requires making certain assumption about design specifications as shown in Sect. 1.3.1.

1.4.1   Selection of Control Charts

The selection of the parameters (factors) to be control charted is an important part of the quality monitoring process. Too many parameters plotted tend to adversely confuse the beneficial effect of control charts, since they will all move in the same direction when the process is out of control. The parameters selected for control charting should be independent from each other and directly related to the overall performance of the product. These same factors can be used in DoE to improve quality.

When introducing control charts into a manufacturing operation, it is beneficial to use factors that are universally recognized such as temperature and relative humidity in a controlled manufacturing environment or readings taken from a process display monitor. Production operators should participate in the charting process to increase their awareness and get them involved in the quality output of their jobs. Several shortcomings have been observed when initially using control charts; some of these to avoid are:

- Improper training of the production operators: collecting a daily sample and calculating the mean and range of the sample data might seem a simple enough task. Some data collection uses complex instruments or measurement systems such as micrometers. If the skill set of some operators is not adequate, then extensive training must be given to make sure the manufacturing operators can successfully perform required data collection and calculations.
- Using software programs for plotting data removes the focus away from data collection and manipulation of control charting. The issues of training and operating software tools become primary factors. Automatic means of plotting control charting should be introduced later in the quality improvement plan for manufacturing.
- Selecting factors that are outside of the production group's direct influence or are difficult or impossible to control could result in a negative perception of the quality effort. An example would be to plot the temperature and humidity of the production floor when there are no adequate environmental controls. Change in seasons brings an "out-of-control" condition that cannot be remedied with operator training. Using DoE to select more environmentally tolerant material in conjunction with control charting of temperature and humidity is more appropriate.

There are many types of control charts, as shown in Fig. 1.23, to be discussed in the next sections. In the latter stage of world-class quality implementation, very low defect rates impact the use of these charts. In many cases, successful implementation of Six Sigma and high process capability has rendered control charts obsolete, and the factory might switch over to TQM tools for maintaining the quality level at $Cpk = 4$ (64 ppm) or Six Sigma 3.4 ppm. Defect rates could become so low that only few defects occur in a single production day. Operators and engineers can pay attention to individual defects, rather than sampling plans of control charts.

1.4.2 Variable Control Charts

Variable processes are those where direct measurements can be made of the quality characteristic in a periodic or daily sample. The daily samples are then compared with a historical record to see if the manufacturing process for the part is in control.

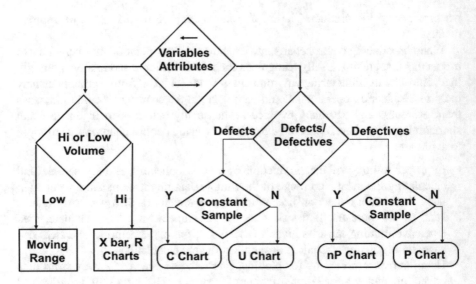

Fig. 1.23 Types of control charts

In $\overline{X}$ and R charts, the sample measurements taken every day are expected to fall within the $\pm 3s$ standard deviations of the distribution of sample means taken in the past. In moving range (MR) charts, the sample is compared with the 3σ of the population standard deviation derived from an $\overline{R}$ estimator of σ. When a sample range falls outside of the $\pm 3s$ limits of $\overline{R}$, the process is declared out of control, and corrective action is needed to restore process control.

Another type of charting for quality in production is the pre-control charts. These charts directly compare the daily measurements to the part specifications, as opposed to historical precedence. They require operators to make periodic measurements, before the start of each shift and then at selected time intervals afterward. They require the operator to adjust the production machines if the measurements fall outside a green zone halfway between the nominal and the specification limits.

Pre-control charts ignore the natural distribution of process or machine variability. Instead, they require a higher level of operator skills and intervention in manufacturing to ensure that production distribution is within halfway of the specification limits. This is in direct opposition to basic quality concepts of analyzing and matching the process distribution to the specification limits in the design phase, thus removing the need to do so every time parts are produced.

In high-volume manufacturing, several measurements can be taken daily for production samples. $\overline{X}$ and R variable control charts are used to monitor sample means and their variability. It is important to note that $\overline{X}$ control charts are derived from the distribution of sample means, which is always normal, regardless of the parent distribution of the population. Population statistics are used for process capability and Six Sigma calculations of the defect rate and might not always be

Table 1.8 Control chart factors

Observations subgroup n	A_2 factor for X_{bar} chart	Lower control R_{bar} limit D_3	Upper control R_{bar} limit D_4	$d_2 = R_{bar}/\sigma$
2	1.88	0	3.27	1.128
3	1.02	0	2.57	1.693
4	0.73	0	2.28	2.059
5	0.58	0	2.11	2.326
6	0.48	0	2.00	2.534
7	0.42	0.08	1.92	2.704
8	0.37	0.14	1.86	2.847
9	0.34	0.18	1.82	2.970
10	0.31	0.22	1.78	3.078
11	0.29	0.26	1.74	3.173
12	0.27	0.28	1.72	3.258
13	0.25	0.31	1.69	3.336
14	0.24	0.33	1.67	3.407
15	0.22	0.35	1.65	3.472
16	0.21	0.36	1.64	3.532
17	0.20	0.38	1.62	3.588
18	0.19	0.39	1.61	3.640
19	0.19	0.40	1.60	3.689
20	0.18	0.41	1.59	3.735

normal. Moving range (*MR*) charts are subsets of control charts used in low-volume applications.

The $\overline{X}$ chart indicates whether current manufacturing process mean is centered around a historical precedent. A moving trend in the plotted daily sample mean $\overline{X}$, whether up or down, indicates a probability of an out-of-control process condition. This could culminate in the daily sample mean being greater or lower than the control chart limits, triggering a corrective action. Possible causes for the $\overline{X}$ chart trends include faulty machine or process settings, improper operator training, or defective materials.

The R chart shows the uniformity or reduced variability of the manufacturing process. If the R chart span within their control limits is narrow, then the product manufacturing variability is uniform. If the R chart span is wide or out of control, then there is a nonuniform effect on the process, such as poor repair or maintenance records, untrained operators, and non-conforming materials.

The variable control charts are generated by taking a historical record of the manufacturing process over time. Walter Shewhart, the father of control charts, recommends, "statistical control cannot be reached until under the same conditions, not less than 25 samples of four each have been taken to satisfy the required criterion." These observations form the historical record of the process and serve as a baseline. All observations from now on are compared to this baseline. From these observations, the sample mean $\overline{X}$ and the sample range R, which are the absolute value of the highest to lowest measurement in the sample, are recorded. At the end of the observation period (minimum 25 samples of 4 each), the grand average of $\overline{X}$s, designated as $\overline{\overline{X}}$, and the mean of Rs, designated as $\overline{R}$, are used as the centerlines for the $\overline{X}$ and R variable control charts.

The $\pm 3s$ limits for the control charts are then calculated using an approximation of standard variation estimators d_2 for control chart factors, shown in Table 1.8. The control chart factors were designated with variables such as A_2, D_3, and D_4 to calculate the control limits of $\overline{X}$ and R control charts. Factor d_2 is important in linking the mean range $\overline{R}$ to the population standard deviation (σ). Control chart limits are $\pm 3s$ limits of the sample mean distribution.

$\overline{X}$ and $\overline{R}$ Chart Control Limits ($\pm 3s$ Limits)

$$\text{Upper control limit } (UCL_X) = \overline{\overline{X}} + 3s = \overline{\overline{X}} + A_2 * \overline{R} \qquad (1.9)$$

$$\text{Lower control limit } (LCL_X) = \overline{\overline{X}} - 3s = \overline{\overline{X}} - A_2 * \overline{R} \qquad (1.10)$$

$\overline{R}$ Control Limits

$$\text{Upper control limit } (UCL_R) = D_4 * \overline{R} \qquad (1.11)$$

$$\text{Lower control limit } (LCL_R) = D_3 * \overline{R} \qquad (1.12)$$

where:

$\overline{X}$ = mean of n observation in a subgroup
$\overline{\overline{X}}$ = grand average of all $\overline{X}$s
$\overline{R}$ = mean range of n observation in a subgroup (highest to lowest value)
A_2 = factor for $\overline{X}$ chart
D_3 = lower control limit factor for R chart
D_4 = upper control limit factor for R chart
d_2 = estimator for σ based on range of samples

1.4.3   Relationship of Control Limits to Specification Limits

Variable control chart limits (UCL, LCL) are independent from specification limits (USL, LSL). Control limits are based on the distribution of sample means (s) related to manufacturing variability, while specification limits are related to population distribution (σ) of parts and based on design parameters. It is desirable to have the specification limits as large as possible when compared to the process control limit.

Multiplying 1/3 the distance from the centerline $\overline{\overline{X}}$ of the $\overline{X}$ and R chart to one of the control limits by $\sqrt{n}$ will determine the population deviation σ, based on Eq. 1.8. A simpler approximation is the use of the formula $\sigma = \overline{R}/d_2$ in control charts to generate σ directly from the control chart data. Factor d_2 can be used as a good estimator for σ when using small number of sample sizes n (less than 20) and their average range $\overline{R}$.

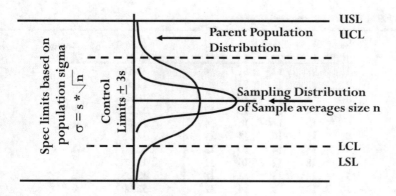

Fig. 1.24 Relationship of control limits to specification limits

The relationship of control limits to specification limits is shown in Fig. 1.24, using the following notations:

1. The variable control charts use the grand sample average $\overline{\overline{X}}$ and sample standard deviation (s), which are the responsibility of the manufacturing engineers and operator teams.
2. Design specifications and their limits (N, USL, LSL) are defined by the design team and could be multiples of σ (4σ or 6σ design). They are not related to the control charts and their limits ($\overline{\overline{X}}$, UCL, LCL).
3. Six Sigma and Cpk, which are management tools, seek to relate the two limits, by forcing design and manufacturing teams to communicate together to achieve overall quality goals. This is accomplished by connecting the design specification to the population (μ) and (σ) through the formulas of Cp and Cpk shown in Eqs. 1.4 and 1.5.
4. Grand average of sample means $\overline{\overline{X}}$ = parent population mean μ of production parts.
5. Standard deviation (s) of production sample means = σ of parent population/$\sqrt{n}$.
6. Variable control chart limits (UCL, LCL) = $\overline{\overline{X}} \pm 3s$ or = $\overline{\overline{X}} \pm A_2 * \overline{R}$ ($d_2 = \overline{R}/\sigma$ from Table 1.8).

$\overline{X}$ and R Variable Control Chart Calculation Example

In this example, a critical dimension for a part is measured as it is being inspected in a machining operation. To set up the variable control chart, four measurements were taken every day for 25 successive days, to approximate the daily production variability. These measurements are then used to calculate $\overline{X}$ and R variable control chart limits. The measurements are shown in Table 1.9. It should be noted that $n = 4$ is the number of observations in each sample. There are 25 sets of subgroups or samples, and if they are taken daily, they represent approximately a 1-month history of production.

Table 1.9 Control chart calculation example

Sample no.	Parts 1	2	3	4	Average $\overline{X}$	Range R
1	9	12	11	14	11.50	5
2	13	16	12	9	12.50	7
3	11	11	10	9	10.25	2
4	14	11	12	12	12.25	3
5	12	14	16	14	14.00	4
6	19	10	13	15	14.25	9
7	13	14	10	13	12.50	4
8	18	11	14	11	13.50	7
9	13	13	11	12	12.25	2
10	12	10	14	12	12.00	4
11	13	10	14	17	13.50	7
12	13	15	10	10	12.00	5
13	16	10	10	11	11.75	6
14	15	15	13	14	14.25	2
15	16	10	14	15	13.75	6
16	12	11	14	9	11.50	5
17	14	10	13	11	12.00	4
18	11	16	13	14	13.50	5
19	12	10	12	13	11.75	3
20	13	10	10	11	11.00	3
21	14	14	10	13	12.75	4
22	13	13	9	10	11.25	4
23	13	13	13	17	14.00	4
24	15	12	15	13	13.75	3
25	15	12	15	13	13.75	3
Totals					315.50	111

During the first day, 4 samples were taken, measuring 9, 12, 11, and 14 thousands of an inch. These were recorded in the top of the four columns of samples. The mean, or $\overline{X}$, was calculated and entered in the fifth column, and the range or R is entered in the sixth column.

$\overline{X}$ sample 1 = (9 + 12 + 11 + 14)/4 = 11.50

The range R is calculated by taking the highest reading for sample one (14) minus the lowest reading (9).

R sample 1 = 14 − 9 = 5

The means of $\overline{X}$ and R are calculated by dividing the column totals of $\overline{X}$ and R by the number of subgroups:

$\overline{\overline{X}}$ = (sum of $\overline{X}$ s)/number of subgroups
$\overline{\overline{X}}$ = 315.50/25 = 12.62
$\overline{R}$ = (sum of Rs)/number of subgroups
$\overline{R}$ = 111/25 = 4.44

Using the control chart in Table 1.8, the control limits can be calculated, using $n = 4$.

$\overline{X}$ Control Limits

$$\text{UCL}_x = \overline{\overline{X}} + A2 * \overline{R} = 12.62 + 0.73 * 4.44 = 15.86$$
$$\text{UCL}_x = \overline{\overline{X}} - A2 * \overline{R} = 12.62 - 0.73 * 4.44 = 9.38$$

$\overline{R}$ Control Limits

$$\text{Upper control limit (UCL}_R) = D_4 * \overline{R} = 2.28 * 4.44 = 9.38$$
$$\text{Lower control limit (LCL}_R) = D_3 * \overline{R} = 0$$

Since the measurements were recorded in thousands of an inch, the centerline of the $\overline{X}$ control chart is $0.01262''$, and the control limits for $\overline{X}$ are $0.01586''$ and $0.00938''$. For the R chart, the centerline is set at 0.00444, and the limits are 0.01012 and 0.

These numbers form the control limits of the $\overline{X}$ and R chart. After the limits have been calculated, the control chart is ready for use in production. Each production day, four readings of the part dimension are taken by the responsible operators, with the mean of the four readings plotted on the $\overline{X}$ chart and the range or difference between the highest and lowest sample readings to be plotted on the R chart. The daily numbers of $\overline{X}$ and R should plot within the control limits. If they plot outside the limits, the production process is out of control, and immediate corrective action should be initiated.

An alternate method for calculating control limits is to use the control limits as $+3s$ ($3*$ standard deviation of the sample distribution s). Sample standard deviation (s) can be calculated from the population standard deviation (σ) using the factor d_2 from Table 1.8 and Eq. 1.8.

$$\sigma = \overline{R}/d_2 = 4.64/2.059 = 2.156$$
$$s = \sigma/\sqrt{n} = 2.156/2 = 1.078$$
$$\pm 3 s = 1.078 * 3 = 3.23 \text{ which is close to } A_2 * \overline{R} \text{ value of } 3.24$$

If the total population of 100 numbers of Table 1.9 is considered, then the standard deviation is $\sigma = 2.156$, which is exactly the one predicted by the $\overline{R}$ estimator. If the specification limits are given, then the Cp, Cpk, and defect rates can be calculated as shown earlier in the chapter.

1.4.4 Variable Control Chart Usage Guidelines

Figures 1.8 and 1.25 are examples of $\overline{X}$ and R charts showing the solder paste thickness deposition process for an SMT (surface-mount technology) soldering process. Several observations can be made from examining these charts:

1. The two charts are measuring process mean and process variability, respectively. While one might be out of control, the other one is not or vice versa. This is due to the independence of the two attributes of mean and variability of the process.

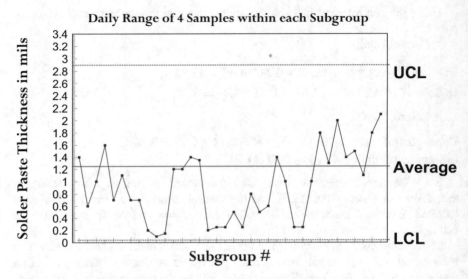

Fig. 1.25 R control chart example of solder paste thickness

2. The two charts are related mathematically since the distance from the $\overline{\overline{X}}$ to one of the control limits is equal to $\pm 3s$ or $A_2 * \overline{R}$. Figure 1.25 indicates $\overline{R} = 1.25$, and it can be multiplied by 0.73 (the A_2 factor for $n = 4$ from Table 1.8), resulting in 0.9125. This is the approximate distance from the $\overline{\overline{X}}$ (sometimes called the centerline or CL) to one of the control limits in the $\overline{X}$ chart in Fig. 1.8 (UCL $-$ $\overline{\overline{X}} = 11.6 - 10.7 = 0.9$).

3. The frequency of taking samples for control charts is left to the manufacturing process quality status. For high-quality processes, a daily sample for each shift is adequate to ensure conformance. For production lines with occasional or frequent quality problems, more sampling might be required, depending on the number of parts being produced or the number of hours since the last sample. This is necessary if reworking of parts is required when an out-of-control condition was detected. In this case, parts produced since the last good $\overline{X}$ plot on the chart must be reworked. In addition, the problem should be investigated by production engineers and possible causes recorded on the chart. Production engineers may require that more frequent samples be taken until the process is more stable. Some of these sampling plans can be found in MIL-STD-105E and its successors MIL-STD-1916 or ANSI/ASQ Z1.4, "Sampling Procedures and Tables for Inspection by Attributes."

4. The control limits should not be recalculated unless there is a change in the manufacturing process. Example could be new materials, machinery, operators, or quality improvement projects by undertaking a successful DoE.

5. When a chart shows an out-of-control condition, the process should be investigated and the reason for the problem identified on the chart. Figure 1.26 shows a

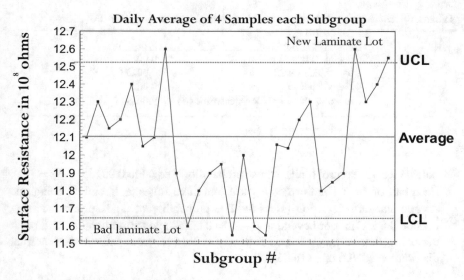

Fig. 1.26 Surface cleanliness measurement example

typical scenario of plotting a parameter such as the surface cleanliness resistance measurements in ohms. A downward trend of surface resistance began, and then successive out-of-control sample means $\overline{X}$ s were recorded. This out-of-control condition triggered a corrective action measure which determined that the cause was due to a defective laminate lot. A new lot resulted in significant increases in the resistance value, which would necessitate recalculating the control limits.

6. Another reason for recalculating control limits is the completion of a successful DoE that helps improve quality. An example is shown in Fig. 1.7, solder process defect rate, where substantial improvements in quality due to a DoE discussed later in the book would require recalculating the control chart limits.

7. In the $\overline{X}$ chart, the upper and lower control limits are usually symmetrical around the $\overline{X}$ or the centerline, such as shown in Fig. 1.8. In the case of only a single specification (USL or LSL), only one corresponding control limit is sufficient. In the $\overline{R}$ chart, symmetry is not necessary when the sample size is less than seven, since D_3, the control factor for the LCL_R, is equal to zero.

8. Many manufacturing sites added additional information such as the specification limits and then calculated the Cp or Cpk on the control charts. This can easily be done by deriving σ either from the $\overline{R}$ or (s) calculations, using the formulas $\sigma = s *$ $\sqrt{n}$ and $\sigma = \overline{R}/d_2$.

9. The most common indicator of out-of-control condition is that one sample mean is plotted outside the $\overline{X}$ chart control limits or one sample range is outside the R chart control limits. If these observations are confined to the upper or lower portion of the chart, then many other indicators of out-of-control condition can be used as well. These indicators have a probability approaching that of one $\overline{X}$ point

Table 1.10 Equivalent probabilities for out-of-control conditions in X_{bar} control charts

Upper half of the chart	Probability
One point above 3 σ line	0.00135
2 out of 3 above 2 σ line	0.00153
4 out of 5 above 1 σ line	0.00277
8 in a row above or below centerline ($\bar{\bar{X}}$)	0.00391

outside the $\pm 3s$ control limits, whose probability is equal to 0.00135 or $f(z = -3)$. Each half of the $\bar{X}$ chart can be divided into three zones (a, b, and c), being one sample standard deviation (s) wide. The probability of an $\bar{X}$ point occurring outside the 2 s limit or beyond is $f(-2) = 0.0228$, and the probability of $\bar{X}$ point occurring outside the 1 s limit is $f(-1) = 0.1587$, from the standard normal distribution (SND) or z (Table 1.7).

The probability of multiple $\bar{X}$ points occurring in succession might be approximately equal to that of the one point outside the $\pm 3s$ control limits. Several conditions might trigger an out-of-control corrective action as shown in Table 1.10. Sample probability calculations are given below:

1. Two out of three sample means above $2 * s$ in zone a = four combinations:
    Three points above + three sets of (above, above, below) + (above, below, above) + (below, above, above).
    $P(>2s)^3 + 3[P(>2s)]^2 * [1 - P(>2s)] = 0.02275^3 + 3 * [0.02275^2 * (1 - 0.02275)] = 0.00153$
2. Four out of five sample means above $1 * s$ in zone b = six combinations:
    Five points above + five sets of (above, above, above, above, below) + (above, above, above, below, above) + (above, above, below, above, above) + (above, below, above, above, above) + (below, above, above, above, above)
    $P(>2s)^5 + 5[P(>2s)]^4 * [1 - P(>2s)] = 0.15866^5 + 5* [0.15866^4 * (1 - 0.15866)] = 0.00277$
3. Eight in row above or below centerline ($\bar{X}$) in zone $c = 0.5^8 = 0.00391$

These three additional triggers can serve as predictors of out-of-control quality conditions, in addition to the single $\bar{X}$ point above or below the control limits.

Example of Variable Control Charts' Relationship to Process Capability

A company's quality team was sent to audit a supplier plant making their parts at 8 ± 3 inches. They read the $\bar{X}$ and R variable part control chart with $n = 4$ and $\bar{X} = 8.1$; $UCL_x = 11.1$ and $LCL_x = 5.1$. What is the process capability for the populations of parts delivered to the company?

Variable Control Chart Solution

From the $\overline{X}$ control chart, $3s = 3$; $s = 1$; $\sigma = s * \sqrt{n} = 2$; mean shift ($\mu$ to N) $= 0.1$
$Cp = 3/(3*2) = 0.5$
$Cpk = $Minimum$\left\{\frac{11-8.1}{3*2}; \frac{8.1-5}{3*2}\right\} = $Minimum $\{0.48; 0.52\} = 0.48$
$Z_1 = (5 - 8.1)/2 = -1.55$; $f(z_1) = 0.06057$
$Z_2 = (11 - 8.1)/2 = 1.45$; $f(-z_2) = 0.07353$
Total rejects $= 0.06057 + 0.07353 = 0.1341 = 13.41\% = 134{,}100$ ppm

1.4.5   Control Charting and Process Capability for Low-Volume Production

Reduced data points can be used successfully to estimate process capability when large data is not feasible to collect, because of cost issues or production volume. While 30 data points are considered statistically significant, a smaller number can be taken, using predetermined error levels and confidence goals to obtain good estimation of process capability.

The moving range methodology allows for a reasonable estimate of process capability for variable control charts. A minimum of ten data points is required to provide a d_2 estimator for σ using moving range method in Table 1.8. It uses individual measurements or defect rates over a representative period. The moving range charts can also be used for attribute charts. Instead of counting defects, the time between defects can be counted and entered as the variable in the chart.

The moving range stands for the difference between successive pairs of numbers in a series of numbers. The absolute value of the difference is used, creating a new set of range numbers, each with two successive numbers. The number of differences or "ranges" is one less than the individual numbers in the series. The chart is built up from the following:

1. The centerline $\overline{MR}$ of the moving range control chart is the mean of all the individual measurements.
2. The individual MR chart control limits can be obtained using the same methodology as variable control charts for individual measurements where the sample standard deviation is assumed to equal the population standard deviations for a sample of one. The standard deviation is determined by the d_2 factor from Table 1.8 for $n = 2$ (1.128). The control limits are calculated from the three standard deviations assuming that the individual measurement control limit is based on the population σ.
   $\overline{X}$ control limits $= \overline{\overline{X}} + 3\sigma$ ($=$ s for $n = 1$) $= \overline{\overline{X}} \pm 3 * [\overline{R}/d_2 \ _{(n=2)} = 1.128] = \overline{\overline{X}} \pm 2.66 * \overline{R}$
3. The mean of the ranges of the successive numbers is called $\overline{MR}$. The control limits are set by the same method for variable control charts: multiplying $\overline{MR}$ with factors D_4 and D_3 for upper and lower control limits for $n = 2$.
   $\overline{MR}$ upper control limits $= \overline{MR} * D_3 \ _{(n \ = \ 2) \ = \ 3.27}$
   $\overline{MR}$ lower control limits $= \overline{MR} * D_4 \ _{(n \ = \ 2) \ = \ 0}$

Examples of Moving Range Calculations

Example 1 Moving Range: Days between defects were counted as a measure of the quality for a manufacturing process. They occurred on the following production calendar days: 23, 45, 98, 123, 154, 167, 189, 232, 287, 311, and 340. Calculate the data for the moving range chart for days between defects.

Days between defects: 22, 53, 25, 31, 13, 22, 43, 55, 24, 29; mean $(\overline{X}) = 31.70$
Moving ranges (Rs): 31, 28, 6, 18, 9, 21, 12, 31, 5; $= 17.89 = \overline{MR}$
$\overline{X}$ control limits $= \overline{\overline{X}} \pm 2.66 * \overline{R} = 31.70 \pm 2.66 * 17.89 = 31.70 \pm 47.58$
UCL $= 79.28$ and LCL $= 0$ since there cannot be a negative control limit.
MR chart control limits: $UCL_{MR} = D_{4\ (n=2)} * \overline{R} = 3.27 * 17.89 = 58.49$; $LCL_{MR} = 0$
Another method to plot this defect data would be defects/month, by dividing the data
   by 30.

Example 2 Moving Range and Resulting Cp/Cpk: Fuses are made in a production line, with specifications of 5 ± 2 ohms. A sample of six fuses' measurements were taken at 3, 6, 6, 4, 5, and 5 ohms. If it is desired to have an $\overline{X}$ and R control charts, what is the quality data (control charts, defect rates, and process capability) for the fuse line?

Moving range method data $= 3\ 0\ 2\ 1\ 0$
Mean $\overline{X} = 4.83$; $\overline{MR} = 1.2$
$UCL_x = 4.83 + 2.66 * \overline{MR} = 8.03$
$LCL_x = 4.83 - 2.66 * \overline{MR} = 1.64$
$UCL_{MR} = D_{4\ (n=2)} * \overline{MR} = 3.27 * 1.2 = 3.92$
$LCL_{MR} = 0$
$\sigma = \frac{1.2}{d_{2(n=2)}} = \frac{1.2}{1.128} = 1.06$
$Cp = \frac{\pm SL}{\pm 3\sigma} = \frac{\pm 2}{3*1.06} = 0.63$

$$Cpk = \text{Minimum}\left\{\frac{USL - \mu}{3\sigma}; \frac{\mu - LSL}{3\sigma}\right\} = \text{Minimum}\left\{\frac{7 - 4.83}{3 * 1.06}; \frac{4.83 - 3}{3 * 1.06}\right\}$$

$$= \text{Minimum}\ \{0.68; 0.57\}$$

$Cpk = 0.57$
$z_1 = \frac{USL - \mu}{\sigma} = \frac{7 - 4.83}{1.06} = 2.04 \rightarrow f(-z_1) = 0.02068$
$z_2 = \frac{\mu - LSL}{\sigma} = \frac{4.83 - 3}{1.06} = 1.72 \rightarrow f(-z_2) = 0.04272$
Defect rate (RR) $= 0.04272 + 0.02068 = 0.0634$ or 6.34% or 63,400 ppm.

1.4.6   *Attribute Control Charts*

Attribute control charts directly measure rejects in production operations, as opposed to measuring a particular value of the quality characteristic as in variable control charts. They were more common in manufacturing because of the following:

1. Attribute or pass-fail test data are easier to measure than actual variable measurement. They can be aided by devices or tools such as go/no-go gages or calibrated for only the specification measurements, as opposed to measuring the full operating spectrum of parts.
2. Attribute data require much less training by operators, since they only must observe a reject indicator or red light, as opposed to making several measurements on gages or test equipment.
3. Attribute data can be directly collected from the manufacturing equipment, especially if there is a high degree of automation.
4. Storage and dissemination of attribute data is also much easier since there is only the reject rate to store versus the actual measurements for variable data.

Attribute charts use different probability distributions than the normal distribution used in variable charts, depending on whether the sample size is constant or changing, as shown in Fig. 1.23. For C and U charts, the Poisson distribution is used, while the P and nP charts use the binomial distribution. Individually different control limits for each variable sample taken must be calculated in P and U charts, while a single control limit value is used in C and nP charts with constant sample size.

All attribute control charts follow the same ($\pm 3s$) sample standard deviation control limits away from centerline methodology used in variable control charts.

For constant samples (C or nP charts)

For Poisson distribution C charts:

$$\text{Centerline} = \text{mean } \lambda \text{ or } \overline{C}; s = \sqrt{\lambda} = \sqrt{\overline{C}} \text{ and } \text{CL}_C = \overline{C} \pm 3\sqrt{\overline{C}} \qquad (1.13)$$

For binomial distribution $n\overline{P}$ charts:

$$\text{Centerline} = \text{mean } n\overline{P}; s = \sqrt{n\overline{P}(1 - \overline{P})} \text{ and } \text{CL}_{nP} = n\overline{P} \pm 3s$$

$$= n\overline{P} \pm 3\sqrt{n\overline{P}(1 - \overline{P})} \qquad (1.14)$$

where:

$\overline{C}$ = mean of defects in time or space
$n\overline{P}$ = number of defectives found in each constant sample n
n = number of units in sample

For changing sample sizes (U or P charts)
For Poisson distribution U charts:

$$\text{Centerline} = \text{mean } \overline{U} = \frac{\sum C_i}{\sum n_i}; s = \sqrt{\frac{\overline{U}}{n_i}} \text{ and } \text{CL}_U = \overline{U} \pm 3s$$

$$= \overline{U} \pm 3 \sqrt{\frac{\overline{U}}{n_i}} \tag{1.15}$$

where n_i is the number of the sample for each U measurement.
For binomial distribution P charts:

$$\overline{P} = \frac{\sum P_i}{\sum n_i}, s = \sqrt{\frac{\overline{P}(1 - \overline{P})}{n_i}}; \text{ and } \text{CL}_P = \overline{P} \pm 3s = \overline{P} \pm 3 \sqrt{\frac{\overline{P}(1 - \overline{P})}{n_i}} \tag{1.16}$$

where:

U = mean number of defects in a sample
P_i = fraction or % defective = number of defective units in sample n_i

The relationship of attribute control charts to process capability is through the defects implied in the charts, with defects being generated by having parts out of specification, though the specifications are not known from the attribute control charts. They are assumed to be two-sided with no mean shift of process mean to specification nominal. The centerline represents the mean defect rate, and the control limits are ($\pm 3s$) distance from the centerline. These defect rates can be translated into an implied Cpk as shown in previous sections.

Example of Attribute Control Chart Calculations and Relationship with Process Capability

100 amp fuses are tested in sample lots of 100 and defectives are found to be 1%. To control the quality, the company takes an hourly sample of 100 fuses and counts the number of defectives. Calculate the Cp, Cpk, population σ, and control limits using an nP chart for constant samples. Assume that the fuses were made in a normally distributed manufacturing process and that the process is centered with no mean shift (process mean μ = specification nominal N) and has a two-sided defect distribution.

Example of Variable Control Chart Solution

One-sided defective rate = $0.01/2 = 0.005 = f(z)$; therefore, $z = 2.575$ from Table 1.7

$Cp = Cpk = z/3 = 0.86 = \pm \text{ SL}/3\sigma$

Using the formula 1.14 for the nP charts:

$$s = \sqrt{n\overline{P}(1 - \overline{P})} = \sqrt{1(0.99)} = 0.995$$
$$\text{UCL}_{nP} = n\overline{P} \pm 3s = 1 + 3*0.995 = 3.985$$
$$\text{LCL}_{nP} = n\overline{P} \pm 3s = 1 - 3*0.995 = 0 \text{ (defects cannot be negative).}$$

In this example, the specification limits were not given, yet the implied Cp and Cpk can be calculated. If a process mean shift to the specification limits is given (such as 1.5σ), it is still possible to calculate Cpk, if it is assumed that the rejects are mostly generated by one side of the distribution. The sample standard deviation (s) for the nP chart is based on the defective rate and has no relationship to the population σ, which is unknown, since no specification limits were given.

Several techniques must be made in the case of the attribute chart connections to quality and process capability:

1. For P and U charts with variable sample sizes, the centerline is measured as defective fraction or percent, and control limits are calculated individually for each sample. For nP and C charts with constant sample sizes, the centerline is measured as integer defects, and control limits are calculated as one value for the chart, ($\pm 3s$) away from the centerline.
2. There is one or a complex set of specifications that are not readily discernable which govern the manufacturing process for the parts and cause defective parts to be made.
3. These specifications are either one- or two-sided, resulting in one- or two-sided defects (defects <LSL and defects >USL).
4. The manufacturing process is assumed to be normally distributed.
5. There is a relationship between the process mean and the specification nominal. In some definitions of process capability, an assumption is made that there is a $\pm 1.5\sigma$ shift from process mean to specification nominal.
6. The control limits of the attribute charts are calculated as ($\pm 3s$), where (s) is the defective sample standard deviation and is not related to the population distribution (σ). The population standard deviation (σ) calculations are different than used in variable control charts. It is indirectly derived from the conversion of defect rates to process capability without any knowledge of the specification limits.

1.4.7   Use of Control Charts in Factories with High Process Capability

The C chart is the most widely used chart in factories that are approaching high process capability such as Six Sigma. Since the defect rates are very low, binomial-based control charts (P and nP) require a very large sample size and hence are impractical to use. For Six Sigma quality, a defect rate of 3.4 PPM would result in an *nP* chart with a centerline probability of 0.0000034. Such a chart would require a very large sample to determine if the process indeed has gone out of control.

Using *C* charts with well-defined areas of opportunity, such as defects per shift or defects per 10,000 units, can be effective for monitoring quality control in high process capability production. In some factories, the discussion has shifted to the number of possibilities of defects or the number of opportunities. Some industries, such as electronics, have defined a new *C* chart metric called the DPMO (defects per million opportunities) chart.

A more realistic way to achieve quality control in factories that approach Six Sigma is to closely couple the total defect reporting to continuous quality improvement actions. The low defect rate produces a small number of total defects per day, even in a large factory. For example, if we assume that a factory produces 5000 ICs per day, and each IC requires 2000 operations, that is a total defect opportunity of 10 million operations per day. For Six Sigma defect rate of 3.4 defects per millions, the total expected defects amount to 34 defects. The management of the factory can review these defects individually each day and then decide what corrective action is needed, whether immediate, short, or long term. They can use the tools of TQM and DMIAC to monitor, organize, and rank defects and initiate a corrective action plan for continuous quality improvements.

1.4.8   DoE Effects on Control Charts

Control charts are methodologies to maintain quality, but they do not improve quality. To have a meaningful control charting program, the principles of quality and its qualitative tools (DMIAC, TQM) must be embedded in the organization and sustained by vigorous management support and employee training. If a monitoring and maintaining system such as control charts constantly indicates out-of-control conditions, then it will be ignored or discontinued.

Two scenarios of the relationship of control charts to DoE can be used:

1. For existing processes, once qualitative quality methods are implemented, a statistical process control (SPC) phase should be launched with control charts to monitor process performance and ensure that quality issues can be detected and corrected promptly. This SPC phase should last a minimum of 6 months and go through at least one quality issue to instill confidence in the process and the team running it, as well as indicating that the process is stable and quality maintained regularly. A DoE project can then be initiated to increase the quality of the process to world-class levels. This sequence is illustrated in Fig. 1.7, solder process defect rate two-year progression.
2. For new processes and design, DoE should be initiated to determine optimum settings (levels) for design and process factors. SPC and control charts can then be used to maintain and ensure process capability parameters indicated by DoE.

1.5 Conclusions

The historical developments of the tools and concepts of qualitative quality methodologies such as TQM, DMIAC, and data charting, as well as quantitative methods of statistical distributions, process capability, Six Sigma, and control charts, were outlined in this chapter. The tools were discussed in terms of understanding how they were developed, relationships to each other, and how does DoE relate to these tools. The proper use of each tool was examined: whether each tool provides comparative analysis or maintains or improves quality. The proper sequence of when to use DoE in conjunction with these tools was also discussed and what role they play in creating more competitive global products. Each tool was discussed individually, including the governing equations and assumption that are commonly made when using the tool.

Many examples were given to demonstrate the use of the tools in typical small- and large-volume manufacturing, as well as sampling techniques and their relationship to populations. The process capability applications in short- versus long-term production were also shown, in conjunction with DoE, Six Sigma, and control charts.

Quantitative and charting analyses also serve as an introduction to common concepts of DoE, to demonstrate DoE and other quality tool outputs, as well as quality tools' triggers to initiate a DoE project. Common misconception of the proper use of these tools and typical errors that quality practitioners make when interpreting these tools are discussed with extensive explanation of the purpose and mathematical derivation basis of quality tools.

References

1. Shina, S., *Green Design and Manufacturing*, McGraw Hill professional series, April 2008

Additional Reading Material

Advanced Product Quality Planning and Control Plan (APQP), Southfield, MI: Automotive Industries Action Group (AIAG), 1995.

Bajaria, H., "Six Sigma-Moving Beyond the Hype," Annual Quality Congress Proceedings, ASQ, 1999.

Box, G., J. Hunter and Hunter, W., *Statistics for Experimenters: Design, Innovation, and Discovery*, 2nd Edition, New York: Wiley-Interscience, 2005.

Harry, M., *The Nature of Six Sigma Quality*, Rolling Meadows, IL: Motorola University, 1988.

Harry, M. and Schroeder R., *Six Sigma*, New York: Doubleday, 2000.

Hahn, G. et al., "The Evolution of Six Sigma," *Quality Engineering*, vol. 12, no. 3, pp. 317–326, 2000.

Iversen, W., "The Six Sigma Shootout," *Assembly Magazine*, June 1993, pp. 20-24.

Kemtovicz, R., *New Product Development: Design and Analysis*, New York: Wiley, 1992.

Liker, J., *The Toyota Way: 14 Management Principles from the World's Greatest Manufacturer*, 1st Edition. New York: McGraw-Hill, 2003.

Shina, S., *Engineering Project Management*, McGraw Hill professional series, December 2014

Shina S., "A Cpk-Based Toolkit for Tolerance Analysis and Design," Engineering Design Conference, UCL, London; July 2002.

Shina, S. and Saigal A. "Manufacturing Costs for electronic Products", Volume 3 of the Encyclopedia of Materials, Elesevier Press, November 2001, pp 2727-2735.

Shina, S. and Saigal A., "Using Cpk as a Design Tool for New System Development", Journal of Quality Engineering, Volume XII, Number 4, 2000, pp. 333-349

Shina S. and Saigal A., "Using Cpk as Design Tool for New System Development," International Conference on Engineering Design (ICED), Vol. 1, pp. 357-360, Munich, August 1999.

Shina S. and Saigal A., "Technology Cost Modeling for the Manufacture of Printed Circuit Boards in New Electronic Products," Journal of Manufacturing Science and Engineering, May 1998, pp 368- 375.

Shina, S. and Saigal, A., "A design Quality Based Cost Model for New Electronic Systems and Products," Journal of Materials, April 1998, pp 29-33.

Shina, S., *Successful Implementation of Concurrent Engineering Products and Processes,* New York: Wiley, 1993.

Shina, S., "The successful use of the Taguchi Method to Increase Manufacturing process Capability", Journal of Quality Engineering, Volume III, Number 3, 1991, pp. 333-349.

Shina, S., *Concurrent Engineering and Design for Manufacture of Electronic Products,* Norwell MA: Kluwer Academic Publishers, 1991.

Smith, B., "Six Sigma Quality, a Must not a Myth", *Machine Design*, February 12, 1993, pp 13-15.

Smith, P. and Reinertsen, D., *Developing Products in Half the Time*, 2nd Edition. New York: John Wiley and Sons, 1998.

Ulrich, K., and Eppinger, S., *Product Design and Development*, 2nd Edition, New York: McGraw-Hill, 2005.

Chapter 2
Samples and Populations: Statistical Tests for Significance of Mean and Variability

The use of statistical tests for significance of mean and variability are the building blocks of Design of Experiment (DoE) analysis. These tests compare the data from two different populations, or samples from populations, to determine whether they are significantly different in mean or variability. These tests assume that samples came from populations that are normally distributed. Testing for normality is discussed in this chapter as well as in Chap. 3.

Many different variants of the statistical tests presented in this chapter will be discussed, including special conditions of testing normal versus non-normal populations; testing the same samples with differing conditions, such as comparing the original samples before and after baking; as well as testing samples from the same or differing population variances.

This chapter is a prelude to Chap. 3, "Regression, Treatments, DoE Design, and Modelling Tools for Data Analysis." Chapter 3 presents multiple factors and levels of significance tests, using the tools presented in this chapter. The progression of significance testing begins with tests outlined in this chapter for comparing two samples from the same or different populations (one factor with two levels), proceeding to treatments comparing multiple levels of the same factor and then on to DoE which is testing significance of multiple factors with multiple levels to achieve desired results. The focus of DoE, in general, and this book, in particular, is to reduce the amount of testing, by using proven techniques of reduction of experiments while maximizing the information obtained from DoE analysis.

Most of these tests are available in common statistical tools, including Microsoft Excel© and Minitab© software packages, and are used to verify the examples in this chapter. Topics to be discussed in this chapter are as follows:

2.1. Sample and Population Statistics. The descriptive statistical terms of mean and variability are discussed with formulas and sample calculations. The distinction between samples and populations needs to be firmly established before any significance testing. The relationship of samples to populations is shown as well as methods to relate them through means and variability.

2.2. Tests for Samples and Populations. These are tests that can be performed to compare means and variability of samples and populations. There are several variants of these tests, depending on whether the parent population is normal or not and whether their variability is unknown but presumed equal or unequal. In addition, test accuracy for comparing sample means can be increased by using the same sample units subjected to different conditions.

These binary tests are called significance tests, since they compare the difference between the samples and populations to those expected according to the standard normal distribution (SND) or its derivatives, the t (Student's), X^2 (chi-square), and F distributions. In DoE terminology, they represent significance testing of two levels of a single factor, while DoE experiments are statistical significance tests for multiple factors and levels.

2.3. Tests for Means. These tests include the z-test for comparing population means to each other and comparing sample means to the parent populations, assuming that the populations are normally distributed. If not, then the Wilcoxon test is used, which uses z-tests for the mean through rank ordering of the two sample data points to extract a common variability.

The t-test is used when population variability is unknown and therefore uses the variability of the test samples. There are three types of t-tests: (1) the d-test, which is a reduced variation t-test, comparing two means when the same sample units are used in subsequent tests under different conditions; (2) when comparing two sample means from populations of unknown variability but presumed equal; and (3) when comparing two sample means from populations of unknown variability but presumed unequal.

Sample size (n) and confidence interval (CI) are used as reverse indicators for significance when an actual population mean is compared to the sample CI, with or without an estimate of the tolerated error. CI provides an easy answer to significance without having to use the SND or t distribution tables every time a sample is being tested for significance.

Tests for Variability. These tests include the X^2 (chi-square) test for comparing a sample variability to its parent population variability, assuming that the population is normally distributed. The F-test is used to compare sample-to-sample variation and is especially important in DoE statistical analysis to verify individual factor significance.

2.4. Conclusions

2.1 Sample and Population Statistics

Calculations based on the properties of samples can provide conclusions about the properties of populations. Some of these relationships between samples' and populations' properties were discussed in Chap. 1. This chapter will explore these relationships further using statistical techniques for testing of pairs of samples or populations using significance. The sample properties can be of two types:

1. Quantitative (variable) as in the diameter of shafts, all produced under replicating conditions in production
2. Qualitative (attribute) as in the set of all solder joints produced under the same soldering conditions

A sample of n shafts can be measured, and its properties of mean and variability can indicate the same properties for the population of all shafts. A solder joint can be either acceptable according to multiple guidelines contained in industry standards (such as IPC 610) or defective for not meeting these guidelines. Sample size (n) is defined as the random choice of n objects from a population, each independent of each other.

One of the relationships between sample and population data is control chart factors shown in Table 1.8, which approximates population standard deviation σ from mean sample range R_{bar} or $\bar{R}$ using the d_2 factor. Observation subgroup n value starts at $n = 2$ in Table 1.8, since one single observation cannot yield a standard deviation. It ends at $n = 20$, since a sample size of $n = 30$ is considered large enough to have the sample mean and standard deviation ($\bar{X}$ and s) approximately equal to population (μ and σ). For the number (n) in the sample size between 20 and 30, the approximation based on d_2 is not considered accurate.

As n approached an infinite (∞) number, sample attributes (mean and standard deviation) become population attributes. Rather than using infinite or very large samples for populations, statisticians prefer $n = 30$ as a definition of sample size (n) large enough to be considered as population. The error between $n = 30$ and $n = \infty$ is less than 5% for the common confidence intervals (CI) of 90% and 95%, as will be shown later in this chapter.

The formulas and definition of common statistical terms for mean (μ) and standard deviation (σ) for samples are shown below:

$$\text{Sample mean (average)} = \bar{X}(X_{bar}) = \sum \frac{X_i}{n} \tag{2.1}$$

$$\text{Sample standard deviation}(s) = \sqrt{\frac{\sum (X_i - \bar{X})^2}{n - 1}} = \sqrt{\frac{\sum X_i^2 - \frac{(\sum X_i)^2}{n}}{n - 1}}$$

$$= \sqrt{\frac{\sum X_i^2 - CF}{n - 1}} \tag{2.2}$$

where X_i is a data point, n is the number of data points in the sample, ($n - 1$) is the degrees of freedom (DOF), $\sum (X_i)^2$ is the sum of the squares (SS), and $\frac{(\sum X_i)^2}{n}$ is the correction factor (CF).

If all data points are symmetrical around 0 (mean $= 0$), then CF $= 0$ and $s = \sqrt{\frac{\sum (X_i)^2}{n-1}}$. The correction factor is simply the adjustment of the SS when the sample mean is not $= 0$. Table 2.1, statistical terms used in sample mean and

Table 2.1 Statistical terms used in sample mean and standard deviation common to DoE

1. $\overline{X} = X_{bar}$ = sample mean (average)
2. s = sample standard deviation
3. n = number of sample data points
4. DOF (degrees of freedom) = $n - 1$
5. MS (mean square) = variance = s^2
6. SS (sum of the squares) = variance (mean square or MS) * DOF
7. CF = correction factor = $\dfrac{\left(\sum X_i\right)^2}{n}$
8. SS = $\sqrt{\dfrac{\sum X_i^2 - CF}{n-1}}$

standard deviation common to DoE, summarizes the terms used for these sample formulas.

When these formulas are applied to the population, then the population mean is called μ, and the population standard deviation is called σ. Their relationships to sample data (X_{bar}, s) were defined in Eq. 1.8, where:

$$\overline{X} = X_{bar} = \mu \text{ and } s = \frac{\sigma}{\sqrt{n}}$$

There is a certain amount of error associated with equating sample mean $\overline{X}$ to population mean μ in Eq. 1.8. Obviously, this error decreases as the sample size (n) increases, and the opposite is also true. Section 2.3.7 "Determination of Sample Size Based on Error" offers a methodology to determine sample size (n) based on the planned margin of error. Inversely, the population mean estimation error from sample mean can be quantified based on the sample size (n).

Another dilemma occurs when using Eq. 2.2 for standard deviation (σ) for populations. The denominator of DOF = $n - 1$ should be changed to n for population standard deviation (σ). As n gets larger, the difference between the two denominators n and $n - 1$ does not significantly alter the value of σ. The most common calculators with statistical capabilities, such as the TI-30X©, show the symbol σ_{n-1} to indicate the formula used to calculate σ is based on DOF = $n - 1$. Similarly, Excel© "STDEV.S" function indicates the standard deviation of a sample using $n - 1$ method.

2.1.1 Standard Deviation Estimation Methodologies and Data Collection

Given the different formulas provided in the previous section, the population standard deviation (σ) can be estimated through several methodologies yielding different results. These methodologies can differ in the way data is collected or

Table 2.2 Summary of methods 1–3 for calculating mean and standard deviation of 32 data points

Subgroup	Measurements								Range (R)	Mean ($\bar{X}$)	s (subgroup)
1	4	5	5	4	8	4	3	7	5	5	1.69
2	2	4	3	7	5	4	2	5	5	4	1.69
3	3	6	6	4	5	4	6	6	3	5	1.20
4	6	2	4	3	8	4	2	3	6	4	2.07
					Mean subgroup				$\bar{R}=4.75$	$\bar{\bar{X}}=4.5$	

Method 1. Total overall variation (32 data points)

s (DOF = 31) = 1.6848; σ (DOF = 32) = 1.6583

Method 2. Within group variation

$\bar{R}=4.75$

$\sigma = \bar{R} / d_2\,(n{=}8) = 4.75 / 2.847 = 1.6684$

Method 3. Between group variation

$s_{samples} = s\,(\bar{X}) = s\,(5, 4, 5, 4) = 0.5774$

$\sigma_{population} = s_{samples} * \sqrt{n} = 0.5774 * \sqrt{8} = 1.633$

grouped. The data collection has to be defined in terms of separating the data into distinct groups in order to reduce the influence or bias due to outside factors, such as the timing of data collection, the effect of temperature or humidity when some of the data is collected, or the influence of the data collection by using separate operators or measurement equipment.

There are four different methods to determine the population standard deviation (σ), some of which are based on data analysis for control charts shown in Chap. 1. They are:

1. Total overall variation. All data is collected into one large group and treated as a single large sample where n is greater than 30. The σ is calculated from Eq. 2.2.
2. Within-group variation. Data is collected into subgroups, and a dispersion statistic R (range) is calculated for each subgroup. All ranges of each subgroup are averaged into an R_{bar} ($\bar{R}$). The σ is then calculated from an estimator (d_2) from Table 1.8. This method is the basis for variable chart control limit calculations and was discussed in Chap. 1.
3. Between-group variation. Data is collected into subgroups, and a sample mean X_{bar} ($\bar{X}$) is calculated for each subgroup. The standard deviation (s) of sample means is calculated. The population (σ) is estimated from the relationship $\sigma = s * \sqrt{n}$ as defined in Eq. 1.8. This method can also be used to obtain σ from control chart limits.

An example is shown in Table 2.2, with a total of 32 data points collected in 4 subgroups of 8 each. Each subgroup is analyzed individually with a range R and a mean $\bar{X}$ for each subgroup. The standard deviation of each subgroup (s) is also shown, as well as the mean range $\bar{R}$ and the grand mean $\bar{\bar{X}}$, which is also the average of the 32 data points. Three different values of σ are calculated using methods 1–3 with intermediate calculations provided.

Table 2.3 Moving range method standard deviation calculations for 32 data points

Subgroup		Measurements							Total	Moving range $\bar{R}$
1	4	5	5	4	8	4	3	7		
Ranges*	1	0	1	4	4	1	4		15	2.1429
2	2	4	3	7	5	4	2	5		
Ranges*	2	1	4	2	1	2	3		15	2.1429
3	3	6	6	4	5	4	6	6		
Ranges*	3	0	2	1	1	2	0		9	1.2857
4	6	2	4	3	8	4	2	3		
Ranges*	4	2	1	5	4	2	1		19	2.7143
									Mean range $\bar{R}$	2.0714

$\sigma = \bar{R} / d_2 = 2.0714 / 1.128$ (n = 2) = 1.8364
*Take pairwise numbers, find the range for each pair, using the middle number twice

4. Moving range (MR) method. This method was discussed in Chap. 1. Data is collected into one small group over time. A range R is calculated from each two successive points. All ranges of each pair are averaged into an R_{bar} ($\bar{R}$). The σ is calculated from an $\bar{R}$ estimator (d_2) for $n = 2$ which is equal to 1.128 (Table 1.8). This is the preferred method for time series data and small data sets from low-volume manufacturing. An example of using moving range method is shown in Table 2.3. Each subgroup is shown with pairwise ranges, and a moving range is calculated for each subgroup. A grand mean moving range $\bar{\bar{R}}$ is calculated and converted to $\sigma = 1.8364$ using $\bar{R}$ estimator (d_2) for $n = 2$ (1.128).

Using the four methods above yielded different values for σ (1.66, 1.67, 1.63, and 1.84). For processes that are in statistical control, these methods should be equivalent over time. For processes not in control, only method 2 (within-group variation) is not sensitive to process mean variations over time. These variations could be due to other factors such as operators, equipment, or data collection environment. The latter may include factors such as temperature, humidity, or pressure. The σ estimate is inflated or deflated with method 1 (total overall variation) and could be severally inflated/deflated with method 3 (between-group variation), since the subgroup sample size would exert a large influence on the conversion between sample (s) and population (σ) standard deviations. If the number of subgroup samples is smaller than $n = 8$, then it could lead to even larger error in σ determination.

The fourth method to calculate σ (moving range (MR) method) is sensitive to individual subgroup ranges. For example, the fourth subgroup has a larger MR value than the other three subgroups, leading to a larger population σ. In conclusion, a particular method for calculating σ should be selected only after considering the circumstances of the process being analyzed and its measurement.

2.1.2 Measurement System Error (GR&R) and Its Impact on Statistical Measurements

The error in the measurement system due to operator or equipment variation can be quantified and isolated from σ using either $\overline{X}$ and R techniques discussed in Chap. 1 or ANOVA to be discussed in Chap. 3 with a methodology called GR&R (gage repeatability and reproducibility). It is quantified by measuring (n) parts, with (k) operators (appraisers), and repeating each measurement in (r) trials. For example, ten parts can be measured by three operators $(k = 3)$, and each measurement is repeated two times $(r = 2)$ using the same equipment for a total of 60 (10 * 3 *2) measurements. The $\overline{R}$ (r) trial grand average $= \overline{\overline{R}}$ from the (k) operators is converted to the standard deviation of equipment variation (σ_{EV}). The maximum difference between operator means X_{diff} is similarly converted to the standard deviation of operator (appraiser) variation (σ_{AV}) by another factor called $d_2{}^*$ (shown in Table 2.4 for $n = 2$–10). The σ_{GRR} is equal to the root mean square of its two components:

$$\sigma_{GRR} = \sqrt{\sigma_{EV}^2 + \sigma_{AV}^2}$$

The total span of the GRR could be calculated as either six sigma $(6 * \sigma_{GRR})$ or $(5.15 * \sigma_{GRR})$. The two spans represent either 99.73% or 99% of parts' measurement. They represent $z = \pm 3$ or $z = \pm 2.575$, respectively, leaving a $f(z)$ of 0.00135 or 0.005 at either tail end of the z distribution.

The measured part variation standard deviation σ_{PV} is obtained by one of the four methods discussed above, preferably using method 2 by converting $\overline{R}$ to σ. It can be added to σ_{GRR} to obtain the total variation (TV) of the parts and the measurement system similar to the σ_{GRR} methodology above:

$$\sigma_{TV} = \sqrt{\sigma_{GRR}^2 + \sigma_{PV}^2} \text{ and TV span} = 6 * \sigma_{TV}$$

The GR&R span can be reduced with better test equipment and training of operators. It is recommended to use a GR&R study if the manufacturing process measurement error (TV span) is suspected to be greater than 10% of the specification

Table 2.4 GR&R example

Operator	A	A	A	A	B	B	B	B	C	C	C	C	Part	$d_2{}^*$
Sample	1st trial	2nd trial	Average	Range	1st trial	2nd trial	Average	Range	1st trial	2nd trial	Average	Range	Means	for 2 = 2-10
N2	0.65	0.6	0.625	0.05	0.55	0.55	0.55	0	0.5	0.55	0.525	0.05	0.5667	1.414
N3	1	1	1	0	1.05	0.95	1	0.1	1.05	1	1.025	0.05	1.0083	1.906
N4	0.85	0.8	0.825	0.05	0.8	0.75	0.775	0.05	0.8	0.8	0.8	0	0.8000	2.237
N5	0.85	0.95	0.9	0.1	0.8	0.75	0.775	0.05	0.8	0.8	0.8	0	0.8250	2.477
N6	0.55	0.45	0.5	0.1	0.4	0.4	0.4	0	0.45	0.5	0.475	0.05	0.4583	2.669
N7	1	1	1	0	1	1.05	1.025	0.05	1	1.05	1.025	0.05	1.0167	2.270
N8	0.95	0.95	0.95	0	0.95	0.9	0.925	0.05	0.95	0.95	0.95	0	0.9417	2.961
N9	0.85	0.8	0.825	0.05	0.75	0.7	0.725	0.05	0.8	0.8	0.8	0	0.7833	3.076
N10	1	1	1	0	1	0.95	0.975	0.05	1.05	1.05	1.05	0	1.0083	3.178
N11	0.6	0.7	0.65	0.1	0.55	0.5	0.525	0.05	0.85	0.8	0.825	0.05	0.6667	
Totals	8.3	8.25	8.275	0.45	7.85	7.5	7.675	0.45	8.25	8.3	8.275	0.25	0.5583	=Parts Xdiff
			Mean Ra =	0.0450			Mean Rb =	0.0450			Mean Rc =	0.0250		

span. While an $\overline{X}$- and R-based study provides percentage of various components of TV span, an ANOVA-based study will provide significance testing, whether in repeating the experiments using the same test equipment or by different operators. This is the advantage of significance testing in that it can apportion the source of variation and whether each is significant or not. Otherwise, acceptable good parts produced might not meet their specifications due to measurement error.

The number of parts, trials, and operators in a GR&R study are usually recommended to be equal to or greater than certain minimums. The number of parts (n) must be greater than or equal to 5. The number of operators (appraisers) k must be greater than 2. The number of trials (r) must be greater than or equal to 2. In addition, $n * k$ should be greater than 15. This gives more confidence in the results since it requires a minimum of 30 measurements, which is close to statistically significant numbers approaching a population.

GR&R studies can be analyzed by several commercially available software statistical packages. An example is given below to show how these calculations can be made by using $\overline{X}$ and R techniques, to better understand the relationships of the different components of a GR&R study.

GR&R Study Example

For this theoretical study, 10 parts were measured twice using the same test equipment by 3 operators, for a total of 60 measurements. The details of the 60 measurements and some of their attributes such as $\overline{\overline{R}}$ and X_{diff} are shown in Table 2.4. The specification tolerance span is 15 or $\pm$ 7.5.

Answer of GR&R Sample Study

$$R_{\text{DoubleBar}} = \frac{R_{\text{BarA}} + R_{\text{BarB}} + R_{\text{BarC}}}{3} = \frac{0.045 + 0.045 + 0.025}{3} = 0.0383$$
$$X_{\text{Diff}} = X_{\text{Operator A}} - X_{\text{Operator B}} = 0.8275 - 0.7675 = 0.060$$

1. Equipment Variation (EV): Repeatability
   $$\sigma_{\text{EV}} = \frac{R_{\text{DoubleBar}}}{d_2} = \frac{0.0383}{1.128} = 0.034$$
   Span of EV $= 6 * \sigma_{\text{EV}} = 6 * 0.034 = 0.204$
   EV% of total variation (TV) $= 0.204/1.089 = 0.1872$ or 18.72% (TV provided later in the answer)
   EV *as % tolerance* $0.187/15 = 1.36\%$
2. Operator (Appraiser) Variation (AV): Reproducibility
   In this case, the operator (appraiser) variability has to be converted from X_{diff} to σ_{AV} using factor $d_2{}^*$. In addition, the equipment variation σ_{EV} has to be subtracted from the σ_{AV} based on one part. Therefore, the $\sigma_{\text{AV}}{}^2$ is divided by 20 (10 parts measured twice) to reduce it to a single part.

$$\sigma_{AV} = \sqrt{\left(\frac{X_{Diff}}{d_2^*}\right)^2 - \frac{\sigma_{EV}^2}{n*r}} = \sqrt{\left(\frac{0.060}{1.906}\right)^2 - \frac{0.034^2}{10*2}} = 0.0305 \ (n = 10 \text{ measurements,}$$

r = two trials)

Span of AV = $6 * \sigma_{AV} = 6 * 0.0305 = 0.183$

%AV of total variation (TV) = $0.183/1.089 = 0.01683 = 16.83\%$

AV as % of tolerance = $0.183/15 = \mathbf{1.22\%}$

3. Total Gage Repeatability and Reproducibility: GR&R

Span of GR&R = $6 * \sigma_{GRR} = 6 * \sqrt{\sigma_{EV}^2 + \sigma_{AV}^2} = 6*$

$\sqrt{0.034^2 + 0.0305^2} = 6 * 0.0457 = 0.274$

%GR&R of total variation (TV) = $0.274/1.089 = 0.2517 = 25.17\%$

GR & R as % tolerance = $0.274/15 = 1.83\%$ (acceptable since it is <10% of total specifications)

4. Part Variation (PV)

σ part variation (PV) = parts X_{diff} (3 operators)/d_2^* (for $n = 10$) = $0.5583/3.178 = 0.1757$

Span of part variation = $6 * \sigma$ part variation = 1.054

%PV of total variation (TV) = $1.54/1.089 = 0.9678 = 96.78\%$

PV as % tolerance = $1.054/15 = 7.03\%$

5. Total Variation (TV)

Total variation (TV) = $6 * \sqrt{\sigma_{GRR}^2 + \sigma_{PV}^2} = 6 * \sqrt{0.0457^2 + 0.1757^2} = 6 * 0.1815 = 1.089$

Total variation as % tolerance = TV/tolerance = $1.089/15 = 7.26\%$

This example demonstrates the power of completing a GR&R study to understand the distribution of the components of variability within data measurements. Performing a GRR study before attempting a variability reduction DoE would be very beneficial in determining whether DoE should be attempted or not.

2.2 Tests for Sample and Population

There are several tests for comparing means as well as variability for samples ($\overline{X}$, s) and populations (μ, σ). These tests are dependent on the confidence percentage desired. Confidence is the complement of significance level $\alpha = 1 - $ confidence. Table 2.5 is a summary of statistical tests for mean and variability. These tests will be explained in detail in the next two Sects. 2.3 and 2.4.

In the case of population tests using SND, the significance level α can be expressed as a given probability of occurrence of a particular event. For example, in a significance test of a medication to lower the level of cholesterol, a certain level of medication is given to one medicated sample group, while a placebo is given to another unmedicated sample or control group. A percentage reduction in cholesterol in the medicated sample does not relate to confidence or significance. It is not quantifiable as to the probability based on normal distribution of data. Significance is a more precise determination of the amount of cholesterol reduction by comparing it to probabilities based on normal distributions using a particular level of significance α or confidence $(1 - \alpha)\%$.

Control charts are examples of significance testing. A sample is taken from production, and its mean $\overline{X}$ is compared to the historical average standard distribution of sample means (s). This sample average is compared to the $\pm 3s$ control limits (LCL, UCL), which has a probability of $f(z) = 0.0027$ (0.27% or 2700 ppm) of randomly occurring instance for two sides (either >UCL or <LCL). When the sample mean exceeds either of the $3s$ limits, then the sample representing the current state of production is declared out of control, and action must be taken to correct the perceived problem. However, there is a remote chance that the out-of-control condition happened randomly by chance and not due to an actual problem. This is usually referred to as a false positive and occurs with the same probability of 0.0027 for two-sided (tails) occurrences.

The purpose of the significance test is to compare the mean or variance difference between sample means to those occurring naturally in normal distributions. When the mean difference is larger, the probability that it is due to chance is lessened. A typical number used in statistical tests is 95% confidence or $\alpha = 0.05$ significance. This represents an outcome occurring in normal distribution only once in 20 trials. If the result of the cholesterol-lowering test is expressed in the calculated z value that is indeed equal to or less than 0.05, then this result of lower cholesterol could be either occurring naturally (probability of 1/20 or 0.05) due to the normal curve behavior or it is specifically due to the medication administered. The 95% confidence or 0.05 significance level α in the SND is given by $f(z_{\alpha} = 0.05)$, corresponding to a critical value of $z_{\alpha} = 0.05 = 1.645$.

If the computed value of z ($z_{computed}$) is $\geq$ critical z value, then the mean difference is significant to 95% confidence or 0.05 significance. If the $z_{computed}$ is < critical z value, then the mean difference is not significant to 95% confidence or 0.05 significance.

Alternately, the probability $f(z_{computed})$ can be extracted from the SND table and compared to the significance level $\alpha = 0.05$. If the $f(z_{computed})$ is ≤ 0.05, then the mean difference is significant. If the $f(z_{computed})$ is >0.05, then the mean difference is not significant. The significance tests can be applied to any confidence level α using the same methodology.

When using other distribution tables such as t, X^2, or F distributions, an additional term called degrees of freedom (DOF) is introduced, usually referred to as $\nu = n - 1$. The same principles used in the SND are applied, by first selecting the desired significance level and the sample degrees of freedom (DOF) $= \nu$. The critical test value is extracted from the table and compared to the computed test value. If the computed value is > critical value, then the test yields a significant difference in mean or variability.

Significance α can be either one-sided, as in the examples above, or two-sided (two-tailed). One-sided tests are those that express whether an attribute of a sample or population (either mean or variability) is only larger or only smaller than the attributes of the second sample or population or as compared to the critical value. In this case, the critical z or t are those corresponding to the significance probability or $f(z)$. As in the example above, 95% confidence one-sided test is equal to significance $\alpha = 0.05$, and the probability of $f(z_{\alpha = 0.05}) = 0.05$ corresponds to critical $z_\alpha = 1.645$.

Two-sided significance tests are those that express whether an attribute of a sample or population (either mean or variability) is either larger or smaller than the attributes of the second sample or population or as compared to the critical value. This two-sided test indicates that the probability is divided equally among the two ends of the distribution: either the computed probability or $f(z_{computed}) \geq 0.025$ or $f(-z_{computed}) \leq -0.025$. As in the example above, 95% confidence two-sided test is equal to significance $\alpha/2 = 0.025$, and the probability of $f(z_{\alpha/2 = 0.025}) = 0.025$ corresponds to critical $z_{\alpha/2 = 0.025} = 1.96$.

2.3 Tests for Means

This section includes the first three tests listed in Table 2.5. They include the z-test for comparing two population means (μ) to each other as well as comparing a sample mean ($\overline{X}$) to the parent population mean (μ), assuming that the parent population is normally distributed. If not, then the Wilcoxon test is used, which uses z-tests for the two means through rank ordering of the two sample data points to extract a common variability. When the parent population σ is known, the z-test can be used to compare two samples of sizes (n) less than 20.

The t-test is used when population variability is unknown and therefore uses the combined variability of the two samples. There are three types of t-tests: (1) the d-test, which is a reduced variation t-test, comparing two means when the same sample units are used in subsequent tests under different conditions; (2) when

Table 2.5 Summary of statistical tests for mean and variability

1. All of the tests are binary comparing two samples and/or populations, assuming that the parent population is normal
2. z-test is used to compare means from two populations (or large sample size $n > 30$) to each other or one sample to population or two samples when the parent population (σ) is known
3. Wilcoxon test is used to compare two means where at least one sample parent population is not normally distributed
4. t-test is used to compare two sample means (s) when parent population (σ) is unknown, with three variants
5. Confidence interval (CI) is a pre-calculated t-test where a sample mean $\overline{X}$ can be shown to be significantly (or not) different to parent population mean (μ)
6. Sample (n) size determination is dependent on the error tolerated, confidence, and population variability (σ)
7. Chi-square test is used to compare a sample variation (s) to parent population standard distribution (σ)
8. F-test is used to compare two sample variances (s), when the parent population (σ) is unknown

comparing two sample means from a population of unknown variability but presumed equal; and (3) when comparing two sample means from a population of unknown variability but presumed unequal. In order to decide which t-test to use when parent population variability is presumed equal or unequal, an F-test for variability significance has to be performed first.

Confidence interval (CI) is a reverse t-test, where the interval of $\pm$ values of the sample is pre-calculated beforehand. If the population mean (μ) falls within the CI of the sample mean ($\overline{X}$), then sample mean is not significantly different to parent population mean (μ). The reverse holds true as well: if the population mean (μ) falls outside of the CI of the sample mean ($\overline{X}$), then sample mean is significantly different than the parent population mean (μ). CI provides for an easy answer to significance testing without having to use the SND or t distribution tables. Sample (n) size determination is dependent on the planned error, confidence, and population variability (σ). A visual presentation of the CI versus population mean is given in Fig. 2.3 as part of the CI discussion.

2.3.1 z-Test for Population Means

The z-test is used for comparing means of populations and samples $X_{bar} = \overline{X}$, if the parent population standard deviation (σ) is known or the sample size (n) $\geq$ 30. In addition, it can be used in the case of comparing small samples when parent population mean (μ) and standard deviation (σ) are known. The equation for z is given as:

$$z = \frac{\overline{X} - \mu}{\sigma / \sqrt{n}}$$ (2.3)

It is important not to confuse this z-test with the z transformation discussed in Chap. 1, given in Eq. 1.3. The z transformation is to convert the normal distribution to SND or z distribution, while the z-test is to compare samples to populations.

A pooled (combined) σ_p has to be calculated when comparing two sample means from a population with known σ or two populations (or very large sample sizes $n > 30$) with two different standard deviations (σ_1 and σ_2). The z Eq. 2.3 is amended as follows:

1. For comparing means ($\overline{X}_1, \overline{X}_2$) of two sample sizes ($n_1$ and n_2) with population σ known:

$$\sigma_p = \sigma * \sqrt{\frac{1}{n_1} + \frac{1}{n_2}} \text{ and } z = \frac{\overline{X}_1 - \overline{X}_2}{\sigma_p}$$ (2.4)

2. For comparing means (μ_1, μ_2) of two populations (or very large sample sizes n_1 and n_2) with two different standard deviations (σ_1 and σ_2):

$$\sigma_p = \sqrt{\frac{\sigma_1^2}{n_1} + \frac{\sigma_2^2}{n_2}} \text{ and } z = \frac{\mu_1 - \mu_2}{\sigma_p} \tag{2.5}$$

z-Test Example 1: Comparing a Large Sample Mean $\overline{X}$ to a Known Parent Population Mean (μ) and σ with Different Significance Results Depending on Test Conditions

The historical record of a normal population of shafts has mean diameter (μ) = 5 mm and standard deviation (σ) = 0.1. A sample of ten shafts was measured with mean $\overline{X}$ = 5.06. Is the sample of ten shafts significantly different than the historical population mean μ = 5 of shafts?

Using Eq. 2.3, the z value of the test of ten shafts = $\frac{5.06-5}{0.1/\sqrt{10}}$ = 1.9.

The probability of $z = 1.9$ extracted from the SND = $f(-z = -1.9) = 0.02872$ or 0.029.

If the z-test is one-sided, the probability (P) = 0.029 < critical value of confidence at 95% or significance (α) = 0.05. Another method is to test using the value of $z = 1.9 >$ critical z one-sided extracted from $f_{(z = 0.05)} = 1.645$. Both tests indicate that there is a significance difference between the sample mean from ten-part sample and the population mean. This one-sided significance difference is based on either sample mean $\overline{X}$ being only greater or only less than the population mean (μ).

If the z-test is two-sided, the probability P for both sides of the SND = $2 \times 0.029 = 0.058 >$ critical value of confidence at 97.5% or significance (α) = 0.025. Another method is to test using the value of $z = 1.9 <$ critical z two-sided extracted from $f_{(z = 0.025)} = 1.96$. Both tests indicate that there is no significance difference between the sample mean from ten-part sample and the population mean based on two-sided (tailed) tests. This two-sided significance difference is based on sample mean being either greater or less than the population mean.

This example demonstrates the importance of significance testing and specifying whether it is one- or two-sided. Close examination of the desired test conditions and goals should determine the test type. For example, in human weight chart tables, there might be different categories of underweight, normal, or overweight. One-sided would be testing a controlled sample mean weight of patients taking diet pills versus those taking placebo for underweight versus normal only or overweight versus normal only. Two-sided would be testing for the same sample for both underweight and overweight at the same time. If the purpose of the experiment is to test weight loss in the sample, then one-sided is the proper test condition.

z-Test Example 2: Comparing Two Sample Means to Known Parent Population Standard Deviation (σ)

Given two samples:

Sample $A =$ 113   116   123   132
Sample $B =$ 113   118   123   126   140

Calculate if the two sample A and B means are significantly different from each other at 95% confidence (based on one- and two-sided), if the standard deviation of the parent population (σ) = 2.5.

If σ is known = 2.5, then z-test is selected for significance testing. Since the two samples are not of the same size, the pooled variance (σ_p) of the two samples should be calculated based on $\sigma = 2.5$; $n_A = 4$ and $n_B = 5$.

The $z_{computed}$ value can be calculated from Eqs. 2.3 and 2.4. The SND or z distribution table (Table 1.7) can be used to generate the computed probability = $f(z_{computed})$ from the z value. The computed probability is then compared to the critical probability value one-sided at significance (α) = 0.05 and critical $z = 1.645$. For two-sided significance test, $\alpha/2 = 0.025$ and critical $z = 1.96$. The difference in sample means = X_{diff} is:

$$\text{Sample } A \text{ mean } (\overline{X_A}) = \frac{113 + 116 + 123 + 132}{4} = 121$$

$$\text{Sample } B \text{ mean } (\overline{X_B}) = \frac{113 + 118 + 123 + 126 + 140}{5} = 124$$

$$X_{diff} = \text{Sample } A \text{ mean} - \text{Sample } B \text{ mean} = 121 - 124 = -3$$

The pooled (combined) standard deviation of the two samples is:

$$\sigma_p = \sigma * \sqrt{\frac{1}{n_A} + \frac{1}{n_B}} = 2.5 * \sqrt{\frac{1}{4} + \frac{1}{5}} = 1.677$$

1. Computed $z = X_{diff}/\sigma_{pooled}$
   $z_{computed} = {}^{x_{diff}}/\sigma_p = \frac{-3}{1.677} = -1.78885$
2. The probability P of the two sample means being significantly different $= f(z)$
   From Z table $\rightarrow f(z_{computed}) = f(-1.78885) = 0.036819$
   One-sided significance test (α) $= f(z_{computed}) < 0.05$ ($\alpha_{critical}$) $\rightarrow$ Sample mean difference is significant.
   Two-sided significance test ($\alpha/2$) $= f(z_{computed}) > 0.025$ ($\alpha/2_{critical}$) $\rightarrow$ Sample mean difference is not significant.
3. The critical value of significance α at 0.05, based on sample means and $\sigma_{combined}$ (one- or two-sided)
   For α at 0.05 one $-$ sided critical $z \rightarrow z_{f(z)} = z_{f(z = 0.05)} = 1.645$.
   One-sided z-test is significant ($Z_{computed} = -1.7885 > Z_{critical} = -1.645$).
   For α at 0.025 two $-$ sided critical $z \rightarrow z_{f(z)} = z_{f(z = 0.025)} = 1.96$.
   Two-sided z-test is not significant ($Z_{computed} = -1.7885 < Z_{critical} = -1.96$).
4. Checking results with Excel© statistical function menu
   NORM.S.DIST($z_{computed}$,TRUE) = 0.0368

2.3.2   The Wilcoxon Test for Non-normal Population Mean Test

The Wilcoxon test is used to compare samples from non-normal distributions. It uses the methodology of dividing the data into ranges and then calculating a z value. The z value is based on the calculated data mean minus the expected data average divided by the data standard deviation, as follows:

1. Take two samples n_1 and n_2. Order the data points from each, keeping note of sample origin.
2. If the sample data values present a ranking tie, then the ranking order is split into ranking tie of dividing the sum of two successive ranks by a factor of 2.
3. Calculated expected ranked mean$_\text{ranked}$ $(\mu) = (n_1 * n_2)/2$.
4. Calculated variable $U = \sum$ ranks of sample 1 or 2.
5. Calculated $\sigma_\text{ranked} = [n_{1*} n_2 * (n_1+n_2+1)/12]$.
6. For large n (>8), $U_x = z_x = $ (calculated U − expected U)/σ.

$$z_\text{ranked} = [(U - (n_1 * n_2/2)]/\sqrt{[n_{1*} n_2 * (n_1+n_2+1)/12]}$$

The Wilcoxon test calculations are shown in Wilcoxon example 1. Rank order can be calculated for both z_rankedA and z_rankedB, since both samples are two sides of the same distribution. The resulting z_ranked should be the same for both samples 1 and 2. A close variant to the Wilcoxon test is sometimes called the Mann-Whitney test.

Wilcoxon Test Example

A Wilcoxon test is performed for the following samples A and B from a non-normally distributed parent population. The difference of the sample means can be calculated as significant or not based on 95% confidence using one- and two-sided. Alternatively, one-sided and two-sided actual probability and critical values can be shown using hand calculations and Excel©.

$A =$ 15 12 18 20 21 23 27 29 30 31
$B =$ 11 13 14 16 17 10 21 22 23 28

The two sample data points were ranked against each other (sorted), and the ranks were summed and then applied to the z_ranked calculations based on Eq. 2.3.

There are two rank ties in the ranking at data points 21 and 23. The ranking were adjusted accordingly. The resultant z_ranked was calculated from samples A and B and was the same as expected.

Number	10	11	12	13	14	15	16	17	18	20
Group	B	B	A	B	B	A	B	B	A	A
Rank	1	2	3	4	5	6	7	8	9	10
Number	21	21	22	23	23	27	28	29	30	31
Group	A	B	B	A	B	A	B	A	A	A
Rank	11.5	11.5	13	14.5	14.5	16	17	18	19	20

Rank of $A = 3 + 6 + 9 + 10 + 11.5 + 14.5 + 16 + 18 + 19 + 20 = 127$

Rank of $B = 1 + 2 + 4 + 5 + 7 + 8 + 11.5 + 13 + 14.5 + 17 = 83$

$U_A = \left(\text{Rank of } A - n_A * \frac{n_A+1}{2}\right) = \left(127 - 10 * \frac{10+1}{2}\right) = 127 - 55 = 72$

$U_B = \left(\text{Rank of } B - n_B * \frac{n_B+1}{2}\right) = \left(83 - 10 * \frac{10+1}{2}\right) = 83 - 55 = 28$

$z_A = \dfrac{U_A - \frac{n_A * n_B}{2}}{\sqrt{n_A * n_A * \frac{n_A + n_A + 1}{12}}} = \dfrac{\left(72 - \frac{10*10}{2}\right)}{\sqrt{10*10*\frac{10+10+1}{12}}} = \dfrac{(72-50)}{13.228} = 1.663$

$z_B = \dfrac{U_B - \frac{n_A * n_B}{2}}{\sqrt{n_B * n_B * \frac{n_B + n_B + 1}{12}}} = \dfrac{\left(28 - \frac{10*10}{2}\right)}{\sqrt{10*10*\frac{10+10+1}{12}}} = \dfrac{(28-50)}{13.228} = -1.663$

The significance tests were summarized as follows:

z-test	$z_{critical}$	$f(z_{critical})$	$z_{computed}$	$f(z_{computed})$ = P-value
Single-Sided	1.645	0.050	1.663	0.04846
Two-Sided	1.960	0.025	1.663	0.04846

For the one-sided z_{ranked} test, the difference between the two samples is significant, because the computed P-value $f(z_{computed})$ is less than the level of significance ($0.04846 < 0.05$) and $z_{computed}$ (1.663) > $z_{critical}$ (1.645). The two-sided test, however, shows that the difference between the two sample is not significant, as the computed P-value is greater than the level of significance ($0.04846 < 0.025$) and $z_{computed}$ (1.663) < $z_{critical}$ (1.96).

2.3.3 t-Tests for Sample Means: Single Sample

When the total number of data points in the samples (n) is small ($n < 20$), very little can be determined by the sampling distribution for small value of n, unless an assumption is made that the sample comes from a normal distribution. The normal distribution assumes an infinite number of occurrences that are represented by the process average μ and standard deviation σ. The Student's t distribution is used when n is small. The data needed to construct this distribution are the sample average $\overline{X}$ (X_{bar}) and sample standard deviation s, as well as the parent normal distribution mean μ.

$$t = \frac{\overline{X} - \mu}{s/\sqrt{n}} \qquad (2.6)$$

where:

t is a random variable having the t distribution with ν = degrees of freedom (DOF) = $n-1$.

Fig. 2.1 Student's
t distribution

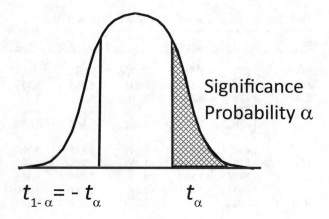

Significance
Probability α

$$t_{1-\alpha} = -t_{\alpha} \qquad t_{\alpha}$$

$\overline{X}$ = sample mean.
μ = population mean.
s = standard deviation of the sample.

Equation 2.6 assumes that the population standard deviation σ is unknown. In this case, the sample standard deviation s is used in the *t*-test. When the population standard deviation σ is known, then it can be used in the *z*-test rather than the sample standard deviation s in the *t*-test. The Student's *t* distribution is used to confirm or refute a particular claim about a sample mean versus the population mean. It is always assumed that the parent distribution of the *t* distribution is normal. It is shown in Fig. 2.1, with the shaded area representing the probability determined by the significance level α. While there is a single t_ν distribution for each DOF (ν), it is common to highlight only the significance level α versus the DOF (ν) for each distinct $t_{\alpha,\nu}$. A selected set of $t_{\alpha,\nu}$ for multiple significance levels of α versus DOF ν is given in Table 2.6. It is not a continuous table for all values of DOF or confidence. For values that are not in the table, extrapolation can be used to determine the approximate $f(t_{\alpha,\nu})$. In addition, more values of $t_{\alpha,\nu}$ for different α and ν can be obtained by searching the Internet or using Excel®. The last row of Table 2.6 gives the complement of significance (α), which is called confidence. It is set to $1 - \alpha$ and expressed as a percent value.

$$\text{Confidence } (\%) = 1 - \text{significance} = 1 - \alpha$$

Figure 2.2 represents the *t* distribution with various DOF (ν) compared with standard normal distribution (SND). The shape of the *t* distribution is similar to the normal distribution. Both are bell shaped and distributed symmetrically around the mean. The *t* distribution mean is equal to zero, and the number of degrees of freedom (DOF) (ν) governs each *t* distribution. The spread of the distribution decreases as the number of degrees of freedom ν increases. The variance of the *t* distribution always exceeds 1, but it approaches 1 when the number n approached infinity (∞). At that time, the *t* distribution becomes the same as the standard normal distribution (SND).

Table 2.6 Selected values of $t_{\alpha,\nu}$ of Student's t distribution

ν	$\alpha = 0.10$	$\alpha = 0.05$	$\alpha = 0.025$	$\alpha = 0.01$	$\alpha = 0.005$	$\alpha = 0.001$	$\alpha = 0.0005$
1	3.078	6.314	12.706	31.821	63.657	318.3	636.6
2	1.886	2.920	4.303	6.965	9.925	22.327	31.600
3	1.638	2.353	3.182	4.541	5.841	10.214	12.922
4	1.533	2.132	2.776	3.747	4.604	7.173	8.610
5	1.476	2.015	2.571	3.365	4.032	5.893	6.869
6	1.440	1.943	2.447	3.143	3.707	5.208	5.959
7	1.415	1.895	2.365	2.998	3.499	4.785	5.408
8	1.397	1.860	2.306	2.896	3.355	4.501	5.041
9	1.383	1.833	2.262	2.821	3.250	4.297	4.781
10	1.372	1.812	2.228	2.764	3.169	4.144	4.587
20	1.325	1.725	2.086	2.528	2.845	3.552	3.849
30	1.310	1.697	2.042	2.457	2.750	3.386	3.646
∞	1.282	1.645	1.960	2.326	2.576	3.090	3.291
Confidence (1−α)	90%	95%	97.5%	99%	99.5%	99.9%	99.95%

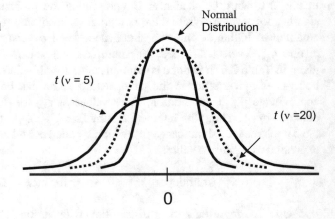

Fig. 2.2 t distribution with various DOFs (ν) compared with standard normal distribution (SND)

Normal Distribution

$t\,(\nu = 5)$

$t\,(\nu = 20)$

0

The t distribution can be used to determine the area under the curve, called significance or α. Certain t-values of α can be considered as critical t_α, such as $t_{\alpha=0.05,\ \nu\ =\ n-1}$ for one-sided or $t_{\alpha/2\ =\ 0.025,\ \nu\ =\ n-1}$ for two-sided t-tests. The t distribution is different than the normal distribution in that the number in the sample or degrees of freedom $\nu = n-1$ has to be considered. Table 2.6 values of critical variables t, called $t_{\alpha, \ \nu \ = \ n-1}$ corresponding to each area under the t distribution curve to the right of α and with $\nu = n-1$ degrees of freedom (DOF).

t-Test Example 1: Single Sample Mean Compared to Population Mean

A sample is collected from a normally distributed population with population mean μ = 12. The sample attributes are sample size $n = 11$, sample mean $\overline{X} = 10$, and sample standard distribution $s = 2.0$. Is the sample mean significantly different than the population mean?

Answer: The t-test for significance is applicable in this problem for small sample size with σ unknown.

$t_{\text{computed}} = (10 - 12)/(2/\sqrt{11}) = -3.32$, and critical $t_{\alpha,\nu}$ ($t_{0.025,10}$) = 2.228 which represents the critical value for t at significance level α and DOF ($\nu = n-1 = 10$)

Since $t_{\text{computed}} = 3.32 > t_{0.025,10} = 2.228$, then the sample mean is significantly different than the population mean with $\alpha/2$ significance (two-sided sample mean significantly greater than or less than population mean).

The actual significance of the sample (3.32) can be determined from Table 2.6 as falling somewhere between $\alpha = 0.005$ (3.169) and $\alpha = 0.001$ (4.144) using DOF (ν) = 10. The actual probability value from an Internet t calculator or Excel© is α = 0.0039 or 99.61% confidence, using the formula =T.DIST(3.32,10,TRUE), which is significant.

2.3.4 t-Tests for Comparing Two Sample Means with Unknown Variance

The t-test for comparing two sample means with population standard deviation (σ) unknown but presumed equal is similar to the basic formula shown in Eq. 2.6. In this case, an F-test (mentioned later in this chapter) for comparing sample standard deviations (s) to each other should be performed first. The F-test can determine whether the unknown population standard deviations are equal or not. The sample standard deviation (s) for a single sample in Eq. 2.6 is substituted by a pooled (combined) standard deviation s_{pooled} (s_p). s_p is made up of the two individual sample deviations (s_1 and s_2), with means X_{bar1} ($\overline{X}_1$) and X_{bar2} ($\overline{X}_2$) and sample sizes n_1 and n_2. This pooled variance (s_p) is used to calculate t according to sample sizes as shown below:

$$s_p = \sqrt{\frac{[(n_1 - 1) * s_1{}^2 + (n_2 - 1)*]s_2{}^2]}{r}} \tag{2.7}$$

and

$$t = \frac{\overline{X}_1 - \overline{X}_2 - d_0}{s_p * \sqrt{\frac{1}{n_1} + \frac{1}{n_2}}} \tag{2.8}$$

where $r = (n_1 + n_2 - 2)$ is the combined degrees of freedom for the test and d_0 is a test difference of the sample means if desired; usually, it is $d_0 = 0$.

Another set of formulas are available in the alternate version of t-tests for comparing two sample means with population standard deviation (σ) unknown but presumed unequal. As discussed above, an F-test can be performed to determine whether to use this t-test version. Microsoft Excel© or any other statistical software can be used for calculating the unequal population deviation t-test version.

t-Test Example 2: Comparing Two Samples of Unknown (Presumed Equal) Variances

Two samples with unknown population variances are given. Determine if the mean difference of the samples is significant or not.

$$A = \quad 113 \quad 116 \quad 123 \quad 132$$
$$B = \quad 113 \quad 118 \quad 123 \quad 126 \quad 140$$

This example is similar to the previous z-test example 2, only with population standard deviation (σ) unknown. This necessitates the use of the t-test for comparing the two sample means. The two samples are not of the same size (n). A pooled variance s_p is calculated based on each sample deviation s_A and s_B; sample size $n_A = 4$ and $n_B = 5$ and the combined r DOF $= 4 + 5 - 2 = 7$. The $t_{computed}$ value can be calculated from Eqs. 2.7 and 2.8 and compared to the $t_{critical}$ value one-sided at significance (α) $= 0.05$ and critical $t_{0.05, \ 7} = 1.895$ from t table (Table 2.6). For two-sided significance test, (α) $= 0.025$; critical $t_{0.025, \ 7} = 2.365$ for combined DOF (ν) $= 7$. Steps for performing significance tests are as follows:

1. Mean and variance of each sample
   Sample A mean $(\overline{X}_A) = \frac{113+116+123+132}{4} = 121$
   Sample B mean $(\overline{X}_B) = \frac{113+118+123+126+140}{5} = 124$

   Sample A variance $= \left(113 - \overline{X}_A\right)^2 + \left(116 - \overline{X}_A\right)^2 + \frac{\left(123-\overline{X}_A\right)+\left(132-\overline{X}_A\right)^2}{4-1=71.3333}$
   Sample B variance $= \left(113 - \overline{X}_B\right)^2 + (118-$
   $\frac{\overline{X}_B)^2+\left(123-\overline{X}_B\right)^2+\left(126-\overline{X}_B\right)^2+\left(140-\overline{X}_B\right)^2}{5-1=104.5}$

2. Number of observations of each sample and degrees of freedom (DOF)
   Sample A observations $(n_A) = 4$
   Sample B observations $(n_B) = 5$
   DOF $=$ (Sample A observations $- 1$) $+$ (Sample
   B observations $- 1$) $= (4 - 1) + (5 - 1) = 7$

3. The difference in sample means $= X_{diff}$
   $X_{diff} =$ Sample A mean $-$ Sample B mean $= 121 - 124 = -3$

4. The pooled standard deviation s_p and variances s_1 and s_2 of the two samples n_A and n_B

$$s_p = \sqrt{\frac{[(n_A-1)*s_A{}^2+(n_B-1)*s_B{}^2]}{r}} = \sqrt{\frac{[(4-1)*71.333+(5-1)*104.5]}{7}} = 9.502$$

Variance $\left(s_p^2\right) = 9.502^2 = 90.2857$

5. The computer t value from items 3 and 4

$$t_{computed} = \frac{\overline{X}_A - \overline{X}_B - do}{s_p * \sqrt{\frac{1}{n_A}+\frac{1}{n_B}}} = \frac{121-124-0}{9.502*\sqrt{\frac{1}{4}+\frac{1}{5}}} = -0.47066$$

6. The critical value for t based on one- or two-sided 5% = significance, degrees of freedom and significance level α at 0.025 or 0.05 from Table 2.6

   t critical one-sided$_{@DOF=7;\ \alpha=0.05}$ =1.895
   t critical two-sided$_{@DOF=7;\ \alpha=0.025}$ =2.365

   The two-sample mean difference $t_{computed}$ is significant at $\alpha = 0.05$ if it is greater than the critical t value for one- or two-sided. The $t_{computed}$ can be positioned in the t table between two levels of α shown:

   t table for DOF $(\nu) = 7$

Computed t		t critical one-sided; α=0.05		t critical two-sided; α=0.025
-0.47066	<	1.895	<	2.365

   One-sided →Since t critical is larger than computed t, t is not significant.
   Two-sided →Since t critical is larger than computed t, t is not significant.

7. The answer can be checked by performing a t-test in Excel©, choosing one or two sides:

   T.TEST(A,B,2,2) = 0.65 {two-sided (tailed), not significant}
   T.TEST(A,B,1,2) = 0.326 (one-sided (tailed), not significant)
   To return the original computed t value:
   T.INV.2T(T.TEST(A,B,2,2), 7) = 0.47066. This is the same magnitude as computed t.

The absolute value of the computed t is the same for both cases, when evaluated directly using Eq. 2.8, and then recalculated using inverse t function in Excel©. This indicates the problem was performed correctly.

2.3.5   d- or Paired t-Test

The d-test is a special consideration of the t-test. It is also called the paired or blocked t-test. It is used when the sample sizes are the same for the two samples being tested for significance, as well as a corresponding relationship between the two samples. This relationship could be in the form of the same parts being subjected to different conditions. In this manner, conditioning of the parts, such as baking, can be evaluated for significance.

The d-test is performed on the mean of the differences between the two samples. The d-test formula is similar to the t-test except that the mean of the mathematical differences (making sure to preserve the mathematical signs of the difference) is used as well as the standard deviation of the difference. The degrees of freedom (DOF) of the d-test are the number of elements in the sample $n-1$. The DOF in the d-test is smaller than the corresponding t-test. The formula for the d-test is as follows:

$$d = \frac{\overline{d}}{s_d/\sqrt{n}} \tag{2.9}$$

where $\overline{d}$ is the mean and s_d is the standard deviations of the differences in the same sample units before and after specific conditioning.

d-Test Example 1

A test was made to evaluate the impact of baking on selected parts. Measurements were made on the same seven parts before and after baking. Does baking make a significant difference for the same parts' mean measurements at 95% confidence (two-sided)?

Baked	3.0	3.7	4.0	3.2	3.6	3.5	4.2	$\overline{X}_1 = 3.6$	s = 0.42
Not baked	3.7	3.6	3.4	3.0	3.6	2.8	3.7	$\overline{X}_2 = 3.4$	s = 0.36
d		−0.7	0.1	0.6	0.2	0.0	0.7	0.5	$\overline{d} = 0.2$	s = 0.476

$t_{paired} = 0.2/(0.476/\sqrt{7}) = 1.11 < $ critical $t_{0.025,\,6} = 2.447$
Computed probability $= 0.155 > $ critical probability $= 0.025$ at 97.5% confidence and two-sided

Conclusion: Baking does not significantly change parts' measurement means using paired d-tests.

d-Test Example 2 and Comparison with t-Test

In certain electronics industries, some devices are turned on (aged) in the stockroom for a period to reduce infant mortality failures before shipping to customers, with a typical period of 100 hours. There were six devices that were initially aged for 100 hours, tested for results, and then aged for another 100 hours and tested again. Is there a need for the extended aging of 100 hours in addition to the standard aging based on the following aging data?

100 hours	22	23	28	19	21	18	$\overline{X}_1 = 21.833$, $s_1 = 3.54$
200 hours	23	14	17	19	16	18	$\overline{X}_2 = 17.833$, $s_2 = 3.06$
Difference	−1	9	11	0	5	0	$\overline{d} = 4$, $s_d = 5.14$

d-Test Calculations

$t_d = \frac{\overline{d}}{s/\sqrt{n}} = \frac{4}{5.14/\sqrt{6}} = 1.907$ which is less than the critical t values for one- or two-sided (tailed) shown below:

$t_{\text{crit one-sided}} = t_{\frac{\alpha}{2}, n-1} = t_{0.05,5} = 2.015$ and $t_{\text{crit two-sided}} = t_{0.025,5} = 2.571$
Probability one-sided = T.TEST(array1,array2, tails = 1, type = 2) = 0.057
Probability two-sided = T.TEST(array1,array2, tails = 2, type = 2) = 0.115

There is no significant difference due to aging either by using the actual one- or two-tailed probability or comparing the resulting d-test to the critical t value for $\alpha = 0.05$ (two-tailed) or 0.025 (one-tailed) and DOF = 5.

The d-test shows that the extended 100 hours (200 hours total) is not needed, as the $t_d < t_{crit}$ (1.907 < 2.015 and 1.907 < 2.571). This shows no significance between the two samples with different aging times.

t-Test Calculations

$\overline{X}_{\text{diff}} = \overline{X}_1 - \overline{X}_2 = 21.8 - 17.8 = 4, \nu = 6 + 6 - 2 = 10$

$S_p = \sqrt{\frac{[(n_A - 1)*s_A{}^2 + (n_B - 1)*s_B{}^2]}{r}} = \sqrt{\frac{[(6-1)*3.54^2 + (6-1)*3.06^2]}{10}} = 3.31$

Computed $t = \frac{21.8 - 17.8}{3.31*\sqrt{\frac{1}{6}+\frac{1}{6}}} = 2.09$.

The one-sided t-test shows that the extra 100 hours is significant ($t_{\text{computed}} > t_{\text{crit one-sided}} = t_{\alpha = 0.05,10}$ or > 2.09 > 1.812), but the two-sided t-test shows that the extra 100 is not significant ($t_{\text{computed}} < t_{\text{crit one-sided}} = t_{\frac{\alpha}{2},10}$ or 2.091 < 2.228). The following are the results using Excel©:

Computed t statistic = 2.09
$P(T < = t)$ one-tailed   0.0315. T.TEST(array1,array2, tails = 1, type = 2) $\rightarrow$ significant
t critical one-tailed = 1.812, from Table 2.6; $r (\nu) = 10$, $\alpha = 0.05$
$P(T < = t)$ two-tailed   0.0629. T.TEST(array1,array2, tails = 2, type = 2) $\rightarrow$ not significant
t critical two-tailed = 2.228, from Table 2.6; $r (\nu) = 10$, $\alpha = 0.025$

The previous two examples of testing for significance of sample means using d- versus t-tests can result in having different conclusions for the same problem. Pairing or d-tests reduce uncontrolled sources of variations by using the same sample for the comparisons of conditions being tested. Pairing should not have interaction between conditions and units. In the baking example 1, the prebake measurements do not affect the post-bake ones. t-tests allow for randomness which can include uncontrolled sources of variations.

These calculations above can be performed with Excel© using Data Analysis Package. The results are shown in Table 2.7, d-test example 2 and comparison with t-test, and match the hand calculations and significance tests above.

2.3.6 Confidence Interval (CI) of the Mean

Confidence intervals (CI) are used to highlight the probability span of data sets within populations of known mean (μ) and variance (σ). It is an inversion of the t-

Table 2.7 *d*-test example 2 and comparison with *t*-test

t-test: paired two samples for means			t-test: two samples assuming equal variances		
	100 hours	200 hours		100 hours	200 hours
Mean	21.833	17.833	Mean	21.83333	17.83333
Variance	12.567	9.367	Variance	12.56667	9.366667
Observations	6.000	6.000	Observations	6	6
Pearson correlation	−0.206		Pooled variance	10.96667	
Hypothesized mean Difference	0.000		Hypothesized mean Difference	0	
df	5.000		df	10	
t stat	1.907		*t* stat	2.092104	
P(T<=t) one-tailed	0.057		*P(T<=t)* one-tailed	0.031453	
t critical one-tailed	2.015		*t* critical one-tailed	1.812461	
P(T<=t) two-tailed	0.115		*P(T<=t)* two-tailed	0.062906	
t critical two-tailed	2.571		*t* critical two-tailed	2.228139	

test, where the sample mean is tested against population mean μ. If a sample mean with its CI span falls outside of the population μ, then it can be considered significantly different than the rest of the samples in the population. The higher the confidence % (or the lower the significance α), the larger the span of the sample CI and its end points, the confidence limits. The reverse is true: lower confidence % results in smaller CI spans and limits. The 95% confidence CI is the most commonly used in calculating sample significance, process capability, and Six Sigma data (C*pk*, defect rates, FTY), while higher confidence numbers (99% and 99.9%) can be used as worst-case condition check on significance of a particular sample.

Figure 2.3, sample confidence interval around the population mean (μ), is an example of the 95% confidence limit CI around each sample mean. It is a ($\pm 2s$) span around sample mean $\overline{X}$, which indicates that 5% (1 out of 20, highlighted third down from the top line shown in Fig. 2.3) of the samples has a span that does not encompass the population mean μ. If a calculated sample CI span falls outside of μ, then the probability of that event occurring is 5%. This is highly unlikely to occur in a normal distribution. It can be considered that this sample mean is significantly different than μ. In other words, it will result in a computed *t*-value > critical *t*-value in the *t*-test comparing the difference of $\overline{X}$ and μ.

The confidence % determines the span of sample confidence interval, through the significance term $\alpha = (1 - \text{confidence})$. The formulas for the CI are given below for high-volume samples (where $n > 30$ and population μ and σ are known) and low-volume samples (n, $\overline{X}$, s), respectively:

$$\mu \pm z_{\alpha/2} * \sigma/\sqrt{n} \tag{2.10}$$

$$\overline{X} \pm t_{\alpha/2,n-1} * s/\sqrt{n} \tag{2.11}$$

Fig. 2.3 Sample confidence interval (CI) versus population mean (μ)

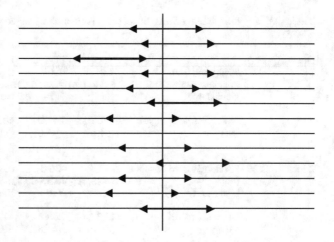

Population Mean μ

Figure 2.3 shows an interpretation of the CI for 13 samples of equal sizes (n) from the same population with a mean (μ) and known σ. The different samples produce different values for $\overline{X}$, and consequently, the interval spans are centered at different points. When σ is known, the CI is the same for all samples, because all their confidence limits are derived from σ according to Eq. 2.10. If the population (σ) is unknown, then the sample standard deviations (s) are used to calculate the CI for each sample from Eq. 2.11, and the CI spans are different for different samples.

When using the CI from two samples ($n < 20$) or populations, the pooled variability formulas for populations (2.5) and samples (2.7) should be used as in the following example. In this case, σ_p and s_p are used instead of σ/n and s/n.

Confidence Interval Example 1

Using the data from d-test example 2 and comparison with t-test, determine the confidence interval for 95% and 90% confidence of the difference in sample averages. Assume that the population standard deviation is unknown but presumed equal.

100 hours   22   23   28   19   21   18 $\overline{X}_1 = 21.833$, $s_1 = 3.54$
200 hours   23   14   17   19   16   18 $\overline{X}_2 = 17.833$, $s_2 = 3.06$

$$S_p = \sqrt{\frac{[(n_A-1)*s_A{}^2+(n_B-1)*s_B{}^2]}{r}} = \sqrt{\frac{[(6-1)*3.54^2+(6-1)*3.06^2]}{10}} = 3.31$$

Critical values of for two-tailed $t_{confidence\%}$ for $r\,(\nu) = 10$, $\alpha = (1 - \text{confidence})$:

$t_{90\%} = t_{\alpha/2,(\,n_1+n_2-2)} = t_{0.05,10} = 1.812$ (T.INV$(0.05, 10)$ from Excel©

$t_{95\%} = t_{\alpha/2,(n_1+n_2-2)} = t_{0.025,10} = 2.228$ (T.INV$(0.025, 10)$ from Excel©

$CI_{90\%} = \overline{X}_{diff} \pm t_{\alpha,(n_1+n_2-2)} * s_p\sqrt{\frac{1}{n1} + \frac{1}{n2}} = 4 \pm 1.812 * 3.31 * \sqrt{\frac{1}{6} + \frac{1}{6}} = 4 \pm 3.46$

$CI_{90\%} = 0.53, 7.47$

Table 2.8 Lipid-lowering effects of Zocor® (simvastatin) in adolescent patients for 24 weeks

Dosage	N		Total–C	LDL–C	HDL–C	TG
Placebo	67	% change from baseline (95% CI)	1.6	1.1	3.6	-3.2
			(–2.2, 5.4)	(–3.4, 5.6)	(–0.7, 8.0)	(–11.8, 5.4)
		significant?	No	No	No	No
		Mean baseline, mg/dL (SD = σ)	278.6 (51.8)	211.9 (49.0)	46.9 (11.9)	90.0 (50.7)
Zocor	106	% change from baseline (95% CI)	–26.5 (–29.6, –23.3)	–36.8 (–40.5, –33.0)	8.3 (4.6, 11.9)	–7.9 (–15.8, 0.0)
		significant?	Yes	Yes	Yes	No
		baseline, mg/dL (SD = σ)	270.2 (44.0)	203.8 (41.5)	47.7 (9.0)	78.3 (46.0)

$$CI_{95\%} = \overline{X}_{diff} \pm t_{\alpha/2,(n_1+n_2-2)} * s_p \sqrt{\frac{1}{n1} + \frac{1}{n2}} = 4 \pm 2.228 * 3.31 * \sqrt{\frac{1}{6} + \frac{1}{6}} = 4 \pm 4.26$$

$$CI_{95\%} = -0.26, 8.26$$

The CI increases with increased confidence % or deceased significance α.

Confidence Interval Example 2: Applications in the Life Sciences Industry
Confidence interval techniques are widely used in the life sciences industry for demonstrating new drug efficacy. An example of drug testing data is taken from the FDA website [1] https://www.accessdata.fda.gov/drugsatfda_docs/label/201 5/019766s093lbl.pdf for Zocor® (simvastatin), specifically for clinical studies in adolescents shown in Table 2.8 of the study. It is summarized in Table 2.8 in the text.

Zocor® is the leading drug for reducing cholesterol. It is effective in reducing the total cholesterol or Total-C; LDL, also known as bad cholesterol (the lower the level, the better); as well as HDL, also known as good cholesterol (the higher the level, the better). Zocor does not affect triglycerides or TG, a test of fatty substances in blood: the lower the level, the better.

The generic name of Zocor® is simvastatin. Table 2.8 shows the reduction of total cholesterol into two separate groups, one treated with placebo and the other with Zocor. For the group of $N = 67$ that was treated with placebo, the % changes for all cholesterol components are small, and all 95% CI values cross the zero level, indicating no significant changes in the measurements, as was indicted by Fig. 2.3. For the group of 106 that was treated with Zocor, the % changes for total cholesterol, LDL (bad) cholesterol, as well as HDL (good) cholesterol are significant at -26.5%, -36.8%, and 8.3% change from baseline, respectively. Their CI does not cross the zero level, indicating significance, as was shown in Fig. 2.3. Triglyceride (TG) level was reduced by -7.9% from baseline, but its CI (-15.8, 0.0) just reaches the zero point, indicating no significant change from baseline due to Zocor®.

The calculations for the 95% confidence interval (CI) for % change of Total-C can be shown as follows. These calculations presented a conversion of the comparisons of average means for placebo versus Zocor® in percentage rather than absolute values from baseline data. In other words, confidence interval CI of the sample was compared against population baseline $\mu = 0$ as was shown in Fig. 2.3. If a sample CI crosses the zero point, then the sample change from baseline is not significant. If a sample CI does not cross the zero point, then the sample change from baseline is significant. A sample calculation from Table 2.8 is shown below for the Zocor total cholesterol group:

Total Zocor cholesterol group actual mean change from baseline $= 270.2 * 26.5\% =$
   -71.603
95% confidence interval for mean (assuming two-sided) $= \pm Z_{0.25}* \sigma/\sqrt{n} = 1.96 *$
   $44/\sqrt{106} = \pm 8.3764$
Upper limit of mean Total-C change at 95% CI $= -71.603 - 8.3764 = -79.9784$
Lower limit of mean Total-C change at 95% CI $= 71.603 + 8.3764 = -63.2256$
% upper limit of mean Total-C change at 95% CI $= -79.9784/270.2 = -29.6\%$
% lower limit of mean Total-C change at 95% CI $= -63.2256/270.2 = -23.3\%$

The CI for the total cholesterol % change from baseline is significant when treated with Zocor® since its CI end point limits are both well above the baseline population mean $\mu = 0$. Zocor was shown to reduce total cholesterol by 26.5%, which is well within the range of -23.3–29.6% CI interval for lowering total cholesterol.

2.3.7   Determination of Sample Size Based on Error

When sample mean is used to estimate population mean, the desired sample size should be at least $n = 30$ to use population statistics (z distribution) rather than the t distribution as shown in Table 2.9, error of $t_{\alpha,30}$ compared with standard normal

Table 2.9 Error of $t_{\alpha,30}$ compared with standard normal distribution (SND)

	α or $f(z)$ = 0.05	α or $f(z)$ = 0.025	α or $f(z)$ = 0.01	α or $f(z)$ = 0.001	α or $f(z)$ = 0.0005
$v = 30$	1.697	2.042	2.457	3.386	3.646
$v = \infty$ or z	1.645	1.960	2.326	3.090	3.291
Error = $t_{\alpha,30} - z$	3.2%	4.2%	5.6%	9.6%	10.8%
Confidence $(1-\alpha)$ or probability for z	95%	97.5%	99%	99.9%	99.95%

distribution (SND). When sample size n approached infinity (∞), then the t distribution becomes equal to the z (normal) distribution. Since an infinite number of samples is impractical, a common $n = 30$ is used in order to use population statistics of mean μ and variability σ, which results in an error % shown in Table 2.9.

The error generated between $t_{\alpha,30}$ and z values varies according to the significance (α) selected. It is lower than 5% for the two important significance levels of $\alpha = 0.05$ for one-sided and 0.025 for two-sided tests. This is important in life science applications, where population-based tests are commonly used. Using a sample $n = 30$ results in an error less than 5% in significance testing rather than using a very large sample size approximating infinity.

To control the resulting error (margin of error), a particular sample (n) size can be specified. This sample size can be expanded to include the general conditions where the standard deviation (σ) is known, either from prior analysis or calculated from the sample when the number of the sample taken is large ($n > 30$). The maximum error E produced when sample mean ($\overline{X}$) is used to estimate μ, the population mean, can be calculated below. The random sample size (n) needed to estimate the mean of a population, with a confidence of $(1 - \alpha)$%, can also be shown as:

$$E = z_{\alpha/2} * \frac{\sigma}{\sqrt{n}} \tag{2.12}$$

$$n = \left[\frac{z_{\alpha/2} * \sigma}{E} \right]^2 \tag{2.13}$$

If the population mean is unknown, the sample standard deviation should not be used for the calculation of sample size n. The t distribution is not a good fit for calculating sample size n. Instead, the noncentral t distribution should be used. Data for the noncentral distribution are available in textbooks and on the Internet.

Sample Size for Specified Error Example

A catheter tested in cow hearts yields a mean $\mu = 20$ (l/m) blood flow and $\sigma = 3.0$. A new catheter is being developed with different materials. How many cows are needed to test the new catheter to ensure (95% confidence) that the new catheter is within 1 (l/m) of the old catheter mean?

Answer: Sample size $(n) = [Z_{\alpha/2} * \sigma/E]^2 = (1.96 * 3/1)^2 = 34.6$ or 35 sample size

2.4   Tests for Variability

These tests are used to compare variability of a sample to population as well as sample to sample. These tests assume that the samples belong to normally distributed populations. If the population standard deviation (σ) is known, then the X^2

(chi-square) test is used in conjunction with the X^2 distribution tables to compare a sample to the parent population or to determine the goodness of fit of a data set to a particular distribution. If σ is unknown, then two separate sample variances (s_1 and s_2) can be compared using the F-test for significance, assuming that their parent population variances are normal.

The X^2 test can be used to determine whether a data set is normal or not. This is an important test prior to using any of the statistical tests listed in this chapter. The F-test can also be used to determine which type of t-test to use in terms of the relationship of the parent populations σ. In addition, the F-test is used to determine the significance of factors that are tested concurrently in a DoE.

The test for significance is performed by comparing the computed value of X^2 or F to the critical values in the corresponding distribution table. The X^2 distribution is not symmetrical, and the significance (α) values at the opposite ends of the x-axis are not equal. Therefore, the X^2 two-sided (two-tailed) significance is determined by performing two separate tests in the high or low end of the x-axis at two values (α/2 and $1-α/2$).

For the F-test, the two sample variances are significantly different if the computed value of F exceeds the critical value in the F table, for the corresponding degrees of freedom for each sample variance. The placement of the samples in the computed F value formula should be made to ensure that computed $F > 1$.

2.4.1 X^2 (Chi-Square) Significance Test

The statistical relationships of the sample and population means have been discussed earlier in this chapter. There is a similar distribution for the sample variance s^2, which can be used to compare population variance or $σ^2$. This distribution is called the chi-square or X^2. The X^2 distribution is not symmetrical but is related to the gamma distribution and is shown in Fig. 2.4. The probability that a random sample

Fig. 2.4 X^2 distribution

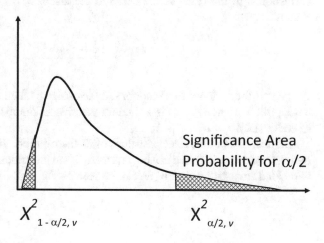

Table 2.10 Selected values of X^2 distribution

ν	$\alpha = 0.995$	0.975	0.95	0.90	0.50	0.10	0.05	0.025	0.005
1	0.0000393	0.000982	0.00393	0.0158	0.455	0.706	3.841	5.024	7.879
2	0.0100	0.0506	0.103	0.211	1.386	4.605	5.991	7.378	10.597
3	0.0717	0.216	0.352	0.584	2.366	6.251	7.815	9.348	12.838
4	0.207	0.484	0.711	1.064	3.357	7.779	9.488	11.143	14.860
5	0.412	0.831	1.145	1.610	4.351	9.236	11.070	12.832	16.750
6	0.676	1.237	1.635	2.204	5.348	10.645	12.592	14.449	18.548
7	0.989	1.690	2.167	2.833	6.346	12.017	14.067	16.013	20.278
8	1.344	2.180	2.733	3.490	7.344	13.362	15.507	17.535	21.955
9	1.735	2.700	3.325	4.168	8.343	14.684	16.919	19.023	23.589
10	2.156	3.247	3.940	4.865	9.342	15.987	18.307	20.483	25.18
15	4.601	6.262	7.261	8.547	14.339	22.307	24.996	27.488	32.801
20	7.434	9.591	10.851	12.443	19.337	28.412	31.410	34.170	39.997
25	10.520	13.120	14.611	16.473	24.337	34.382	37.652	40.646	46.928
30	13.787	16.791	18.493	20.559	29.336	40.256	43.773	46.979	53.67

produces a X^2 value greater than some specified value is equal to the area of the curve to the right of the value. In Fig. 2.4, the probability or the shaded area $(1-\alpha/2)$ corresponds to the value of the curve to the left of $X^2_{(1-\alpha/2)}$. The formula for X^2 is shown below:

$$X^2 = \frac{(n-1) * s^2}{\sigma^2} \tag{2.14}$$

where s^2 is the variance of a random sample of size n taken from a normal population having the variance σ^2. X^2 is a random variable in the distribution with degrees of freedom (DOF) $= \nu = n - 1$.

Table 2.10 contains selected values of X^2 distribution. Since it is not symmetrical, two X^2 values will have to be returned when considering 95% confidence for two-sided (two-tailed) limits, as can be seen in Fig. 2.4.

X^2 (Chi-Square) Test Example

Five samples were taken from a normal population of parts from a factory with historical data showing mean $= 3$ and $\sigma = 1$. The samples' measurements were 2.0, 2.5, 3.0, 3.5, and 4.0. Does this sample of parts support the belief that the sample came from the factory with σ equal to 1 at 95% two-sided (two-tailed) confidence?

Answer. Sample mean $\overline{X} = 3$ and sample standard deviation $(s) = 0.79$

$\text{SS}_{\text{sample}} = \sum x^2 - \frac{(\sum x)^2}{n} = 47.5 - 15^2/5 = 2.5$

$\text{Variance} = s^2 = \text{ss}/(n - 1) = 2.5/4 = 0.625$

$s = 0.79$

$X^2 = (4 * 0.79^2)/1 = 2.50$

The computed value of X^2 (2.50) with $\text{DOF} = 4$ is close to 50% confidence critical value in Table 2.10 ($\alpha = 0.025$ and $v = 4$ at 3.357). It is greater than $\alpha = 0.975$ and less than $\alpha = 0.025$ (0.484 and 11.143) critical significance X^2 values. Therefore, based on sample variance, it is highly likely that the sample was made at that factory.

2.4.2 X^2 Goodness of Fit Test and Checking for Normality

When using the sampling tests outlined in this chapter, such as z, t, F, and X^2, it is important to determine that the samples come from a normal distribution. The X^2 distribution can be used to determine if the observed data follows a particular distribution by comparing the observed data to the expected data from the distribution being analyzed. This is called goodness of fit test. The equation for this test is as follows:

$$X^2 = \sum \frac{(\text{Observed} - \text{Expected})^2}{\text{Expected}} \qquad (2.15)$$

The technique of goodness of fit to a particular distribution starts by dividing the observed data into groupings of interval values or histogram (a column chart that shows frequency data) matching the proposed distribution. Usually five intervals (bins) are used to distribute the data.

For each interval or bin, the number of actual data elements is collected as well as the number of expected data elements based on the proposed distribution. Limits are set for each interval, and two subsets of data are collected for each interval:

1. Observed data values—actual number of collected data points in each interval or bin

2. Expected data values—theoretical probability of number of data points in each interval or bin that would result if the data values were based on the distribution being tested within each interval limit

A X^2 term is calculated for each interval. The total X^2 of terms is collected from all intervals and then compared against the critical value of significance (most likely 0.05) in the X^2 Table 2.10. If the calculated X^2 is within the critical chi-square limits for 0.05 significance, then the proposed distribution is a good fit.

The X^2 test can be used for checking normality of the data, which is a very important attribute to verify before attempting to use a DoE to design or improve a process or a product. Alternate methods for testing for normality will also be shown in Chap. 3.

X^2 (Chi-Square) Goodness of Fit Test Example 1: Linear Distribution

In an experiment, 100 data points were collected and separated into 5 bins of equal intervals. The range of each interval, starting from minimum to maximum, was set similar to five histogram interval (bin) ranges. This is accomplished by dividing the entire range of values into a series of five intervals and then counting how many values fall within each interval. The intervals are specified as consecutive (adjacent to each other) and are usually of equal size.

After completing these operations on the 100 data points, the calculations are shown below:

Interval	Observed (O)	Expected (E)	$(O-E)^2/E$
1	14	20	1.8
2	20	20	0
3	21	20	0.05
4	27	20	2.45
5	18	20	0.2
Total	100	100	4.5

The total X^2 terms add up to 4.5, which is compared to the critical chi-square term for DOF $= 4$ and 0.05 significance (9.488). Since $4.5 < 9.488$, the linear distribution is a good fit for the observed data. The total X^2 terms are also greater than the $1 - 0.05 = 0.95$ significance at DOF $= 4$ (0.711) indicating non-significance at the lower end of the axis or $0.711 < 4.5 < 9.488$.

X^2 (Chi-Square) Goodness of Fit Test Example 2: Normal Distribution

Thirty (30) data measurements were taken from parts produced in a production line. It was decided to check if the parts produced by the line are normally distributed or not before a DoE was to be performed to enhance line quality. The data was sorted prior to checking for normality using X^2 goodness of fit test. The data set is shown below. The 30 data set values' mean μ was 8843.43, and the standard deviation σ was 742.97. The 30 numbers are as follows in ascending order:

7739 7796 7797 7922 8012 8113 8146 8149 8288 8319
8354 8457 **8570** **8787** **8889** **8919** **8956** **8979** **8984** 9095
9326 9380 9450 9534 9565 9619 9820 9858 10170 10310

The 30 data points were divided into 5 histogram bins (intervals) with the ranges as follows: <8000, 8000–8500, 8500–9000, 9000–9500, and > 9500. The number of data points in each bin is observed, and the matching z distribution probability is calculated. The X^2 term for each interval is calculated according to Eq. 2.15 and summed up for the total. A detailed explanation of the calculations for one interval and the summary for all intervals is given below:

Interval 8500–9000 (highlighted in the 30 data set table above):

Observed number of data points within the interval range = 7

z (Lower Interval Range Limit − LIL) $= \frac{x-\mu}{\sigma} = \frac{8500-8843.43}{742.97} = -0.462$

Expected number of values < LIL based on 30 data points $= 30 * f(z = -0.462) =$ 30 * 0.32195 = 9.6586

z (Upper Interval Range Limit UIL) $= \frac{x-\mu}{\sigma} = \frac{9000-8843.43}{742.97} = 0.211$

Expected number of values < UIL based on 30 data points $= 30 * f(z = -0.211) =$ 30 * 0.58345 = 17.5035

Expected number of data points within the interval range = 17.5035 − 9.6586 = 7.845

X^2 term for interval range 8500–9000 = (observed − expected)^2/expected = (7 − 7.845)^2/7.845 = 0.091

This interval analysis summary is highlighted in the total X^2 term table below:

Interval	Observed (O)	Expected (E)	$(O-E)^2/E$
<8000	4	3.844	0.006
8000–8500	8	5.814	0.822
8500–9000	**7**	**7.845**	**0.091**
9000–9500	4	6.844	1.182
>9500	7	5.653	0.321
Total	30	30	2.422

The total X^2 terms add up to 2.422, which is compared to the critical chi-square term for DOF = 4 and the two tails of Table 2.10 at 0.05 and 0.95 significance (9.488 and 0.711). Since $0.711 < 2.422 < 9.488$, the normal distribution is a good fit for the observed data.

2.4.3   F-Test

The t distribution is used to compare sample means, while F distribution is used to compare sample variances, when populations' variances are unknown. Selected F distribution values are given in Table 2.11, for 95% confidence or $\alpha = 0.05$ significance and various degrees of freedom (DOF) of the two samples. It is a representation of the one-sided (right-tailed) area to the right of the computed $F_{0.05,\ v1,\ v2}$ value. Table 2.11 returns the one-sided $\alpha = 0.05$ significance used in DoE analysis. The F.Test in Excel© returns the two-sided values, which is not used in DoE analysis.

The labels for the two DOFs (factors for the horizontal and error for the vertical) correspond to the use of the F tables in determining factor significance in DoE. Other

Table 2.11   Selected F table values for 95% confidence or $\alpha = 0.05$ significance

DOF error	DOF factors								
	1	2	3	4	5	6	10	20	30
1	161.4	199.5	215.7	224.6	230.2	234.0	241.9	248.0	250.1
2	18.51	19.00	19.16	19.25	19.30	19.33	19.40	19.45	19.46
3	10.13	9.55	9.28	9.12	9.01	8.94	8.79	8.66	8.62
4	7.71	6.94	6.59	6.39	6.26	6.16	5.96	5.80	5.75
5	6.61	5.79	5.41	5.19	5.05	4.95	4.74	4.56	4.50
6	5.99	5.14	4.76	4.53	4.39	4.28	4.06	3.87	3.81
7	5.59	4.74	4.35	4.12	3.97	3.87	3.64	3.44	3.38
8	5.32	4.46	4.07	3.84	3.69	3.58	3.35	3.15	3.08
9	5.12	4.26	3.86	3.63	3.48	3.37	3.14	2.94	2.86
10	4.96	4.10	3.71	3.48	3.33	3.22	2.98	2.77	2.70
15	4.54	3.68	3.29	3.06	2.90	2.79	2.54	2.33	2.25
20	4.35	3.49	3.10	2.87	2.71	2.60	2.35	2.12	2.04
25	4.24	3.39	2.99	2.76	2.60	2.49	2.24	2.01	1.92
30	4.17	3.32	2.92	2.69	2.53	2.42	2.16	1.93	1.84
60	4.00	3.15	2.76	2.53	2.37	2.25	1.99	1.75	1.65
∞	3.84	3.00	2.60	2.37	2.21	2.10	1.91	1.57	1.46

values of the F tables for different sample sizes and confidence intervals can be found in the Internet or extracted from the Excel© functions F.INV.RT.

F distribution is the ratio of sample variances (s_1 and s_2) with DOF values ($v_1 = n_1-1$ and $v_2 = n_2-1$) each. This computed F ratio is compared to the critical F table ratio, based on their degrees of freedom. The formula for calculating the variable F is as follows:

$$F = \frac{s_1^2}{s_2^2} \text{ with } v_1 = n_1 - 1. \text{ and } v_2 = n_2 - 1 \qquad (2.16)$$

F should always be > than 1, so the ratio of numerator to denominator should be arranged accordingly.

F-Test Example 1
Two samples are taken from the same normal population: $n_1 = 7$ and $n_2 = 13$. What is the P that the variance of sample $n_1 = 3$ * variance of sample n_2?

Answer: $F_{0.05,6,12} = 3$ for $r_1 = 6$ and $r_2 = 12$
Probability $= 5\%$ approximately from Table 2.11 ($F_{0.05,6,10} = 3.22$ and $F_{0.05,6,15} = 2.79$). From Excel© function $=$ F.INV.RT(0.05,6,12), it returns 2.9961. From the Internet, $F_{0.05,6,12} = 2.9961 = 3$ (http://socr.ucla.edu/Applets.dir/F_Table.html)

F-Test Example 2
A test was performed on two samples of materials, one made in nitrogen (N) environment and the other in air (less expensive). Perform an F-test using hand calculations and Excel© to indicate whether the sample variances are significant to each other or not and which t-test type to use in the future (variances are unknown but presumed equal/not equal).

$N =$ 35 36 42 31
Air $=$ 27 25 26

Answer:
$$F = \frac{S_N^2}{S_{air}^2} = \frac{4.55^2}{1^2} = 20.67$$

They are two degrees of freedom (DOF) associated with the F-test. The placement of the variances in the equation for the F-test is considered by ensuring F-test ratio is greater than 1. The critical F value in Table 2.11 at DOF (3, 2) $= 19.16$.

To determine the critical value in Excel©, the function F.INV.RT(0.05,3,2) is used, since the F-test used for DoE is only one-sided (right-tailed) and returns the same value as Table 2.10 at 19.16. The data analysis function in Excel® can be used to check on the hand calculations above.

	Nitrogen	Air
Mean	36	26
Variance	20.67	1
Observations	4	3
df	3	2
F	20.67	
P(F<=f) one-tailed	0.05	
F critical one-tailed	19.16	

The computed F value is larger than the critical F value (20.7 > 19.16), and the sample variances are significant to each other; therefore, the t-test for the two variables should be performed assuming that variances are unknown but presumed not equal.

2.5   Conclusions

It was shown in this chapter how to perform multiple analysis conditions of applying statistical significance tests to small as well as large samples with population variances known or unknown. Statistical tools such as moving range and the z, t, F, and X^2 tests can be used to compare samples and populations and determine whether their mean or variability differences are significant or not to a particular level.

Variants of z or t-tests of two means were included to determine sample mean confidence intervals for quick visual check of significance to population means and the number required in each sample to determine the error from sample to population mean. In addition, variants of the X^2 test can be used to determine whether a data set follows a particular distribution. This is very important, since many of these tests, including the F-test used for factor significance in DoE, are contingent on the parent population being normally distributed.

Many examples were given to demonstrate these tests in industrial contexts, with detailed step-by-step analysis, and checked with results from commercial statistical software packages. These tests can be used as precursors to DoE analysis by reducing the amount of experiments to be performed prior to the DoE project start. They represent the basic comparative analysis of binary tests of single factor with two levels. In Chap. 3, treatments will be discussed for testing a single factor with multiple levels. The remaining chapters in this book discuss DoE which is design and analysis of experiments with multiple factors and levels. In addition, techniques to reduce the number of experiments while extracting the maximum information from a DoE analysis will be demonstrated in future chapters.

References

1. Highlights of Prescribing Information, Zocor (simvastatin) Tablets, by MERCK SHARP & DOHME LTD. Cramlington, Northumberland, UK NE23 3JU. Copyright © 1999–2015.

Additional Reading Material

Burr, I., *Engineering Statistics and Quality Control*, NY: McGraw Hill, 1953.

Bronshtein, I., *Handbook of Mathematics,* Leipzig: Verlag Press, 1985.

Ducan, J., *Quality Control and Industrial Statistics*, 4th Edition. Homewood IL: Richard D. Irwin. 1995

Guenther W., *Concepts of Statistical Interference*, New York: McGraw Hill, 1973.

Johnson, R., *Probability and Statistics for Engineers*, fifth edition, Englewood Cliffs, NJ, Prentice-Hall, 1994.

Kuehl, R., *Design of Experiments: Statistical Principles of Research Design and Analysis,* 2nd Edition, New York: Duxbury Press, 1991.

Lawson, J., *Design and Analysis of Experiments with R*, Boca Raton, FL Chapman & Hall/CRC, 2014.

Russell, K., *Design of Experiments for Generalized Linear Models*, Boca Raton, FL: Chapman & Hall/CRC, 2019.

Selvamuthu, D. and Das, D., *Introduction to Statistical Methods, Design of Experiments and Statistical Quality Control*, New York: Springer, 2020.

Toutenburg, H.*, Statistical Analysis of Designed Experiments,* 3rd Edition, New York: Springer, 2002.

Walpole R. and Myers, R., *Probability and Statistics for Engineers and Scientists*, New York: Macmillan Publishing Company, 1993.

Wu, J. and Hamada, M., *Experiments: Planning, Analysis, and Optimization,* 3rd Edition, New York: Wiley, 2021.

Young, H., *Statistical Treatment of Experimental Data*, New York: McGraw Hill, 1962.

Chapter 3
Regression, Treatments, DoE Design, and Modelling Tools

In this chapter, the building blocks of DoE experimental design and analysis will be discussed in a step-by-step approach. The first step is regression data analysis for two related variables using modelling to determine their relationships and separating the model coefficients from noise or error in the data. The mean and variability testing tools discussed in the previous chapter will be used to determine significance of the model and its coefficients, using two separate methodologies: one is based on the *t*-test for significance testing of each model coefficient and the other is the F-test for comparing each coefficient variance to the error variance. Every analysis term in the two methodologies will be derived and explained with formulas and examples, including those from statistical software packages.

This approach of using regression analysis for two related variables' coefficient mean and variance will be used for the next statistical analysis of data generated by treatments. It consists of separating the two variables into distinct groupings (levels) from each other and then comparing their respective data outcomes. Treatments' use of multiple levels in designing experiments allows for more in-depth understanding of the relationship of the two related variables: whether the effect of all treatments' level variations is significant or not using the F-test and comparing pairwise treatments' level outcomes for significance using *t*-tests.

Full factorial designed experiments are then introduced to separate the effects of different variables (called factors in DoE) with multiple levels. The outcome data is analyzed by modelling individual factor coefficients versus noise (error) of the experiments. DoE is a form of multiple regression, with the advantage of measuring the effects of all factors and levels concurrently, not one a time. This allows for quantifying factors' effects on each other, called interactions, and including them in the model. A process of model reduction, whereby non-significant factors are removed from the model and added to the error, called pooling, will be shown, resulting in a significant model where the error is expressed as a confidence interval.

Topics to be discussed in this chapter are as follows:

3.1. Regression Analysis. Regression modelling and analysis are introduced with the basic function of linear regression, using the method of least squares to relate two levels of one factor. The two levels consist of one explanatory or input variable, and the other is a dependent (observed) or output variable of a design or process to be evaluated. Linear regression analysis tools presented include (1) linear regression equation, (2) model coefficient estimates and significance using t-tests as well as model summary giving estimates of model variability and significance, and (3) analysis of variance (ANOVA) examining sources of variation and using the F-test to determine significance of the model coefficients (terms). The model terms are converted to the standard error (SE) for each coefficient yielding the 95% confidence interval as well as the t-value and its associated probability for each coefficient. ANOVA divides the sources of model variability into model terms' estimates and the error associated with the design or process. The regression model terms are compared against the error using F-tests, and the resulting probability indicates whether each model term is significant or not.

3.2. Treatment Design and Analysis. Treatments are presented as controlled designed experiments to further enhance the modelling and analysis of one factor with multiple levels. Experiments are grouped in certain repeated levels and outcomes observed to determine the total factor effect. Modelling of the treatments' outcomes is performed by separating each experiment result into three basic components including the total treatments' mean, the effect of each treatment level grouping, and the error associated with each individual experiment. Significance of the total treatments' effect can be quantified by ANOVA using F-test. Significance of pairwise comparison of levels can be quantified by using t-tests. An alternate to the variance F-test, called percentage contribution ($p\%$), is discussed and contrasted with the F-test.

3.3. Full Factorial DoE Design and Analysis. The DoE full factorial model of designing and analyzing multiple factors with multiple levels is presented using the design and analysis tools used for linear regression and treatments. This methodology offers several advantages over the one-factor-at-a-time (OFAT) analysis: (1) the interaction or effect of one factor upon another can be measured; (2) an individual factor can be deemed not significant, while its interaction with another factor can be significant, possibly resulting in a different conclusion about its effect; and (3) certain groupings of factors and levels can be compared to each other to find the best performance of the design or process to be investigated by DoE.

Full factorial implies using all possible combinations of factors and levels. In later chapters, experiment reduction will be presented by limiting all factors to the same number of levels in the DoE. This is accomplished by introducing a fixed set of experiments, called orthogonal arrays (OAs). These OAs cannot be altered for DoE use unless strict relationships are observed between factors and levels.

3.4. Full Factorial DoE Design and Analysis Case Study. An industrial case study is used to illustrate the steps taken in designing and analyzing a full factorial DoE. It is the second DoE performed by a consortium of companies headed by the author to investigate the conversion of electronics manufacturing from leaded to green lead-free soldering technology. It is a good example of using project team approach to DoE by pooling the team members' experience to design and analyze successive DoE phases as knowledge is in increased with each phase.

The first phase was initiated with a feasibility study to identify green soldering technology factors (design space) in controlled settings. The second phase was launched using the knowledge gained from the first DoE to investigate the use of green technology in wider ranges of materials and process parameters (process map) for low-cost high-quality green manufacturing. The third phase was completed as a complement to the second phase for reliability testing in successfully comparing green materials with legacy lead-based soldering using t-tests. The last phase DoE consisted of applying the conclusions of the previous phases toward implementing green technology into the different consortium companies' designs and processes. The first phase DoE using factors with three levels will be discussed in Chap. 5.

3.5. Conclusions

3.1 Regression Analysis

Regression is a methodology to fit certain relationships of observed data that adhere to a particular polynomial equation. The most basic is linear regression which attempts to model the relationship between two variables by fitting a linear equation for the relationship of an input to observed (output) data. One variable is considered as an explanatory or independent input variable, and the other is a dependent or observed output variable. An example would be to model the relationship of new product burn-in temperature duration in the factory on its reliability or failure rate in the field.

The model implies a relationship between the two variables of interest. This does not necessarily imply that one variable causes the other (e.g., burn-in does not cause higher reliability) but that there is some significant relationship between the two variables.

A scatterplot is a useful tool to determine the strength of the relationship between two variables. If there appears to be no association between the proposed explanatory and dependent variables (i.e., the scatterplot does not indicate any increasing or decreasing trends), then a linear regression model to the data is probably not appropriate. This ratio of model prediction to data variation can be seen in Fig. 3.1, data variation types versus model prediction, with three data presentation plots. In the plot on the left, the relationship of the two variables has increasing or decreasing trends and can be relatively predicted by a linear model. Too much noise or error in the data leads to poor model prediction as shown in the plot on the right,

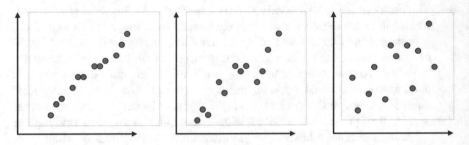

Fig. 3.1 Data variation types versus model prediction

while smaller errors in the data leads to successful model prediction as shown in the middle plot that can be quantified with significance testing.

A valuable numerical measure of association between two variables is the coefficient of correlation R and its square called the coefficient of determinations or R-squared (R^2). R^2 is a value between 0 and 1 indicating the strength of the observed data for the two variables. A value of 0 indicates no relationship, and a value of 1 indicates an absolute correlation between the two variables.

R^2 indicates the proportion of variance in the dependent variable that can be predicted from the model of its relationship to the independent variable. Typically, a value of $R^2 = 0.90$ or 90% indicates strong correlation to determine that the model is significantly applicable to the relationship.

A linear regression line has an equation of the form $y = a + b\,x$, where x is the independent variable and y is the dependent variable. The slope of the line is b, and a is the intercept (the value of y when $x = 0$).

3.1.1 Least Squares Regression

The most common method for fitting a linear regression approximation is the method of least squares. The best-fitted line for the observed data is created by minimizing the sum of the squares of the vertical deviations from each data point to the least squares line. When a point that lies on the fitted line exactly, its vertical deviation = 0. The mean of the dependent variable $\overline{Y}$ is equal to $a + b\,\overline{X}$ where $\overline{X}$ is the mean of the independent variable and should sit directly on the fitted line.

The vertical deviation of each data point y is the difference between the fitted value of y in the model and the observed value, called the residual for that data point. The mathematical sum of the residuals should be close to zero, while the sum of the squares of the residuals is equal to the sum of the squares of the error. Figure 3.2 is a visual illustration of least squares method, showing a fitted data point lying directly on the least squares line, separated from the observed data point by the residual in the vertical axis.

Fig. 3.2 Visual illustration
of least squares method

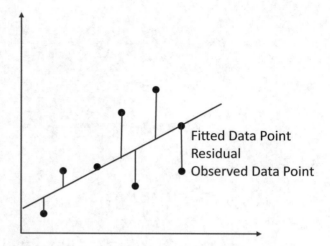

Figure 3.2 is an example of explaining data in terms of conformance and variation. If the data conforms to a particular equation model, then it can be divided into conformance data (data that can be explained) and variation (noise or error) data, which cannot be explained because it originates from uncontrolled factors such differences due to four Ms in the cause-and-effect diagram: Machines, Materials, "Men," and Methods. Significance of dependent (output) data is achieved when variation data is of much smaller value than the conformance data. This ratio of conformance or explained data versus total data (conformance + error variation) is indicated by R^2.

Linear regression analysis uses the method of least squares for analysis of data from a set of measurements. The model regression equation chooses the values that minimize the sum of the squares of deviations of these measurements (residuals). In other words, it minimizes variance or the square standard deviation of the error.

3.1.2   Linear Regression Analysis Using Model Coefficients Estimates

In its simplest form, linear regression analysis can be presented in conforming dependent data to a straight-line model. Taking a set of data of n values of x and corresponding y values, a linear equation of estimates of model coefficients is shown below:

$$\hat{y} = a + b\,\hat{x}, \text{ where } \hat{y} \text{ and } \hat{x} \text{ are model coeffect estimates} \qquad (3.1)$$

$$b = \frac{S_{xy}}{S_{xx}} = \frac{\sum(x - \bar{x})(y - \bar{y})}{\sum(x - \bar{x})^2} = \frac{n\sum xy - \sum x\sum y}{n\sum x^2 - (\sum x)^2} \tag{3.2}$$

and

$$a = \bar{y} - b\bar{x} = \frac{\sum y - b*\sum x}{n} \tag{3.3}$$

Or

$$a = \frac{\sum y - \frac{n\sum xy - \sum x\sum y}{n\sum x^2 - (\sum x)^2}\sum x}{n}$$

$$= \left(\frac{1}{n}\right)\frac{n\sum y\sum x^2 - \sum y(\sum x)^2 - n\sum xy\sum x + \sum y(\sum x)^2}{n\sum x^2 - (\sum x)^2}$$

$$= \frac{\sum y\sum x^2 - \sum xy\sum x}{n\sum x^2 - (\sum x)^2} \tag{3.4}$$

where S_{xx} and S_{yy} are the sum of the squares of the variables x and y, while S_{xy} is the sum of the squares of the variable xy, called the covariance of xy.

The significance of the model coefficient estimates of a (constant) and b (slope) can be calculated using the sum of the square of the error (SSE). There are two methods for calculating SSE: it is equal to the sum of the squares of model residuals or equal to $S_{yy} - b*S_{xy}$:

$$SSE = \sum(\text{Observed data} - \text{Fitted data})^2 = \sum \text{Residuals}^2 \tag{3.5}$$

$$SSE = S_{yy} - bS_{xy} \tag{3.6}$$

where S_{yy} and S_{xy} are sum of the squares of y and xy.

Significance can be evaluated from either the confidence interval (CI) or the calculated t-value of the model coefficients. An intermediate term from the CI Eq. 2.11 called the standard error (SE) can be used to simplify the significance evaluation as follows:

$$\text{Stadard Error} = SE = s_e/\sqrt{n} \tag{3.7}$$

$$CI = \pm t_{\frac{\alpha}{2}, n-2} s_e/\sqrt{n} = \pm t_{\frac{\alpha}{2}, n-2} SE \tag{3.8}$$

$$t_{\text{Coefficient}} = \frac{\bar{x} - \mu}{s/\sqrt{n}} = \frac{\text{Coefficient}}{SE} \tag{3.9}$$

$$s_e = \sqrt{\frac{\text{SSE}}{\text{DoF}_e}} = \sqrt{\frac{\text{SSE}}{n-2}} \qquad (3.10)$$

Equation 3.9 is modified from the original t-value Eq. 2.6 since the coefficient is the shift of the sample mean $(\overline{X})$ from the population mean (μ). Equation 3.10 is a definition of s_e (standard deviation of the error) calculated from the SSE using the general sample standard deviation Eq. 2.2. The regression degrees of freedom (DOF) for s_e is $n-2$, based on the total DOF $(n-1)$ – (regression DOF $= 1$).

The values of SE for the two predicted coefficients in the linear regression model are as follows:

$$SE_a = s_e \sqrt{\sum x^2 / n S_{xx}} \qquad (3.11)$$

$$SE_b = s_e / \sqrt{S_{xx}} \qquad (3.12)$$

where S_{xx} is the sum of the squares of x.

Linear Regression Analysis Example

An example of using linear regression to model the relationship between input variable (X) and dependent variable output (Y) of an electronic amplifier is given in Table 3.1. Model fitted versus residual data for linear regression example is shown in Table 3.2. The calculations for the model including predictions, standard error (SE), 95% confidence interval (CI), t-value, and resultant probability for each

Table 3.1 Linear regression example

	X	Y	XY
	1	10	10
	2	35	70
	3	40	120
	4	70	280
	5	90	450
	6	95	570
	7	110	770
	8	120	960
	9	130	1170
	10	140	1400
	11	160	1760
	12	180	2160
Total	78	1180	9720
$Sxx/Syy/Sxy$	143	29916.67	2050

Table 3.2 Model fitted versus residual data for linear regression example

X	Y	Fitted	Residual	SS (Residual)
1	10	19.49	−9.49	90.01
2	35	33.82	1.18	1.39
3	40	48.16	−8.16	66.56
4	70	62.49	7.51	56.34
5	90	76.83	13.17	173.45
6	95	91.17	3.83	14.70
7	110	105.50	4.50	20.24
8	120	119.84	0.16	0.03
9	130	134.17	−4.17	17.41
10	140	148.51	−8.51	72.39
11	160	162.84	−2.84	8.09
12	180	177.18	2.82	7.96
Total 78	1180	1180	1.818E−06	528.55

coefficient are given below, including two different methods for calculating coefficients a and b:

$$\sum x = 78,\ \sum y = 1180,\ \sum xy = 9720,\ \sum x^2 = 650,\ \sum y^2 = 145950$$

$$S_{xx} = \sum x^2 - \frac{\left(\sum x\right)^2}{n} = 650 - \frac{78^2}{12} = 143,\ s_x = \sqrt{\frac{S_{xx}}{n-1}} = \sqrt{\frac{143}{11}} = 3.61$$

$$S_{yy} = \sum y^2 - \frac{\left(\sum y\right)^2}{n} = 152138 - \frac{(1180)^2}{12} = 29916.67,\ s_y = \sqrt{\frac{S_{yy}}{n-1}} = \sqrt{\frac{29916.67}{11}} = 52.15$$

$$S_{xy} = \sum (x - \bar{x})(y - \bar{y}) = 2050$$

$$b = \frac{S_{xy}}{S_{xx}} = \frac{2050}{143} = 14.336,\ \text{or}$$

$$b = \frac{n\sum xy - \sum x \sum y}{n\sum x^2 - \left(\sum x\right)^2} = \frac{12*9720 - 78*1180}{12*650 - 78^2} = 14.336$$

$$a = \frac{\sum y - b*\sum x}{n} = \frac{1180 - 14.336*78}{12} = 5.1515,\ \text{or}$$

$$a = \frac{\sum y \sum x^2 - \sum xy \sum x}{n\sum x^2 - \left(\sum x\right)^2} = \frac{1180*650 - 9720*78}{12*650 - 78^2} = 5.1515$$

$y = 5.1515 + 14.336\ x$ (model predicted linear regression equation)

SSE = 528.55 (SS residuals from the bottom of Table 3.2), or

SSE = $S_{yy} - bS_{xy}$ = 29916.67 − 14.336 ∗ 2050 = 528.55

$$SE_a = s_e\sqrt{\frac{\sum x^2}{nS_{xx}}} = \sqrt{\frac{SSE}{n-2}}\sqrt{\frac{\sum x^2}{nS_{xx}}} = \sqrt{\frac{528.55}{10}}\sqrt{\frac{650}{12*143}} = 4.47$$

$$SE_b = s_e/\sqrt{S_{xx}} = \sqrt{\left(\frac{SSE}{n-2}\right)/S_{xx}} = \sqrt{\left(\frac{528.55}{10}\right)/143} = 0.608$$

$CI_a = 5.15 \pm t_{a/2,n-2}SE_a$　　　　$= \pm t_{0.025,10} * SE_a = 5.15 \pm 2.228 * 4.47 = 5.15 \pm 9.97 = (-4.82, 15.12)$

$CI_b = 14.336 \pm t_{a/2,n-2}SE_b$　　　$= 14.336 \pm 2.228 * 0.608 = 14.336 \pm 1.35 = (12.98, 15.69)$

Table 3.3 Model coefficient estimates of linear regression example

Predictor	Coefficient	SE Coef.	95% CI	t value	$P > t$
Constant	5.15	4.47	(−4.82, 15.12)	1.15	0.2769
X	14.336	0.608	(12.98, 15.69)	23.58	0.000

$$t_a = \frac{\bar{x} - \mu}{s/\sqrt{n}} = \frac{a}{SE_a} = \frac{5.15}{4.47} = 1.15 \ (P > t = 0.2769 \quad \text{[from Excel}^\copyright \text{ function TDIST}$$
$$(1.15,10,2)]$$
$$t_b = \frac{\bar{x} - \mu}{s/\sqrt{n}} = \frac{b}{SE_b} = \frac{14.336}{0.608} = 23.58 \ (P > t = 0.00, \text{ since } t_b \text{ is very large.}$$

A summary of the above calculations is shown in Table 3.3, model coefficient estimates of linear regression example. The equation for a straight line is determined from the data and linear approximation of $y = 14.336 \, x + 5.15$. This linear equation is the best approximation of a line that goes through all sample data points. The residual from each individual point to the line is a minimum. Statistical analysis was performed to determine if this approximation is significant. Significance in statistical terms implies a degree of certainty about the assumption of a straight-line approximation of the relationship between the two variables x and y.

The significance analysis summarized in Table 3.3 was determined by both confidence interval (CI) and probability $(P > t)$ based on two-sided (tails) t-test. The model constant estimate value of $a = 5.15$ is not significant since the CI of $\pm$ 9.97 exceeds the value of the constant (a) and the CI crosses the zero axis. In addition, the probability of the constant t_a value (1.15) is lower than the critical value of $t_{\alpha/2, \, n-2}$ or $t_{0.025, \, 10} = 2.228$ indicating non-significance. On the other hand, the model slope estimated value of $b = 14.336$ is significant using both the CI (not crossing zero axis) and t-test (t_b value is greater than the critical value of $t_{\alpha/2, \, n-2}$ or $t_{0.025, \, 10} = 2.228$).

The fitted column values shown in Table 3.2 correspond to the use of the model equation $y = 14.336 + 5.15 \, x$ for the set of all 12 X and Y data points. The total of the fitted values equals the total of the actual observed values of Y (1180). The residual column is the difference between individual observed and fitted Y values for each pair of input x and dependent variable output y, as shown in Fig. 3.2. The residual total is close to 0 (1.818E-06), and the sum of the squares of the residuals (SS$_{Residuals}$) = 528.55 is the same as the SS of the error (SSE) calculated above from Eq. 3.6 and shown in Table 3.2.

If the model was reduced by eliminating the non-significant constant a and using only the slope in the equation $y = 14.336 \, x$, then the fitted values of y are reduced, and the residual totals are not equal to zero. In this case, a polynomial regression model will have to be used to increase the accuracy of the model.

3.1.3   Linear Regression Analysis Using ANOVA

Analysis of variance (ANOVA) can be performed on the dependent variable y to divide the variance of the model into explained variance based on the regression model terms and unexplained error or noise of the design or process being analyzed. For this section of linear regression, as well as the next "Treatments" section, these are the only two components of ANOVA being analyzed for significance. For the DoE analysis in Sect. 3.3, the model coefficients will appear in both the model coefficient estimates and the ANOVA. DoE ANOVA can be used to reduce the model by merging or pooling non-significant coefficients or factors into the error.

Table 3.4, regression analysis terms, using ANOVA of the dependent variable y, illustrates the division of conformance versus unexplained variation. Three types of sum of squares (SS) are used in the ANOVA table:

1. SST is a total sum of squares which is the sum of conformance and variation. It is equal to the actual total squares of the dependent Y variable minus a correction factor (CF). CF is needed to compensate for the fact that the mean of all dependent variable numbers Y is not equal to zero. When the mean of Y is zero, the $CF = 0$. This will be shown in Fig. 3.2 as a horizontal least squares line.
2. SSE is the error sum of squares which is a total sum of unexplained variation. It is equal to the error based on the sum of square of residuals, which is the observed (Y) dependent variables minus fitted (predicted) values. For each individual y variable, there is an error or residual between the predicted value of the variable (as it is located exactly on the estimated least squares line) and its actual value y as was shown in Fig. 3.2 and Eqs. 3.5 and 3.6.
3. SSR is the sum of squares of the regression or the sum of squares of explained data conformance. It can be calculated independently or by subtracting error SSE from total SST.

The ANOVA table format is shown in Table 3.5, with formulas attached. Several terms are introduced including the sources of variation which are regression, error, and total. The other columns in Table 3.5 are given as follows:

1. Degrees of freedom is referred to as $DOF = \nu = n - 1$. The DOF of the SST is equal to $n - 1$. Regression has DOF of 1, and the error has DOF equal to the SST DOF minus the regression DOF, or $n - 2$.
2. Sum of squares (SS) are given by the three types discussed earlier (SSR, SSE, and SST).

Table 3.4 Regression analysis terms

Total sum of squares $= SST = \sum y^2 - \left(\sum y\right)^2 \big/_n = \sum y^2 - CF$
Error sum of squares $= SSE = \sum (\text{Observed data} - \text{Fitted data})^2 = \sum \text{Residuals}^2$ or $SSE = S_{yy} - bS_{xy}$
Regression sum of the squares $= SSR = SST - SSE$

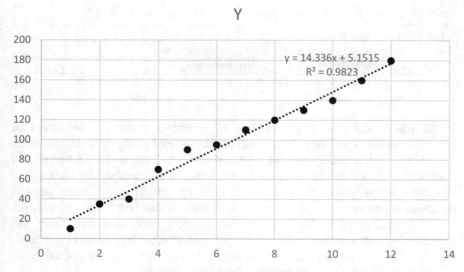

Fig. 3.3 Scatter diagram of linear regression sample data with regression equation and R^2

$$y = b_0 + b_1 * x_1 + b_2 * x_2 + b_3 * (x_{1*}x_2) + b_4 * x_1^2 + b_5 * x_2^2 \qquad (3.13)$$

$$R^2 = \text{SSR/SST} = (1 - \text{SSE/SST}) \qquad (3.14)$$

$R^2 = 0$ denotes no correlation and $R^2 = 1$ is a very positive correlation. Usually, an $R^2 \geq 0.90$ is desired for significance. In the linear regression example, $R^2 = \text{SSR/SST} = 29288.11/299916.67 = 0.9823$ or 98.23%, which is very close to $p\%$ of 98.06%. The difference is that the SSR' subtracts the MSE from the SSR. The two values are close when the error is small but diverge when the error is large. It is normally recommended that a DoE is declared not significant if the error $p\%$ value is greater than 30%.

Figure 3.3, scatter diagram of linear regression sample data with regression equation and R^2, is an Excel© plot of the data in the linear regression example. The regression equation and the R^2 match the calculated results in this section.

3.1.5 Using Linear Regression for Normality Checking

Linear regression analysis can be performed on variable data by transforming them into a new $y = f(x)$ equation. The regression as well as other statistical analysis can be performed on the transformed data, especially with DoE and binary tests discussed in Chap. 2. Another common method of transforming data is to reduce the data range by multiplying the y variables with a power factor <1, to reduce variability and determine coefficient significance of the transformed data. Examples

Table 3.7 Examples of data transformation for improved statistical analysis

Exponential Transformation : $y - ae^{-bx}$, $y = LN(x)$, $y = LN(a) + bx$
Power Transformation : $y = ax^b$, $y = LOG(x)$, $y = LOG(a) + bx$
Reciprocal Transformation : $y = a + b/x$, $y = 1/x$, $y = a + 1/bx$

of data transformation are given in Table 3.7, examples of data transformation for improved statistical analysis. These methods can be used singly or in combination.

Checking for normality was demonstrated in Chap. 2 using the chi-square (X^2) distribution. It can also be achieved by the normal score method by plotting the equivalent normally distributed variable data on the y-axis. The transformed data can be modelled into an estimated linear regression and evaluated for normality using R^2.

The normal score method consists of ranking data in ascending order and transforming it into normal distribution equivalent scores. Once the data is ranked, then a normal probability $f(z)$ is calculated for each rank and the normal score z obtained from the z distribution using the formula below:

$$\text{Probability} f(z) = \frac{i - 0.5}{n} \tag{3.15}$$

where i is the rank number from 1, 2, 3,..., n and n is the number of observed data points. A normal score (NS) is calculated for each data point based on the value of z corresponding to the function of z, $f(z)$.

An example of using normal score method for checking normality is given in Table 3.8, example of observed data and checking for normality. In this example, each of 30 observed data independent variables $y(x)$ is sorted, ranked, and then converted to a normal probability $f(z)$ using Equation 3.15. A normal score is calculated using the z distribution. In this example, the resulting normal score or the z value corresponding to the probability $P = f(z)$ was obtained using the Excel© formula NORM.S.INV(P).

The observed data y was plotted versus the rank order i in a scatter diagram and shown in Fig. 3.4, linear regression plot and histogram of observed versus ranked data example. The linear regression equation model estimate of data versus its rank is shown in the plot as well as the $R^2 = 0.8773$, indicating the linear model is not statistically significant. A histogram of the ranked data with five bins in Figure 3.4 shows almost a lognormal distribution of the 30 data points. This indicated a non-linear relationship between dependent variable y and rank i based on the resulting R^2 value.

The sample data was then converted to a normal score in the last column of Table 3.8 and tested using regression analysis as shown in Fig. 3.5. In this case, the regression model estimate of the normal score is shown in the plot as well as the $R^2 = 0.8882$ indicating the normal model as being closer to statistically significant. A histogram of the data with five bins shows almost a normal distribution, with a center bin range that is higher than the rest of the ranges and data distributed on

Table 3.8 Example of observed data and checking for normality

Observed data y	Sorted data y	Observed rank (i)	Probability $f(z)$	Normal score
110	52	8	0.250	−0.67
120	55	9	0.283	−0.57
257	78	24	0.783	0.78
254	88	23	0.750	0.67
155	88	11	0.350	−0.39
52	99	1	0.017	−2.13
78	99	3	0.083	−1.38
340	110	25	0.817	0.90
221	120	19	0.617	0.30
178	143	17	0.550	0.13
221	155	19	0.617	0.30
143	164	10	0.317	−0.48
165	164	14	0.450	−0.13
99	165	6	0.183	−0.90
348	168	26	0.850	1.04
480	168	29	0.950	1.64
168	178	15	0.483	−0.04
231	185	21	0.683	0.48
88	221	4	0.117	−1.19
164	221	12	0.383	−0.30
55	231	2	0.050	−1.64
450	231	28	0.917	1.38
185	254	18	0.583	0.21
99	257	6	0.183	−0.90
348	340	26	0.850	1.04
480	348	29	0.950	1.64
168	348	15	0.483	−0.04
231	450	21	0.683	0.48
88	480	4	0.117	−1.19
164	480	12	0.383	−0.30

either side of the middle range or mean. However, the data was not symmetrical around the mean as expected.

The observed data in Table 3.8 was transformed by multiplying each dependent variable y by the function $y = LOG\sqrt{x}$. The same analysis process shown for observed data was performed using transformed data, as shown in Table 3.9. The range of transformed data is much smaller and therefore reduces the span of the data ranges in the vertical axis y. The transformed data in the example is shown in Fig. 3.6, normal score regression plot and histogram. In this case, the regression equation estimate is shown in the plot as well as $R^2 = 0.9871$, indicating the model approximation of a normally distributed transformed data is highly significant. A histogram of the data with five bins shows a normal function, with a center bin range that is higher than the rest of the ranges and data distributed almost equally on either

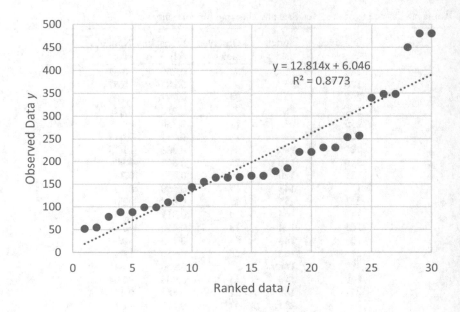

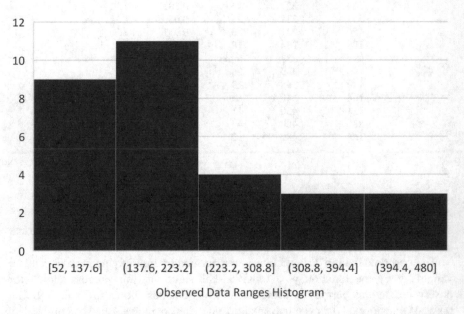

Fig. 3.4 Linear regression plot and histogram of observed versus ranked data example

side of the middle range or mean, in a symmetrical pattern as expected in a normal distribution.

In summary, the observed data in Table 3.8 was analyzed for a linear regression model based on Eq. 3.15 and proved to be non-significant in Fig. 3.4. Then it was

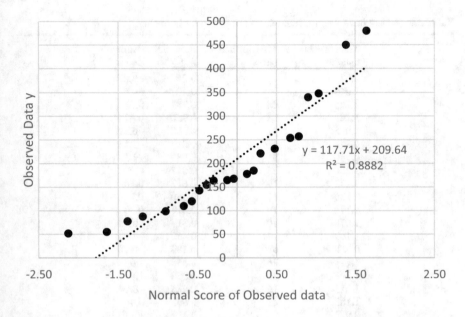

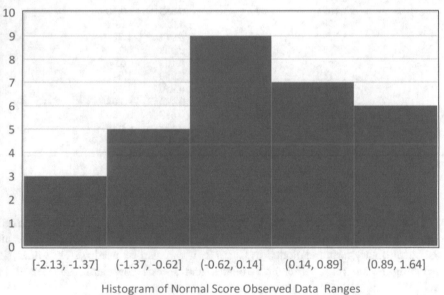

Fig. 3.5   Normal score regression plot and histogram for example of observed data

analyzed for normality and converted to a normal score, and a normal model was not significant in Fig. 3.4. The observed data was transformed using the function $LOG\sqrt{x}$ and checked for normality using Table 3.9 and Fig. 3.6 and proved that transformed data was significantly normal with an $R^2 = 0.9871$.

Table 3.9 Transformed data for example and checking for normality

Transformed data y	Transformed rank (i)	Probability $f(z)$	Normal score
1.021	8	0.250	−0.67
1.040	9	0.283	−0.57
1.205	24	0.783	0.78
1.202	23	0.750	0.67
1.095	11	0.350	−0.39
0.858	1	0.017	−2.13
0.946	3	0.083	−1.38
1.266	25	0.817	0.90
1.172	19	0.617	0.30
1.125	17	0.550	0.13
1.172	19	0.617	0.30
1.078	10	0.317	−0.48
1.109	14	0.450	−0.13
0.998	6	0.183	−0.90
1.271	26	0.850	1.04
1.341	29	0.950	1.64
1.113	15	0.483	−0.04
1.182	21	0.683	0.48
0.972	4	0.117	−1.19
1.107	12	0.383	−0.30
0.870	2	0.050	−1.64
1.327	28	0.917	1.38
1.134	18	0.583	0.21
0.998	6	0.183	−0.90
1.271	26	0.850	1.04
1.341	29	0.950	1.64
1.113	15	0.483	−0.04
1.182	21	0.683	0.48
0.972	4	0.117	−1.19
1.107	12	0.383	−0.30

This methodology can be used to perform statistical testing on the transformed data to satisfy the requirement for using normally distributed populations for all statistical binary tests mentioned in Chap. 2. It can also be used for DoE analysis including t-tests for coefficient estimates and F-tests for significance used in ANOVA.

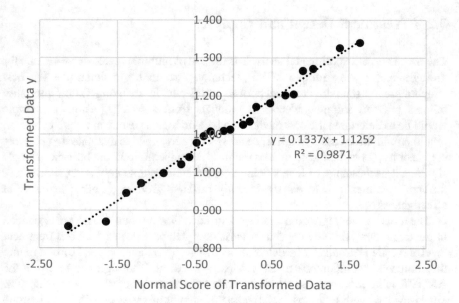

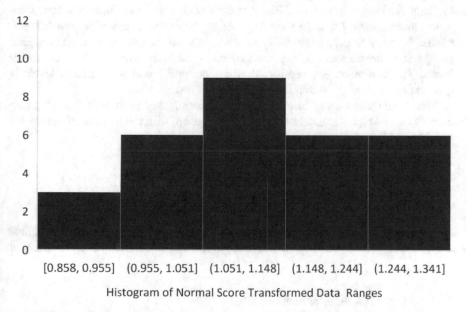

Fig. 3.6 Normal score regression plot and histogram of transformed data

3.2 Treatment Design and Analysis

The next level of experimental analysis is the use of multiple levels or treatments for a single factor. An example of this technique would be to determine the best temperature or time to increase burn-in efficiency in reducing infant mortality failures prior to shipping electronic products from a factory. Another example would be to determine the best medication level to treat a particular health condition or ailment. In these cases, significance analysis is performed to evaluate two possible conclusions: (1) is the factor such as the time or temperature in the burn-in example significant in the range of levels being considered and separately and (2) are these different treatments or levels significantly different from each other in terms of effectiveness?

The tools used in regression analysis disued in the previous section are expanded in treatment analysis. The dependent variable y data is repeated for each treatment level and called "groups." The outcome of each group is then analyzed by examining the sources of variation due to the groups or the error. The regression term in the ANOVA table is substituted by a term called "between groups" of treatments. The error is called "within groups," relating to the amount of variation within each group versus variation group to group. The F-test, which is the ratio of between- to within-group mean square (variance), can be used to determine the significance of treatments. In other words, treatments are not significant when the variability within groups is more substantial than the variability between groups due to treatment levels. The treatments are significant when variability due to treatment levels is greater than variability within the levels themselves.

Treatments can be expressed in terms of identifying each level by its group number and its repetition numbers within each group, with i = number of treatments and j = number of repetitions, as follows:

Treatment group $1 = y_{11} + y_{12} + y_{13} + \ldots + y_{1j}$
Treatment group $2 = y_{21} + y_{22} + y_{23} + \ldots + y_{2j}$
Treatment group $3 = y_{31} + y_{32} + y_{33} + \ldots + y_{3j}$
Treatment group $i = y_{i1} + y_{i2} + y_{i3} + \ldots + y_{ij}$

Table 3.10, ANOVA testing of treatments, shows the arrangement for statistical analysis using ANOVA for determining significance of treatments of i = number of

Table 3.10 ANOVA testing of treatments

Source of variation	DoF	Sum of squares	Mean square (Variance)	F-Test
Between Group	$i-1$	SSR	$V(R) = SSR/i\,(-1)$	$V(R)/V(E)$
Within Groups	$i(j-1)$	SSE	$V(E) = SSE/i\,(j-1)$	
Total	$ij-1$	SST	$V(T) = SST/(ij-1)$	

treatments and j = number of repetitions. The formulas for calculating the sum of squares terms in Table 3.10 are modified from the previous section as follows:

$$SST = \sum y_{ij}^2 - \left(\sum y_{ij}\right)^2 \big/_{ij} \tag{3.16}$$

$$SSR = \left(\sum y_i^2\right)\big/_j - \left(\sum y_{ij}\right)^2 \big/_{ij} \tag{3.17}$$

$$SSE = SST - SSR \tag{3.18}$$

where $\left(\sum y_{ij}\right)^2 \big/_{ij}$ is the correction factor (CF).

3.2.1 Treatment Design and Analysis Example

In this example, treatment analysis will be shown to illustrate the use of fitted versus residual modelling and testing for significance based on treatment ANOVA. A set of five treatments was used to determine the efficacy of a hypothetical drug to treat low high-density lipoprotein (HDL) cholesterol, where the higher value is better and minimum desired HDL level is 40 mg/dL. The five treatments are called A–E corresponding perhaps to placebo, 10, 20, 40, and 80 mg doses with varying sample sizes adding to a total of 17 patients treated. Table 3.11, treatment decomposition into groups, treatment deviations, and residuals, shows the observed HDL level data and a model of its decomposition into model elements.

Table 3.11 breaks down the original 17 data points into the grand mean, the treatment deviation which is the fitted mean shift due to each treatment, and the residuals within each treatment. The residuals show the variation (noise or error) within each treatment deviation.

The sum of the residuals for each treatment is equal to zero. If the same position data point is added from each table, the original treatment number is reconstituted.

Table 3.11 Treatment decomposition into groups, deviations, and residuals

Treatment groups					Treatment deviations					Residuals				
A	B	C	D	E	A	B	C	D	E	A	B	C	D	E
31	32	37	41	41	−3	1	1	−3.33	5	1	−2	3	11.33	3
31	35	36	23	37	−3	1	1	−3.33	5	1	1	2	−6.67	−1
28	38	29	25	36	−3	1	1	−3.33	5	−2	4	−5	−4.67	−2
30	31				−3	1				0	−3			

30	34	34	29.7	38 = Deviation mean
33 = Grand mean				

Table 3.12   Treatment example of ANOVA

Summary

Groups	Count	Sum	Average	Variance
A	4	120	30	2
B	4	136	34	10
C	3	102	34	19
D	3	89	29.67	97.33
E	3	114	38	7

Anova

Source of variation	SS	df	MS	F	P-value	F critical
Between groups	151.33	4	37.83	1.61	0.24	3.26
Within groups	282.67	12	23.56			
Total	434	16				

For example, the first number in treatment A is $y_{11} = 31$ and can be decomposed into the following:

y_{11} = grand mean + treatment A deviation + residual
$y_{11} = 31 = 33 - 3 + 1 = 31$

The calculations for the ANOVA table from the data for the 17 patients in Table 3.11 are as follows:

Grand mean = 33; grand $\sum y^2$ = 18,947, grand total $\sum y$= 561, CF = $561^2/17$ = 18,513
SST = $\sum y^2$ − CF = 18,947 − 18,513 = 434
Treatment total $(\sum Column)^2/i = (120)^2/4 + (136)^2/4 + (102)^2/3 + (89)^2/3 + (114)^2/3$
   = 18,664.33
SSR = total Col^2/n − CF = 18,664.33 − 18,513 = 151.33
SSE = SST − SSR = 434 − 151.33 = 282.67

The full treatment data summary and ANOVA is shown in Table 3.12, treatment example of ANOVA. This table was generated using the Excel$^©$ Analysis ToolPAK, extracted from the add-ins available for data analysis using the "ANOVA single factor" function in the Data Analysis drop-down menu. The data from Excel$^©$ matched the hand calculations above.

The DOF for between groups is $i-1 = 4$ based on the five groups. Since there is a total of 17 patients, the total DOF is 16. The within-group (error) DOF is calculated as DOF (total) − DOF (between) = 16−4 = 12. Another way to interpret the DOF (error) is to think of the original 5 treatments with 1 patient each and then consider that the rest of the 12 patients' data constitute the replications of the original 5 treatments. If the theoretical drug was perfect treatments for all patients, the

Table 3.13   Treatment example of pairwise *t*-tests

Comparison tests	Probability @ 95% 2-tails
t-test A*B	0.06
t-test A*C	0.14
t-test A*D	0.95
*t-test A*E*	*0.00 (significant)*
t-test B*C	1.00
t-test B*D	0.44
t-test B*E	0.14
t-test C*D	0.52
t-test C*E	0.25
t-test D*E	0.23

12 replications would be equal to each corresponding treatment, and the SSE would be zero. The computed F-value is calculated to be 1.61, which is lower than the critical $F_{0.05,\ 4,\ 12} = 3.26$, and its probability is 0.24 >> 0.025 critical probability of 95% one-tailed. Therefore, the overall treatment using the theoretical drug is not significant (effective) in treating low HDL.

An important outcome of treatments is the ability to compare each treatment level to other levels, to investigate whether the mean differences are significant, even if the total treatment experiment is not significant, by performing pairwise *t*-tests. If the treatment experiments are significant, then the pairwise testing of the levels can determine which levels should be selected for significant results.

Table 3.13, treatment example of pairwise *t*-tests, shows comparing the means of different levels of treatments. Only one pairwise *t*-test (levels A*E) was significant, while the total treatment experiment was not. The number (*n*) of *t*-tests to be performed comparing pairwise means of number of levels *i* is equal to the following formula:

$$\text{Number of pairwise } t - \text{tests for treatments}$$
$$= (i - 1) + (i - 2) + \ldots + 1 \tag{3.19}$$

where *i* is the number of treatment levels. For this example of five treatments, 4 + 3 + 2 + 1 =10 *t*-tests have to be performed.

3.2.2   Significance Determination Techniques and p% Contribution

The significance testing in the ANOVA of SS$'$ and $p\%$, shown in Table 3.5, presents an alternate method to test for significance of linear regression and treatments, as well as DoE factorial analysis in the next section and chapters. The $p\%$ is used to indicate significance when it is $\geq 5\%$, which is roughly equivalent to 95% confidence, and is sometimes used as an alternative to the F-test. It includes the subtraction of the MSE from the SSR (between) using the following equations in treatment ANOVA:

$$SSR' = SSR - MSE * DOF \text{ (between)} \tag{3.20}$$

$$SSE' = SSE + MSE * DOF \text{ (between)} \tag{3.21}$$

$$SST' = SST \tag{3.22}$$

The $p\%$ value is a percentage distribution of the sources of variation. The rationale for the modification of the SSR and SSE is assuming that SSR includes the variability due to the error and therefore must be reduced by subtracting the error MSE times the between-group DOF. To keep the SST constant, the SSE must be increased by the same amount. Obviously, if the SSE is small, the effect of the modification of sum of the squares is correspondingly small, and $p\%$ for between groups approaches the coefficient of determination R^2 discussed earlier in the chapter.

If the DoE error is large, as will be seen in this treatment example, then the two methods can yield different indications of significance, as shown below:

$SSR' = SSR - MSE * DOF(\text{between})$
$SSR' = 151.33 - 4 * 23.56 = 57.11$
$SSR\ p\% = 57.11/434 = 13.16\%$
$SSE\ p\% = 100 - SSR\ p\% = 86.84\%$
$SSE' = 282.67 + 4 * 23.56 = 376.89$
$SSE\ p\% = 376.89/434 = 86.84\%$

These two methods of significance testing in ANOVA using F-test or $p\%$ could result in different conclusions, as the F-test is a binary outcome, while the $p\%$ is a variable which can be used to quantify the levels of significance. When the modified SSE (SSE') error in DoE analysis is $\leq 30\%$, then the two methodologies converge and yield almost the same conclusions using the $p\% = 5\%$ threshold for significance.

For DoE's ANOVA of multiple factors and levels, the use of the $p\%$ threshold for significance could have negative implications in model reduction, possibly leading to different regression equations and predicted (fitted) values of the model. For this reason, many popular DoE and statistical software analysis packages such as Excel$^{\copyright}$ and Minitab$^{\copyright}$ do not provide $p\%$ calculations in their ANOVA.

3.3   Full Factorial DoE Design and Analysis

Full factorial DoE design and analysis is the next link in the chain of regression analysis discussed in this chapter, allowing for the opportunity to investigate or optimize a design or a process using multiple factors and levels. While it might be called multiple regression for using similar tools and techniques of regression and treatments, it offers great advantage over separate one-factor-at-a time (OFAT) experiments in determining the significance of each factor and its relationships with other factors, called "interactions."

The traditional approach to DoE assumes the designer is interested in building a model of a product or process and then using DoE to verify or reduce the model. This model could be in the form of a regression equation indicating the relationship of the design to the factors used in the DoE. Prior knowledge or experience of the designer about the relevance of possible factor candidates can help reduce the scope of the DoE into a manageable size for analysis. The tools of TQM and DMAIC can help in reducing the number of factors to be used in full factorial DoE. The use of design teams for DoE projects is recommended to pool knowledge of team members for selecting a proper number of factors and levels and hence DoE size reductions.

3.3.1   Limiting DoE Scope with Design Space and Process Map

One of the major concerns in full factorial DoE is reducing the size and scope of the experiments. Large DoE can be difficult to justify to management, considering the time and effort undertaken versus the perceived benefits. In addition, keeping track of large amounts of data pertaining to each experiment performed could be vulnerable to errors of labeling, tracking, and recording. The use of critical test equipment and measuring instruments for performing DoE can compete with other users lengthening the time the experiments can be completed and analyzed. This might render DoE results obsolete by the time the DoE project is concluded.

The design team can use several options to reduce the number of experiments in full factorial DoE, by limiting the numbers of factors and levels. These techniques can be summarized as follows:

1. Reducing the number of factors by pooling the collective knowledge of the team using TQM and DMAIC tools and methodologies. These were discussed in Chap. 1 and included tools such as brainstorming, cause-and-effect diagrams, as well as Pareto charts to better understand the design or process relationships and select the most probable factors for DoE analysis.
2. Reducing the number of factors or levels by performing smaller experiments prior to a full factorial DoE. Experiments for selecting proper levels for each factor include the statistical testing tools disused in Chap. 2 or treatments discussed

earlier in this chapter. Examples of prior testing for DoE are those seeking a reduction in material, technology, or supplier selection as factors or levels.

3. Limiting the investigation scope of DoE by specifying an operational range for the design output. This range can be specified by considering safe operation of the design or analyzing the intended customer use of the product being designed. This range is called the design space, which is determined by the selection of factor levels where the conclusions of the DoE analysis are valid. DoE analysis does not predict the behavior of the design model outside of the design space.

4. Like the design space, a process map could be developed for the process being analyzed, limiting each factor to a recommended range of levels based on the prior testing or analysis. For example, if two factors interact to produce additive damaging result, such as the heat setting and belt speed in a conveyorized oven, care must be taken not to exceed the allowable maximum temperature of parts curing in a DoE.

5. For both design space and process map, careful consideration of DoE level selections should be undertaken. Factor levels that are too close such as $< \pm 5\%$ difference in input (independent) variable are not recommended since the effect of that factor on DoE outcomes can be overwhelmed with DoE error. On the other hand, wide differences in factor levels such $> \pm 20\%$ cannot demonstrate the presence of non-linear behavior of the dependent variable or output.

More discussion of factor and level selection will be forthcoming in the next chapters when using orthogonal arrays (OAs).

3.3.2 Full Factorial DoE Design Analysis Using Interactions

The first step in performing a full factorial (FF) DoE is completing the suggested steps for limiting DoE scope in Sect. 3.3.1. The design team should agree on an appropriate set of dependent variables of factors and levels and the suggested dependent variable (DoE outcome) definitions. The total number of experiments to be performed should equal the following equation for (n) factors:

$$\#\text{FF experiments} = \#\text{Levels}_{\text{factor 1}} * \#\text{Levels}_{\text{factor 2}} * \ldots * \#\text{Levels}_{\text{factor } n} \quad (3.23)$$

An example of determining the number of experiments for a three-factor (A, B, and C) DoE with five, three, and two levels is $5 * 3 * 2 = 30$ experiments, shown in Fig. 3.7. It is a visual presentation of the combination of all 30 experiments in the following sequential manner:

1. $A_1 \, B_1 \, C_1$
2. $A_1 \, B_1 \, C_2$
3. $A_1 \, B_2 \, C_1$
4. $A_1 \, B_2 \, C_2$
5. $A_1 \, B_3 \, C_1$

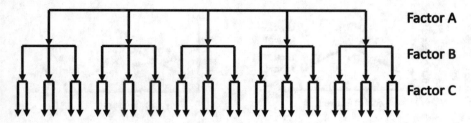

Fig. 3.7 Visual presentation of 30 experiments for 3 factors at 5, 3, and 2 levels

6. $A_1 B_3 C_2$
7. $A_2 B_1 C_1$
8. $A_2 B_1 C_2$
9. $A_2 B_2 C_1$
10. $A_2 B_2 C_2$
11. $A_2 B_3 C_1$
12. $A_2 B_3 C_2$
13–24. There are six experiments with A_3 and six experiments with A_4
25. $A_5 B_1 C_1$
26. $A_5 B_1 C_2$
27. $A_5 B_2 C_1$
28. $A_5 B_2 C_2$
29. $A_5 B_3 C_1$
30. $A_5 B_3 C_2$

To determine significance, more experiments must be performed to generate error to test the significance of the model as well as each factor. The most common method is to replicate the total set of experiments. For the previous example, one replication represents 30 additional experiments for a total of 60 experiments. In Chap. 6, methodologies for reducing the number of replications for generating error will be discussed.

The FF model coefficients as well as those for each factor can be estimated and tested for significance as was shown in Table 3.3. The FF model ANOVA is like regression analysis shown in Table 3.5, except that each factor can be analyzed independently as well as its relationship to other factors, called interactions.

Interactions are defined by the number of factors and could be calculated in multiple ways. In the example of three factors A, B, and C of five, three, and two levels, the number of interactions would be three two-way and one three-way. For each interaction, there are several groupings of levels from one factor interacting with levels from another factor. Interaction grouping can be plotted against the dependent variable y (DoE outcome) to determine the best settings among the interacting factors. For the example of three factors A, B, and C of five, three, and two levels, interactions and their groupings are as follows:

A*B with 15 pairwise level interactions (5*3 levels = 15) = [A_1B_1, A_1B_2, A_1B_3, A_2B_1, A_2B_2, A_2B_3,...,A_5B_1, A_5B_2, A_5B_3]

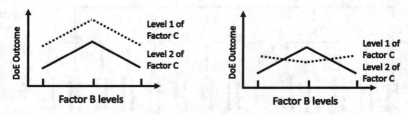

Fig. 3.8 Visual presentation of interaction for factors B and C

A*C with ten pairwise level interactions (5*2 levels = 10) = [A_1C_1, A_1C_2, A_2C_1, A_2C_2,..., A_5C_1, A_5C_2]
B*C with six pairwise level interactions (3*2 levels = 6) = [B_1C_1, B_1C_2, B_2C_1, B_2C_2, B_3C_1, B_3C_2]
A*B*C with all 30 levels (5*3*2 = 30), usually indicated as [(A*B) * C]

These interactions are usually plotted with each factor level used as multiple linear plots in the x-axis versus DoE outcome in the y-axis as shown in Fig. 3.8 for pairwise interactions. The pairwise plots show the two-factor interactions only while averaging out all other factors. Each interaction plot point represents a division of the 30 experiment outcomes' (y variable) data points into groupings of similar factors/levels. For example, each of the 6 interaction B*C plot points is the mean of 5 experiments, such as the group B_1C_1 that appears 5 times in the 30 experiments corresponding to the 5 levels of factor A. The B*C interaction plot is independent from factor A levels since they are averaged out for each of the six plot points.

Figure 3.8 is a visual presentation of factor B (three levels) and C (two levels) interactions. It shows two possible interaction plots of six points [B_1C_1, B_1C_2, B_2C_1, B_2C_2, B_3C_1, B_3C_2] versus DoE outcomes. The plot to the left shows an additive relation between the levels of factors B and C, indicating that the B*C interaction is linear and probably not significant. The plot in Fig. 3.8 to the right shows a non-linear relationship between the two factors, and the interaction B*C is probably significant. The significance of an interaction can be statistically determined using ANOVA, and if true, the regression equation and predicted value of the DoE outcome should include the interaction term.

ANOVA for full factorial DoE is very similar to treatments, with regression being substituted by factors and interactions. Table 3.14 shows the allocation of the degrees of freedom (DOF) in the ANOVA table for a three-factor model of DoE. The DOF of each factor is (number of levels−1). The total DOF for the model is the multiplication of (factor levels) − 1, and the error DOF is derived from the replication of the entire DoE.

Each factor and interaction model coefficient can be calculated and tested for significance. A non-significant factor or interaction can be removed from the model and added to the error. This process of reducing the model, called pooling, should continue until only significant factors or interactions remain. If there are no significant factors after pooling, then the DoE is deemed a failure, and a new DoE should be initiated with alternate factors and levels.

Table 3.14 Full factorial DoE distribution of degrees of freedom (DOF)

Source	Degrees of Freedom (DOF)
Factor A (x levels)	$x-1$
Factor B (y levels)	$y-1$
Factor C (z levels)	$z-1$
Interaction factors (A*B)	$(x-1)*(y-1)$
Interaction factors (A*C)	$(x-1)*(z-1)$
Interaction factors (B*C)	$(y-1)*(z-1)$
Interaction factors (A*B*C)	$(x-1)*(y-1)*(z-1)$
Model total	$(x*y*z)-1$
Error	# of replications (x*y*z)
DoE total	Total # of experiments-1

A regression equation and predicted outcome should be calculated from the reduced model of only significant factors and interactions. There are three possible alternative goals for the DoE outcome or the dependent variable y: larger is better, smaller is better, and a target value is desired. The corresponding levels from the factor and interaction plots can be selected and plugged in the regression equation for the predicted value of the DoE.

In some models, there might be non-significant factors, while their interactions with other factors are significant. In this situation, the regression equation should contain the significant interactions, and the DoE predicted values can reflect these contributions from interactions. While it is not possible to set interaction levels, using the interaction plots such as the ones demonstrated in Fig. 3.8 can clearly show a desired minimum or maximum outcome depending on the selected levels of the two interacting factors.

The methodology described above can result in achieving the desired outcome, whether it is maximum or minimum. For a desired target value, there might be several paths in the model by alternate selection of factors and interaction levels. In this case, other methods can influence which path is the most appropriate, such the reduced variability plots discussed in Chap. 9 "Variability Reduction Techniques and Combining with Mean Analysis."

For multiple desired outcomes, where there are several different outcomes to be analyzed from the same set of DoE experiments, a balance of factor and level selection should be undertaken to best satisfy all outcomes at the same time. In many cases, one factor level would improve one outcome, while a different level of the same factor would improve a second outcome. The decision of which level to select might be using a ratio of the two levels if there are variables or selecting a particular level (such as material A or material B) that influences one outcome more than another outcome. The strength of this influence can be measured either by the percent contribution $p\%$ or the direct ratio of SSF (factor) versus SST computed in the ANOVA table. Multiple outcome analysis is discussed in Chap. 10 "Strategies for Multiple Outcome Analysis and Summary of DoE Case Studies and Techniques."

When DoE analysis is combined with other conditions such as the cost of the design, the materials, or the process being investigated, different combinations of factors and levels could be tested to see if they compare favorably by using multiple pairwise t-tests as was performed in Table 3.13. For full factorial DoE with large number of experiments, multiple comparison procedures can be performed. One of the step-down procedures, starting with the most significant differences in the model and grouping the factor/level combinations into equally significant ranges, is called the Ryan-Einot-Gabriel-Welsch (REGW) multiple range procedure. It is used in the case study of full factorial DoE in the next section.

3.4 Full Factorial DoE Design and Analysis Case Study: Green Electronics Manufacturing

The worldwide movement to phase out lead from electronic products presented many challenges for companies throughout the electronics supply chain. The author brought together eight local firms to collaborate using DoE methodologies to investigate successful high-quality and reliability manufacture of lead-free printed circuit boards (PCBs), using the legacy tin-lead (Sn-Pb) soldering technology as baseline. A series of DoE phases were undertaken to build up the knowledge from one DoE project to the next, starting with a feasibility study of green lead-free technology.

The results of the phase I DoE showed that zero-defect soldering is achievable with green materials. DoE factors included solder alloys (Sn/Ag/Cu, Sn/Ag, Sn/Bi), PCB surface finishes (OSP and ENIG), thermal profiles (soak and linear with various times above liquidus), and reflow environments (air and nitrogen). Following thermal cycling, PCBs were visually inspected for quality performance outcomes, and their electronic components' leads were pull tested for reliability outcomes. This phase I DoE will be outlined in detail in Chap. 5 "Three-Level Factorial Design and Analysis Techniques."

Phase II DoE was initiated, and a second set of test PCBs with a wide variety of materials and processes was manufactured. Several solder pastes based on Sn/Ag/Cu (SAC) solder alloys were tested with a wide variety of surface finishes and reflowed using either air or nitrogen. Quality and reliability outcome assessments using visual inspection as well as component lead pull test analysis were performed and data analyzed in this section to determine significance of factors and levels as well as comparison with legacy (Sn-Pb) solder technology.

3.4.1   Summary of Phase I Green Electronics DoE Case Study

Phase I DoE for green soldering compared five factors and some of their interactions with other factors:

1. Green Lead-Free Solders with three levels (95.5Sn-3.8Ag-0.7Cu (SAC), 96.5Sn-3.5Ag, and 57Sn-43Bi)
2. PCB Surface Finishes with two levels: ENIG (Electroless Nickel/Immersion Gold) and OSP (Organic Solderability Preservative)
3. Reflow Atmosphere with two levels (air versus nitrogen)
4. Reflow Profile (ramp and soak versus linear ramp)
5. Oven Time Above Liquidus (TAL) (60, 90, and 120 s)

Quality and reliability outcomes were measured using visual defects per industry standards and solder joint pull test strength as reflowed and after thermal cycling. Optimal results were obtained using these two outcomes with the Sn-Ag-Cu (SAC) solder alloy. It was determined that the linear ramp profile approach decreases the process map requiring tighter controls but used less energy and thus is more cost-effective. Table 3.15 shows several DoE phase I combinations that yielded visual defect-free results which was extremely encouraging for the project team. These results were incorporated into phase II DoE design.

3.4.2   Phase II of Green Electronics DoE Case Study

The project team decided to use the SAC alloy technology proven by phase I results as the most significant green soldering material and measure its interactions with a wider range of PCB surface finishes and soldering environments. The team was reluctant to use nitrogen recommended in phase I because of higher costs. It was decided to repeat the use of atmosphere as a third factor in hope of finding a combination of factors and levels that would result in successful green soldering in air. In addition, a fourth factor was added which was component technology for the PCBs using different component types and finishes for reliability testing. It was

Table 3.15 Visual defect-free results of phase I Green Electronics DoE

Paste	Surface finish	TAL(secs)	Soak	Nitrogen
Sn/Ag/Cu	ENIG	90	No	Yes
Sn/Ag/Cu	ENIG	120	Yes	Yes
Sn/Ag	ENIG	60	Yes	Yes
Sn/Ag	ENIG	90	No	Yes

Fig. 3.9 Phase II DoE test vehicle

decided not to include it in phase II, but rather test this factor independently and compare it to legacy (Sn-Pb) soldering using *t*-tests.

Prior to phase II DoE launch, the team evaluated six SAC green solder (paste) candidates from various suppliers and narrowed the choice to three levels. The team used separate tests in the selection process including:

- Smell.
- Paste ability to roll when printing.
- Print quality consisting of measuring total height and height variance across the deposited solder pad. Paste heights were measured on the same pad locations for each print and paste. These locations include a fine pitch pad, a standard IC pad, a BGA pad, and a discrete component pad.
- Clogging using three triggers (after each print, after 30, and after 60). Three PCB panels were printed for each SAC paste at each time frame.
- Sticking characteristics, to squeegee and/or stencil.
- Ease of cleaning the stencil.

A new PCB test vehicle was manufactured for phase II DoE with the different combinations of factors and levels. It is shown in Fig. 3.9.

Phase II DoE was designed with the following factors and levels, and the design matrix is shown in Table 3.16. The 30-experiment matrix was replicated once for determining the error for a total of 60 test vehicles:

1. PCB Surface Finishes with five levels (Solder Mask Over Bare Copper with Hot Air Solder Leveling (SMOBC/HASL), Matte Finish Tin (Sn) Electroplate,

Table 3.16 Green Electronics Phase II DoE design matrix

Experiment no.	Surface Finish	Solder Paste	Atmosphere
1	(1) SMOBC/HASL	"A"	Air
2	(1) SMOBC/HASL	"A"	Nitrogen
3	(1) SMOBC/HASL	"B"	Air
4	(1) SMOBC/HASL	"B"	Nitrogen
5	(1) SMOBC/HASL	"C"	Air
6	(1) SMOBC/HASL	"C"	Nitrogen
7	(2) OSP	"A"	Air
8	(2) OSP	"A"	Air
9	(2) OSP	"B"	Nitrogen
10	(2) OSP	"B"	Air
11	(2) OSP	"C"	Nitrogen
12	(2) OSP	"C"	Air
13	(3) ENIG	"A"	Nitrogen
14	(3) ENIG	"A"	Air
15	(3) ENIG	"B"	Air
16	(3) ENIG	"B"	Nitrogen
17	(3) ENIG	"C"	Air
18	(3) ENIG	"C"	Nitrogen
19	(4) Matte Sn	"A"	Air
20	(4) Matte Sn	"A"	Nitrogen
21	(4) Matte Sn	"B"	Air
22	(4) Matte Sn	"B"	Air
23	(4) Matte Sn	"C"	Nitrogen
24	(4) Matte Sn	"C"	Air
25	(5) Imm. AG	"A"	Nitrogen
26	(5) Imm. AG	"A"	Air
27	(5) Imm. AG	"B"	Nitrogen
28	(5) Imm. AG	"B"	Air
29	(5) Imm. AG	"C"	Air
30	(5) Imm. AG	"C"	Nitrogen

Immersion Silver (Ag), Organic Solderability Preservative (OSP), and Electroless Nickel/Immersion Gold (ENIG).

2. Lead-free SAC solders from three different suppliers (A, B, and C), all incorporating no-clean fluxes. The same reflow profile was used for all three levels.
3. Soldering Reflow Atmosphere with two levels (Air versus Nitrogen).

There were two other factors not used in the phase II DoE but analyzed separately using *t*-tests.

4. Component lead finishes with four levels (matte Sn plating, tin/silver/copper, nickel/palladium/gold, and nickel/gold). These were pulled after thermal cycling

to measure reliability outcomes and compared to legacy (Sn-Pb) solder using pairwise *t*-tests.
5. Legacy (Sn-Pb) soldered PCBs were made as control cohorts to compare against the green solder technology using pairwise *t*-tests.

The 60 test vehicles (PCBs) were manufactured according to the design matrix in Table 3.16. All 60 PCBs were tested for several soldering defect types as well as the total defects for each experiment run.

3.4.3 Analysis of Phase II DoE

Nine main categories of common defects were selected, and all PCBs were inspected using criteria from industry standards. Those observed defects were photographed and recorded into a spreadsheet. ANOVA using Minitab® is shown in Table 3.17 using 0.35 power transformed for total defect data for reduced data ranges.

Statistically significant results are indicated in bold in Table 3.17. The overall DoE model analysis of variance was statistically significant. All three main factors (Surface Finish, Solder Paste, and Atmosphere) were found to be significant. In terms of their relative importance, Atmosphere explained the most variation, Paste was the next most important, and Finish explained the least. A significant main factor in the ANOVA only indicates that one or more of the factor's levels differ significantly from one another, not necessarily that all levels differ.

Multiple comparison tests are provided to identify which of the levels differ significantly from one another. The two most important factors' interaction (Paste*Atmosphere) also had a statistically significant interaction. A nearly statistically significant interaction was Finish*Atmosphere, shown in bold and italic.

Total defect is the sum of the nine different types of defects measured in this DoE and shown in Table 3.18. The significance is indicated by either Y (Yes) or N (No). When a defect type is significant by factor, their significance rank is identified by a second digit after the letter Y. For example, Paste is the second significant factor for

Table 3.17 Green Electronics Phase II DoE ANOVA

Source	DOF	Sum of squares	Mean square	F-value	Probability > F
Finish	4	44.6816022	11.1704005	**7.33**	**0.0003**
Paste	2	78.9621665	39.4810833	**25.91**	**<0.0001**
Atmosphere	1	132.3624551	132.3624551	**86.88**	**<0.0001**
Finish*Paste	8	16.0395976	2.0049497	1.32	0.2735
Finish*Atmosphere	4	15.2827444	3.8206861	*2.51*	*0.0629*
Paste*Atmosphere	2	54.3289039	27.1644519	**17.83**	**<0.0001**
Finish*Paste*Atmosphere	8	21.8263762	2.7282970	1.79	0.1184
Model	29	363.4838458	12.5339257	**8.23**	**<0.0001**
Error	30	45.7076735	1.5235891		
Total	59	409.1915194			

Table 3.18 Green electronics statistically significant outcome effects' summary

Property	Main effects			2 Factor interactions			3 Factor interactions
	Finish	Paste	Atm.	Finish*Paste	Finish*Atm.	Paste*Atm.	Finish*Paste*Atm.
Total defects	Y-3	Y-2	Y-1	N	N	Y	N
Cold solder joining	Y-3	Y-2	Y-1	N	N	Y	N
Nonwetting	N	Y-2	Y-1	N	N	N	N
Solder balls	Y	N	N	N	N	N	N
Dewetting	N	N	N	N	N	N	N
Bridging	Y	N	N	N	N	N	N
Pin/Blow holes[a]	N	N	N	N	N	N	N
Shiny[b]	Y-1	Y-2	N	N	N	N	N
Residue[b]	N	Y-1	Y-2	N	N	N	N
Smooth[b]	N	N	N	N	N	N	N

[a]Pin/blow holes: no defects of this type were observed
[b]Qualitatively measured properties

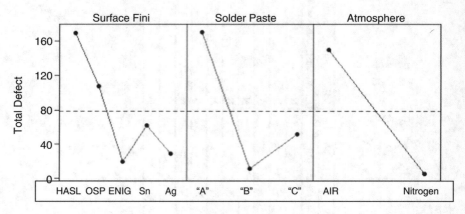

Fig. 3.10 Phase II DoE plot of total defects versus factor coefficients

nonwetting defects after Atmosphere in Table 3.18. A potential risk in performing a statistical analysis on such a sum is that differing mechanisms may come into play for differing defect types. This combined variable may blur such effects. In terms of the defect types in the total, total defect was most strongly correlated to cold solder joining defects.

Statistical analyses are subsequently provided for each specific defect type, as summarized in Table 3.18, statistically significant outcome effects' summary. Statistically significant outcomes are indicated with a Y and are bolded. When more than one statistically significant main outcome is indicated, they are ordered in importance (1=greatest impact).

Figure 3.10 shows phase II DoE plot of total defects' outcome versus independent factor coefficients. The mean total defects for the 60 test vehicles were 78.17, while the main factors have statistically significant effect on the outcome. The most important factor (both in ANOVA and factor plots) is Atmosphere with Nitrogen significantly outperforming Air in total defects by 6 to 150.33. The next factor was Solder Paste with A, B, and C scoring 171.25, 11.35, and 51.90 total defects, respectively. The least important significant factor was Surface Finish with five levels scoring total defect means of 170.25, 107.58, 20.42, 62.92, and 29.67. All 3 factors have the same mean as the total 60 experiments' mean, which is a good check on the analysis. Significance tests were performed for each defect type and each factor with results similar to the total defect significance with slight differences.

It is apparent that the best outcome would include the levels that correspond to the lowest value of total defects in each plot of Fig. 3.10, which are Surface Finish level 3 (ENIG), Solder Paste level 2 (B), and Atmosphere level 2 (Nitrogen). Both Nitrogen and ENIG represent the most expensive choice of levels. It was decided to test all possible combinations of the factors and interactions to search for significance groupings using the REGW multiple range test for total defects discussed in the previous section to determine if there are other combinations that might not be significantly different than the best combination of ENIG, B, and Nitrogen. This

Table 3.19 Green electronics phase II DoE factor multiple range test for total defects

	Groupings	Mean	N	Level
Surface finish	A	170.25	12	SMOBC/HASL
	B	107.58	12	OSP
	B	20.42	12	ENIG
	B	62.92	12	Matte Sn
	B	29.67	12	Imm. Ag
Paste	A	171.25	20	A
	B	11.35	20	B
	C	51.90	20	C
Atmosphere	A	150.33	30	Air
	B	6.00	30	Nitrogen
Paste x atmosphere	A	337.10	10	"A" Solder Paste*Air
	B	98.70	10	"C" Solder Paste*Air
	C	15.20	10	"B" Solder Paste*Air
	C	7.50	10	"B" Solder Paste*Nitrogen
	C	5.40	10	"A" Solder Paste*Nitrogen
	C	5.10	10	"C" Solder Paste*Nitrogen

REGW test replaces the pairwise tests for treatment levels shown in Table 3.13, for each phase II DoE factor and interaction.

Table 3.19 shows the REGW grouping letters (A, B, and C) for levels that are not significantly different from each other. For factor Surface Finish, the finish level SMOBC/HASL significantly differs (poorly) from all other finishes. No other finishes were found to be statistically different from one another at the significance 0.05 level. For factor Paste, all paste levels were found to differ significantly from all other pastes. Solder Paste "B" performed best. For factor Atmosphere, Nitrogen performed significantly better than Air.

For the only significant interaction, Paste*Atmosphere, Solder Paste A*Air combination was significantly worse than all other combinations. Solder Paste C*Air combination was significantly worse than all other remaining combinations. The bottom four combinations could not be statistically differentiated from each other in terms of reduced total defects. This was good news for the project team, since it would suggest that using Solder Paste B with Air would work in any of the last four levels of surface finish (all levels except for SMOBC/HASL).

Interaction Plot - Data Means for Total Defect

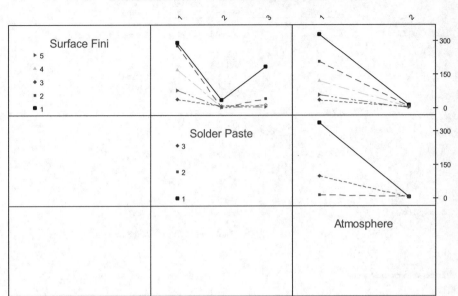

Fig. 3.11 Phase II DoE interaction plots

Figure 3.11, phase II DoE interaction plots, is a typical output of Minitab® which shows all three two-way interactions of the three main factors, though only the interaction Paste*Atmosphere is significant (the lower interaction plot to the right). Level 2 of Solder Paste (B solder), versus both Atmosphere levels 1 (Air) and 2 (Nitrogen), represented by the bottom dashed line, shows almost the same value of total defects, as was indicated by the multiple range test above. For the non-significant interaction Surface Finish*Atmosphere, ENIG (level 3) shows the best performance versus the two Atmosphere levels of Air or nitrogen represented by the dashed line with diamond or rhombus ends, while the next best performance is Silver (immersion Ag, level 5) represented by the dashed line with forward arrow-head ends. These two Surface Finish levels correspond to the best performers in Fig. 3.11. The same can be said for the non-significant interaction Surface Finish*Solder Paste for the same best performers (ENIG and Ag) for low total defects. Solder Paste B (level 2 of solder paste) performs the best for all five levels of Surface Finish.

In conclusion, the phase II DoE analysis above indicated that green electronics soldering can be performed with lower cost and high quality over a wide range of surface finishes using solder B and air. The reliability tests were performed using *t*-tests of the different combination for each surface finish and solder B with air.

For reliability testing, pull tests were performed after thermal cycling to simulate aging on different types of components and their finishes. Since the pull test results were influenced by factors such as the PCB component pad size footprint as well as component finishes, it was decided not to include these factors as part of the second phase DoE. Instead, different combinations of component types and finishes were tested using pairwise t-tests versus legacy (Sn-Pb) soldered PCBs used as baseline. Only Solder Paste B and Air were used in the t-tests with all four best performing levels of Surface Finish and compared with legacy Sn-Pb for each component types and finish. All tests showed non-significance mean differences between green solder B and legacy Sn-Pb baseline. The consortium labeled this phase of pairwise testing DoE project phase III.

The project team concluded that phase II DoE demonstrated the feasibility of low-cost, high-quality, and reliability green electronics soldering and to proceed to implement the next phase IV of testing using consortium companies' specific designs, materials, and processes.

3.5 Conclusions

This chapter introduced the concepts of regression and modelling to investigate the relationship of dependent variable outcomes versus independent input factor(s). The relationship was discussed in three steps, starting with linear regression to fit a least squares line model to data and check for significance. The next step is treatments, which allows significance model testing for multiple levels of one factor for significance as well as t-tests for pairwise comparing levels against each other. The third step is full factorial DoE for model testing of multiple factors with multiple levels.

All these steps were analyzed using several methodologies. These include model coefficient estimates and their individual confidence interval, t-values, and probability coefficient computed t versus critical t. ANOVA was demonstrated to distribute the sources of variation among regression or error, allowing for significance testing of the model using F-tests. For the multiple factor DoE, the factors and their interactions were added to both the model coefficient estimates and ANOVA. Techniques for model reduction by pooling factor coefficients and generating regression equation with predicted outputs were also shown.

A full factorial DoE design and analysis case study was used to demonstrate some of the techniques of developing a DoE project. They include pre-project analysis for reducing the number of experiments and scope of the DoE project, as well as post-project experiments' design and analysis of related factors that are difficult to control in the project. Special considerations for multiple outcome analysis and successive DoE projects were demonstrated and will be discussed with more depth in later chapters. Methods for using DoE to optimize dependent variables' outcomes (minimum, maximum, and target) were shown, with additional methodologies to select the best combinations of factors and levels for target outcomes.

The use of statistical software packages such as Excel$^{©}$ and Minitab$^{®}$ was demonstrated as data analysis tools to reduce hand calculations. In addition, using software for plotting the analysis results in graphical form indicating the relationship of factors and levels can be helpful in selecting proper factor levels for model output prediction.

Additional Reading Material

Aktar I. and Shina, S. "The Development of Test Methodologies for Determining the Formulations and Mechanical Properties of Lead Free Nano Solder", PAN PAC conference of the SMTA, Kona, HI, February 2017.

Faraway, J., *Linear Models with R*, Boca Raton, FL: Chapman & Hall/CRC, 2014.

Gao, F., Gu, Z. and Shina, S., "Lead-Free Nanosolder Based Nanomaterials Assembly and Integration, APEX conference, Las Vegas, NV, March 2014.

Goos, P. and Jones, B., *Optimal Design of Experiments: A Case Study*, New York: Wiley, 2011.

Hwang J., *Environment Friendly Electronics: Lead-Free Technology,* Electrochemical Publications Ltd, 2001.

Madhyastha, N., Ravi, S. and Praveena A., *A First Course in Linear Models and Design of Experiments*, New York: Springer, 2020.

Morris, M., *Design of Experiments: An Introduction Based on Linear Models,* New York: Chapman & Hall/CRC, 2017a.

Morose, G., Shina, S., Anderson, R. and Pasquino, H., "Failure Analysis Results for Evaluating the Long Term Reliability of Test Vehicles Assembled with Lead Free Materials", IMAPS New England Conference, Boxborough MA, May, 2010a

Morose, G., Shina, S. and Farrell, R.," Supply chain collaboration to achieve toxics use reduction", Journal of Cleaner Production, (2010b), doi:10.1016/j.jclepro.2010.04.004.

Morose G., Shina, S., et al.," Quality Evaluation of Lead-free Solders, Halogen-free Laminates, and Nanomaterial Surface Finishes for Assembly of Printed Circuit Boards", Journal of Surface Mount Technology, (2010c) v22 i4; pp 13-21).

Morose G. and Shina, S., "Transitioning to Lead-Free Electronics: Now a Business Necessity", New England Environmental Journal, September 2005.

Morris, M., *Design of Experiments: An Introduction Based on Linear Models*, Boca Raton, FL: Chapman & Hall/CRC, 2017b.

Shina, S., *Green Design and Manufacturing*, McGraw Hill professional series, April 2008.

Shina, S., et al; "Quality and Reliability Testing of Circuit Boards Assembled With Lead Free Components, Finishes, Soldering Materials And Processes In Simulated Production Conditions", PAN PAC conference; January 2009.

Wernicki, E., Gu, Z. and Shina S., "Novel Nanocomposite Solders for High Reliability Electronics Applications", Partnerships for Innovation (PFI) Grantee Conference. Atlanta, GA, June 2018.

Chapter 4
Two-Level Factorial Design and Analysis Techniques

The concepts of Design of Experiments (DoE), alternately known as "robust design" or "variability reduction," have been used to reduce some of the sources of manufacturing variation or manipulate a design toward its intended performance. In addition, DoE is an excellent methodology to investigate new materials and process alternatives.

In this chapter, the general model of full factorial DoE experiments discussed in the previous chapter will be explored further by outlining a formal approach to conducting a DoE project. One of the issues of full factorial DoE projects is that the number of experiments increases sharply as the number of factors and levels increase. By limiting the DoE design to the same number of levels for each factor, such as two levels, experiment reduction can be realized. The objective is to reduce the number of experiments without sacrificing the accuracy of DoE results. The DoE design team can pool their collective knowledge of the design or process to manipulate the DoE methodology to produce maximum results with minimum experimental effort.

The two-level factorial design tasks include DoE project goal setting in identifying single or multiple output variables (outcomes) and how to measure and classify them. It is followed by modelling the design or process to be investigated by selecting input factors and levels and fitting them into predetermined orthogonal arrays (OAs). The OA selection is dependent on the team's goals of exploring the interactions of the factors, as well as the financial, legal, and time constraints of the project. In addition, the team must plan for quantifying DoE error through different methodologies discussed in this chapter as well as in Chap. 6.

Once the experiments are completed and results recorded, DoE analysis tasks begin with verifying the design model by calculating factor coefficients as well as ANOVA discussed in previous chapters to determine significant factors and interactions. The regression equation and predicted outcome(s) can be calculated and compared with the original project goals. The team can achieve the goal of satisfying a single or multiple outcome(s) by selecting proper factor levels for each one. In addition, the team can balance factor and level selection in a compromise to meet the

S. Shina, *Industrial Design of Experiments*, https://doi.org/10.1007/978-3-030-86267-1_4

goals of multiple conflicting outcomes. Finally, the team can validate the predicted DoE outcome(s) by conducting a confirming experiment with the selected factors and levels.

Two different approaches to DoE design and analysis will be considered with differences explained and recommendations provided for the proper use of each method or a combination of both. One is the Taguchi method, which emphasizes the use of team knowledge for reducing experiments through the use of specialized tools and pre-selected combinations of factors and interactions. The other is the classical DoE methodologies as outlined in the book *Statistics for Experimenters*, by Box, Hunter, and Hunter. Its approach is to achieve DoE goals with no initial bias in factor and level selection. It provides for sequential DoE to build knowledge about the design to be investigated while reducing the number of experiments needed to verify the model. The intent of this chapter is to help the design team select the proper approach of which tools and methods to use for their DoE project from either methodology. DoE design and analysis should focus on the same goals which are reduction of experiments and verification of design models. DoE projects should reach the same conclusions in selecting proper factors and levels to meet desired outcomes.

If design teams are not satisfied with DoE conclusions either due to lack of achieving goals of the project or the desire to refine the model, a follow-on DoE can be designed to investigate the model further based on the results of the initial DoE. Different factors or levels can be considered as well as selecting newly available materials, equipment, and processes. These DoE sequencing techniques will be discussed in Chap. 8.

Topics to be discussed in this chapter are as follows:

4.1. DoE Definitions, Expectations, and Processes

A general DoE definition is given, as well as expectations of proactive improvement of the product and process design. The reasons for DoE are discussed, including the effects of error and other external and internal conditions that contribute to the variability of products and processes. An algorithm for conducting a DoE project is presented including model construction, selection of input factors and levels, as well as output quality characteristics and verifying outcomes.

4.2. Two-Level Factorial DoE Design

The set of DoE alternatives using predetermined orthogonal arrays (OAs) with two-level factors is presented with emphasis on factor interactions and confounding. Different modes of using OAs for screening (saturated), partial, or full factorial are explained. The use of two-level OA in experiment reduction and sequential approach to DoE projects is examined. Different DoE methodologies are presented with examples for each.

4.3. Two-Level OA Analysis and Model Reductions

The analysis approach of the previous chapter is extended to include two-level factors and their interactions. ANOVA is shown with assigned degree of freedom (DOF) for each factor. Various types of handling error are listed including pooling of factors to reduce the model.

4.4. Two-Level DoE Case Studies

These case studies are drawn from various uses of two-level DoE in design and manufacturing in different industries. They illustrate the use of the different types of two-level arrays in full and partial factorials and saturated (screening) design modes using OA such as L8, L16, and L32. Error is handled either in reducing the model by pooling the smallest factor(s) or by experiment replications. Other modes such as screening design, half factorial, and reduced error by using CenterPoint or replicating some lines are discussed in Chap. 6 and summarized in the list of industrial experiments in Chap. 10.

4.5. Conclusions

4.1 DoE Definitions, Expectations, and Processes

Evolving quality definitions include understanding of the cost of quality or its inverse, quality loss. It is presented visually in Fig. 4.1. The classical definition of quality is "conformance to specifications," where quality loss is considered binary: if a design or process meets specifications, then it is acceptable, and no quality loss is observed, no matter how close the design or process outcome is to its specifications. Quality loss is applied to design as well as manufacturing: any attribute of a product that is outside of the specification limits results in a quality loss, but there is no loss if the attribute is within specifications.

New quality techniques such as Six Sigma attempt to coordinate design activities with manufacturing variability. Specification limits, set by design, are considered against reducing manufacturing variability to produce robust products. Ideal design and manufacturing goals for shipped products are to meet specification nominals. The evolving quality definitions shifted from meeting specification limits to

Fig. 4.1 Evolving quality definitions

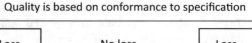

Quality is based on conformance to specification

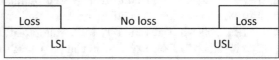

| Loss | No loss | Loss |

LSL   USL

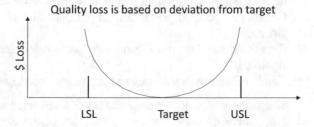

Quality loss is based on deviation from target

$ Loss

LSL   Target   USL

Fig. 4.2 Arrangement of
DoE elements

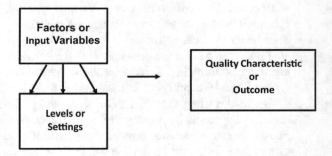

achieving nominal performance targets. In this manner, products shipped to customers inherently carry quality loss if it does not meet intended performance target. Reducing manufacturing variability or increasing design robustness through DoE is the best method to minimize quality loss. The quality loss function (QLF) is quantified by reducing both process average shift from specification nominal and variability. It will be discussed further with examples in Chaps. 8 and 9.

DoE is a systematic approach for determining the effect of factors and their possible interactions toward achieving particular outcome(s). It is used to quantify the source and resolution of variation and magnitude of error when comparing DoE outcomes to desired target goals. Using DoE, a design or process can be modelled and then manipulated to provide a target or minimal/maximum performance outcome or reduce its variability or both. This is accomplished by setting factors that affect DoE outcomes to predetermined input levels and analyzing the output using a set of predetermined OA, as shown in Fig. 4.2, arrangement of DoE elements. These OAs can be used in screening, partial, or full factorial modes.

The objectives of DoE are to adjust the model to the desired outcome by properly choosing the best combination of factors and levels. Increased knowledge about the model can be achieved by having project teams decide on the relevant factors that influence the model and assigning different values to each factor. This is accomplished by collecting maximum information from the DoE results using minimum resources. The factors can be categorized to determine which factors affect the average, variability, and both average and variability or have no effect on DoE quality characteristics, depending on the result of varying factor levels on DoE outcomes. This was shown in Fig. 1.22, example of DoE outcomes with four factors.

The results of a DoE experiment can be one of the following:

1. Identify the most important factors that influence DoE outcome(s).
2. Determine factor(s) levels for significant factors which optimizes desired outcome.
3. Determine best or most economic setting for factors' levels that are not significant.
4. Validate (confirm) responses and implement in design or production.

The success of a DoE experiment is not determined solely by just achieving the desired outcome. Important information about the design or the manufacturing process can be gleaned from any DoE project. This information can be used in future DoE or through using more traditional quality improvement processes such as TQM or DMAIC. Information gained from DoE can be listed as follows:

1. Factors that are significant for influencing design or process mean, reducing variability or both and which factors are not significant. If none of the factors are found to be significant, then the design of the experiment must be repeated to include factors or levels not previously considered.
2. Proper balance between mean shift from target versus variability reduction by choosing proper factor levels. The choices of certain factor levels can shift process mean, while others can reduce the variability, or both. While good results can be obtained by moving process mean to the maximum or minimum possible or to achieve a design target, this action can be tempered by selecting alternate factors and levels to achieve greater robustness in reducing variability. Quality loss can be used to make decisions based on economic considerations of mean shift versus variability.
3. The predicted DoE outcome can be determined when design factors are set to specified levels. Expected error and CI can also be calculated for predicted outcome(s).
4. Predicted outcome(s) can be validated with confirming experiments.

4.1.1 DoE Lifecycle Process

DoE is best characterized as making several assumptions in a model about the design or the process being studied, quantifying these assumptions by the choice of factors and levels, and then running experiments to determine if these model assumptions are valid. It is a mix of several tools that have been developed to optimize performance based on statistical analysis, significance tests, and error calculations.

The steps in conducting a successful DoE are as follows:

Conducting industrial DoE experiments involves using many of the tools of quality common to TQM or DMAIC. It is advantageous to form a team to perform the tasks of designing DoE and interpreting the results. Teams have shared experiences in design and can achieve broad consensus on different approaches to DoE and designs being analyzed.

The success of a DoE project is dependent on selecting the proper team members, identifying the correct factors and levels, focusing on optimizing and measuring the quality characteristics, and analyzing the results. Steps in performing a successful DoE project are as follows, as outlined in Table 4.1:

Table 4.1 DoE project process steps

DoE project definition
Select project goals
Design space/process map
Select DoE project team
Select and quantify outcomes
Identify model factors and levels
Use TQM/DMAIC to select independent factors
Factor level selection
When to use two, three, or multiple levels
Design the experiment
Select interactions (all, some, or none)
Select a predetermined OA
Select DoE error source
Select experiment sequence and dividing (blocking)
Perform the DoE and collect outcomes based on selected OA
Data analysis for each outcome
Analyze response using model coefficient estimates and ANOVA
Graphical analysis for best level setting of significant factors
Reduce model if necessary, by pooling non-significant factors
Predict outcomes using regression equation
Combine multiple outcome response by balancing selected factors and levels for each outcome
Predict reduced model desired outcomes using selected levels to match responses
Confirm (verify) the predicted outcome before implementation

- DoE Project Definition. The first task in performing a DoE project is to outline the goals of the project and to select a design or process to be investigated. While only one design or process outcome can be optimized at a time, many character-istics can be measured while performing experiments and analyzed separately, from the same DoE matrix. The final factor and level selections can be a mix of those recommended by analyzing each characteristic, by balancing the objectives for each.

  - Design Space or Process Map. This task focuses on creating the boundaries for designing a product or a process operating condition. DoE should not be constrained to a small part of the design, hence not having the opportunity to study the interactions of the different parts of the total design. On the other hand, DoE should not be all encompassing in attempting to optimize a wide span of product design use by customers or extreme operating conditions of processes. Ideally, the total design use by customers should be analyzed and a compromise made in developing a plan for a succession of DoE. Each additional DoE provides more knowledge about the design or process to be optimized.

- DoE Project Team Selection. A DoE project team should be selected to conduct the experiments and perform the analysis. The team should be composed of those knowledgeable in the materials, product, and process and should solicit inputs from all stakeholders involved in the design to be optimized. It is not necessary to have in-depth technical understanding of the science or technology of the design, but team members should have experience in similar or previous designs. Knowledge in statistical methods, and in particular DoE techniques, should be available within the team, either through a statistician or someone having received training or experience in DoE.
- Select and Quantify Outcomes. Outcomes are DoE output responses, sometimes called the quality characteristic(s). One of the first tasks of the team is to select the design or process outcome to be optimized, either as a maximum, minimum, or set to a desired target. This is the dependent variable in the regression analysis discussed in Chap. 2. Multiple outcomes could be analyzed using the same set of DoE as input. Techniques for combining the knowledge formed by multiple response analysis into a recommended set of factor and level selections will be discussed in Chap. 10.
- Identify Model Factors and Levels. DoE projects can be performed in two approaches: one method is to select multiple factors and use a screening DoE experiment, usually a saturated design (to be explained in the next section), to narrow down factor selections. A follow-on DoE, preferably a full factorial experiment, is used to complete the selection of the optimal factors and their levels. The second method is to have the team members consider the DoE project as a single opportunity to try out as many as possible factors and levels, because of the lack of time or resources available. In this case, partial factorial experiments are used, with some assumptions as to the relationships of factors (interactions), to maximize the benefits and minimize the resources and time spent on a single experiment.

  - TQM/DMAIC tools such as brainstorming and cause-and-effect diagram techniques should be used to select the number of factors and the different levels for each factor. The selection process should outline factors that are as independent as possible from other factors and hence are additive in controlling the quality characteristic(s) to be optimized. This is important in reducing the interactions of factors, which are difficult to quantify statistically. Interactions can be considered for all factors and levels for full factorial experiments or some of the factors in partial factorial experiments. Factor interactions are ignored in screening experiments.
  - Factor level selection. Good selection of levels for each factor used in DoE is important in achieving the proper design space. Levels that are either too close together or too far apart in value should not be selected, because they do not represent a continuum of the impact of the factor on the quality characteristic.
  - Two-level factors could be selected for new designs or if there is little confidence in the adequacy of the current design, based on the collective judgment of the DoE team. By choosing two levels, more factors can be tested

within a small number of experiments as will be demonstrated later. In addition, the direction of better design performance can be ascertained for future DoE. Two-level DoEs offer only linear relationships of factors versus DoE outcomes. They cannot indicate any non-linear or polynomial relationships. They are disused in this chapter.

– Three-level factors could be chosen if the project team is confident in that the current design is performing adequately but needs to be improved upon. The current level should be in the center of a 10–20% span represented by the other two levels. In this manner, a DoE can help in finding an optimum set of factor levels in the design space or process map. Three-level DoE are discussed in Chap. 5.

– Multiple-level factors should be chosen for survey DoE. The team can select many new technologies or materials within one factor to identify which one can perform best within DoE design space. The number of multiple levels can be equal to squares of two or three levels, such as four, eight, or nine levels. They are easier to perform since they fit easily into the set of predetermined two- or three-level OA. Multiple-level factors are discussed in Chap. 8.

– The selected levels should be well within the operating range of the factors within the design space or process map. In the green soldering full factorial experiment mentioned in Chap. 3, the combination of soldering temperature factors and levels should not result in having components soldered beyond their maximum temperature and time exposure specifications.

• Design the Experiment

– Deciding on factor interactions and their levels is very critical to DoE project size. In the case of new designs, or if there is no consensus within a DoE team as to the relationship of factors, full factorial design should be selected to gain maximum knowledge. When project teams feel confident about using their experience in ignoring some interactions, partial factorial DoEs are recommended. Ignoring all interactions, as in the case of screening designs, could be used to reduce the number of factors to a manageable significant few that can then be further investigated in a full factorial DoE. This screening option represents a possible two-step alternative to large single full factorial DoE projects by reducing the numbers of factors to be fully investigated. A variant technique of the two-step screening DoE, called "foldover," can sequentially examine each significant factor at a time. Screening-based DoEs will be discussed in Chap. 8.

– Select a predetermined orthogonal array (OA). Most DoE experiments use a set of predetermined OA available to conduct the experiment, with two, three, or mixed levels. There are only certain combinations of factors and their levels available in OA to perform experiments. Compromise might be necessary to achieve DoE project economy by selecting a given number of factors and levels that can fit within one of the predetermined OAs. There are only a small number of these arrays for two and three levels, and their size increases geometrically with the number of factors selected.

- Select DoE error source. DoE regression statistical analysis requires a source of error to test for the significance of the model coefficients. Section 4.4 discusses the types of error handling techniques in DoE design and analysis. More details on error handling are given in Chap. 6.
- Select experiment sequence and dividing (blocking). OAs are arranged in terms of the number of experiments, factors, and levels. The experiments should be conducted in a random order from the array matrix. If possible, they should be completed within one shift or 1 day to reduce the effect of uncontrolled variables such as weather or operators conducting the experiments. If that is not possible, DoE can be divided into blocks to be performed separately while minimizing the effect of the uncontrolled variables. This will be explained in Chap. 6.

- Perform the DoE and Collect Outcomes Based on Selected OAs. The physical specimen manufactured using DoE factor levels or subjected to process parameters or materials as prescribed by the OA should be labeled and preserved for later DoE analysis. Measurements should be made by the same operators and using the same equipment, if possible, to minimize variability due to GR&R.

  - Data analysis for each outcome. Once DoEs are performed, outcome data can be analyzed graphically to determine the optimal settings of significant factor levels. In addition, statistical analysis can be performed to determine the significance of each factor's effect on each outcome, using model coefficient estimates and analysis of variance (ANOVA). Significant factors can be set to the proper level, and non-significant factors can be set to the most economic conditions. If they are part of significant interactions, then they can be set according to the interaction analysis.
  - Graphical analysis of the data is sufficient to determine the best level setting for each significant factor to adjust outcomes to target and reduce design variability.
  - Statistical analysis provides more details on the effect of each factor on DoE outcome and its significance. Model reduction can be achieved with pooling of non-significant factors and adding them to the error. The reduced model after statistical analysis can be used to create a regression equation for predicting DoE outcome(s).
  - Combine multiple outcomes' response by balancing selected factors and levels for each outcome into a set of universal factor and level recommendations. This is further explained in Chap. 10.

- Prediction with Multiple Outcomes. Once the graphical and statistical analysis are completed, multiple outcomes can be predicted based on the reduced model of factor levels. These choices could be a compromise between setting the DoE outcome to target value and reducing variability. A recommended factor level might cause variability to be reduced, yet at the same time, the process mean could be shifted from target. Another case is when multiple characteristics are to be optimized using one experiment with many separate outcomes. For example, a DoE could be performed to design a new plastic material to be injected molded. The material and process design can have several desired characteristics including modulus of elasticity, density, amount of flash after injection, gel time, flow rate, and free rise volume. A DoE could be designed using an OA that determines what ratios and composition of raw materials are to be used, as well as the injection molding machine process parameters. Measurement of all the desired character-istics can be performed and then the data analyzed to determine the best set of raw material ratios for each characteristic. A compromise of all recommended factor levels will have to be made to achieve the best overall plastic product.

- Confirmation of Predicted Outcomes. Once all choices and predictions of DoE project have been agreed upon, a confirmation (verification) experiment run should be made before final adoption of DoE recommendations, to verify the outcome analysis. This confirmation will test out the entire "new" design process before full implementation takes place. In manufacturing, the newly adjusted process should continue to be monitored through statistical quality control methods for a 6-month minimum period before any attempts should be made to further increase the robustness of the process by launching another DoE.

4.1.2   DoE Project Timing and Error Source

Reducing production variability is one of the most common methods used to increase process capability index and attain Six Sigma quality as was discussed in Chap. 1. Manipulation of means and variability can be addressed by using a combination of two strategies:

1. Traditional quality control, where the focus is to maintain the current production processes or new designs within a specified area of variability through control charts, regular maintenance, and calibration of production processes and equip-ment. The use of TQM and DMAIC principles is critical to permeate the quality culture within the organization.
2. Proactive quality control, where the goal is reducing process variability or increasing design robustness through defect analysis and DoE to achieve high quality through targeting of specific process operations or designs. DoE projects should be initiated before a new design or process are implemented to optimize their performance. It could also be used once existing processes matured to a steady level with defect reduction and maintenance through traditional quality control. This effort can be guided by many of the tools of TQM and corrective action processes. This timing of a DoE project for soldering processes was shown in Fig. 1.7 for a 2-year project to achieve world-class quality. The sequence of

using quality tools was as follows: TQM phase to reverse poor quality due to increased production, followed by a statistical process control phase to maintain quality, and finally a DoE project to bring world-class quality using new materials and processes.

Traditional quality should be well instituted before attempting DoE projects. No amount of DoE and defect analysis can rectify poor-quality designs or operations that are out of control. DoE-based improvement can only be felt temporarily, before being negated by a manufacturing operation that is out of control where production personnel, materials, and processes are constantly changing. The sources of defects, as outlined in Chap. 1, are due to the interaction between product specifications and process variability. This interaction originates from one of two sources: either the process is not centered, where the process mean (μ) is not equal to the target value (specification nominal N), or the product design and/or process variability is too high. Either one or a combination of both can influence product defect levels.

It is much easier to identify and collect data and rectify the first situation: a process mean not equal to target. Incoming materials, equipment settings, and performance can be measured and, if not equal to target, can be corrected by strict adherence to specifications. Materials' properties such as geometrical tolerances, density, tensile strength, hardness, etc. can be easily measured and rectified by working with production personnel and suppliers. Equipment factors such as temperature, pressure, speeds and feeds, and motion accuracy can be measured against original purchase specifications and readjusted as necessary with regular maintenance.

The calibration of production and measurement equipment is usually achieved by using instruments or gages which are inherently more accurate than the equipment to be calibrated. In addition, instruments' accuracy must be certified, through traceability to the National Institute of Standards and Technology (NIST). It is common to use calibration equipment whose accuracy and resolution are at least ten times than of the equipment being calibrated. An investigation of equipment, operator, or measurement error can be quantified through the use of statistical analysis tools such as gage repeatability and reproducibility (GR&R), discussed in chap. 2.

Reducing variability of production or design measurements is more difficult and requires a thorough examination of the sources of variation. Some of these causes are uncontrolled factors or noise. They can be generated from the following:

- External conditions, imposed by the environment under which the product is manufactured or used, such as temperature, humidity, dirt, dust, shock, vibration, human error, etc. These conditions are beyond the control of design and manufacturing engineers. They are difficult to predict, and it is expensive to design specific characteristics to satisfy all possible conditions under which the product is expected to operate.
- Internal conditions under which the product is stored or used, such as friction, fatigue, creep, rust, corrosion, thermal stress, etc. These conditions must be specified correctly within the normal use of the product. Many customers will overuse the product, and they expect that the product will continue to operate

even beyond its normal range. The design could be robust to ensure proper operations beyond advertised specifications.

DoE is focused on improving the robustness of the product functionality in external and internal conditions of operation. It seeks to determine the best set of process materials and factors to ensure that product characteristics' mean is equal to the specified nominal and the variability is as small as possible.

4.2   Two-Level Factorial DoE Design

Orthogonal arrays (OAs) are most used in DoE since they are balanced. There are an equal number of levels for each factor in the experiment. The outcome of each factor can be studied, while other factors are changing their levels, versus each single level of the factor being investigated.

OAs are different from one-factor-at-a-time (OFAT) experiments, as shown in Table 4.2 with four factors and two levels. The first experiment consists of all factors set to level (1) as a baseline. Additional experiments are added where each factor level is varied to the second level (2) individually, while the rest of the factors are kept constant. There is a deficiency in OFAT techniques: it does not allow for measuring the effect of varying other factors at the same time as one factor being changed. This mathematical relationship of factors, called factor interaction, needs to be analyzed to take full advantage of DoE.

The same four factors at two levels can be presented in Table 4.3, shown as successive OA arrangements. Two of the four factors (A and B) at two levels can be investigated first in $4 = 2^2$ experiments while keeping other factors at the same levels. More factors can be added in terms of perceived importance while keeping the results of the first experiment as part of the full DoE analysis. If a third factor (C) is added to DoE, then a second level of this factor is selected with the original four experiments for factors A and B for a total of $8 = 2^3$ experiments. The fourth factor (D) can also be added with a second level for an additional eight OA experiments for a total of $16 = 2^4$ experiments for four factors with two levels.

Table 4.2 One-factor-at-a-time (OFAT) DoE

Expt.		Factors		
#	A	B	C	D
1	1	1	1	1
2	2	1	1	1
3	1	2	1	1
4	1	1	2	1
5	1	1	1	2

Table 4.3 Successive two-level orthogonal array arrangements

Expt. #	Factors D	C	B	A	
1	1	1	1	1	
2	1	1	1	2.	Start with two Factors
3	1	1	2	1	of Primary Interest
4	1	1	2	2	
5	1	2	1	1	
6	1	2	1	2.	Add 3rd Factor if
7	1	2	2	1	Results Inconclusive
8	1	2	2	2	
9	2	1	1	1	
10	2	1	1	2	
11	2	1	2	1	
12	2	1	2	2	Add 4th factor of
13	2	2	1	1.	Least Importance
14	2	2	1	2	
15	2	2	2	1	
16	2	2	2	2	

OFAT experiments for (n) factors result in $n + 1$ total experiments, while OAs with (n) factors and two levels result in 2^n experiments with the added advantage of being able to measure the effect of factor interactions. Interactions for n factors range from two-way to n-way. The number of two-way interactions for (n) factors is represented by the product formula below:

$$\text{Two} - \text{way interactions with } n \text{ factors} = \prod_{i=1}^{n-1} (n - i) \qquad (4.1)$$

The number of columns in a two-level OA is equal to the number of experiments -1. For example, if four-factor DoEs are considered (A, B, C, and D) for 2 levels, there is a total of 15 OA columns, which is usually called L16 for 16 total experiments $= 2^4$. They include 4 columns for the main factors and their 11 interactions. They are:

- Six two-way interactions: $(n - 1) * (n - 2) * (n - 3) = $ AB, AC, AD, BC, BD, CD
- Four three-way interactions: ABC, ABD, ACD, BCD
- One four-way interaction $=$ ABCD

Two-level OA can measure interactions, through several techniques:

Table 4.4 Two-way interactions for two-level assignments

A	B	A * B	A	B	A * B
1	1	1	-1	-1	1
1	2	2	-1	1	-1
2	1	2	1	-1	-1
2	2	1	1	1	1

1. Full factorial DoEs are used to evaluate the effects of all factors and their interactions. For every number of factors (n), there are multiple two- to n-way interactions to be considered, as in the example for L16 above. The levels in the interaction columns are derived from the multiplication of the levels of the originating factors, using either as an exclusive OR (XOR) logical relationship for levels 1 and 2 or simple multiplication of levels -1 and 1 as shown in Table 4.4. Levels 1 and 2 are favored by the Taguchi method and its associated analysis software, while classical statistics use levels -1 and 1 as outlined by DoE analysis software such as Minitab©.

   While an assumption can be made that those levels 1 and 2 are equivalent to levels -1 and 1, their two-way interactions are not equivalent, since levels (2) * (2) = 1 and levels (1) * (1) = (1) in Table 4.4. This will result in opposite algebraic signs for two-way and other even numbered interaction coefficients and in regression equations. Use of graphical analysis of factor plots to determine which factor level affects DoE outcomes will alleviate any possible confusion associated with factor level designations. Single factor and odd level (such as three-way) interactions do not have this algebraic sign problem.

2. Fractional factorial DoEs provide a cost-effective way of determining the significance of selected factors and interactions while increasing the chance of DoE analysis error. A fractional factorial DoE uses a portion of the full factorial columns to estimate main factor coefficients and their interactions. The unused interaction columns are then assigned to other additional factors. This results in the condition called "confounding," where the assigned factor to an interaction column could be confounded with the interaction coefficient that is normally found in the column. By selectively choosing where to confound, a fractional factorial DoE could be used to study more factors with fewer experiments. Good DoE planning could minimize confounding problems, especially with higher-order interactions. There are several levels of confounding methodologies called "resolutions," which will be discussed in Chap. 7.

   Interaction columns to be confounded might be significant, counteracting the additional factor effect, making its coefficient either higher or lower in value. Confounding might result in faulty DoE analysis, by assuming that some interactions (especially higher-order interactions) are not significant. If a confounded column is shown to be significant, additional experiments, discussed in Chap. 7, should be performed to separate the additional factor from its confounding interaction.

3. A special variant of partial factorial experiments is called half fraction factorial. In this case, only a selected half of the full factorial experiments are performed, and an additional factor is assigned to the highest-order interaction. For the L16 mentioned above, it is a full factorial for four factors (A–D), and if a fifth factor (E) is assigned to the highest interaction column 15 which is interaction ABCD, then the DoE can be performed in an L16 rather than an L32 for 5 factors. The assumption is that the three- and four-way interaction coefficients are small and confounding is minimal. The 15 interactions between the fifth factor E and the other 4 factors are not considered in this design (AE, BE, CE, DE, ABE, ACE, BCE, ADE, BDE, CDE, ABCE, ABDE, ACDE, BCDE, ABCDE). The four two-way interactions between the five factors are placed in columns assigned to the three-way interactions in the L16, thus confounding them. This technique is known as resolution V in classical DoE and is referred to as $L16 = 2^{5-1}$, which indicates a half fraction design for five factors. For the smaller L8 OA consisting of three main factors (A, B, and C) and four interactions (AB, AC, BC, and ABC), a fourth factor (D) can be added to column ABC confounding it. This half fraction L8 DoE use is called resolution IV with the symbol 2^{4-1}.

4. Screening design DoEs are resolution III designs that allow all OA columns to be assigned to different factors. All factors are confounded by two-way interactions. An example is using three factors in L4 or $L4 = 2^{3-1}$. They represent a minimum set of experiments for the number of factors considered. They are called "saturated designs," because they are commonly used to whittle down the number of factors quickly through smaller DoE. Follow-on full factorial DoE can then be performed on the top 3 or 4 significant factors from the screening DoE. The assumption in screening DoE is that all interactions are small and can be ignored compared to the main factor effects.

4.2.1 Commonly Used Two-Level Orthogonal Arrays

The most used two-level OAs are full factorial implementations of $n = 2, 3, 4$, and 5 factors, commonly known as L4, L8, L16, and L32 in full factorial mode. The Lx nomenclature is the (x) number of experiments corresponding to (n) factors with two levels ($x = 2^n$). These four OAs are listed in Tables 4.5, 4.6, 4.7 and 4.8 in the level (1) and (2) configurations. In general, the factor sequence labels are either alphabetic (A, B, C,...) or numeric (1, 2, 3,...).

For Table 4.5 (L4) and Table 4.6 (L8), the top rows show the factors and interactions with their alphabetical and numerical labels. The interaction columns are labeled as two or three ways. For Table 4.7 (L16), two types of interactions are shown: the three-way interactions confounded with the two-way interactions of the four main factors (A, B, C, and D) versus the fifth factor E (half fraction mode)

Table 4.5 Orthogonal array L4 with two main factors (1/A and 2/B) and their interaction 3/C = 1*2/A*B

Factor	1	2	3 = 12
Expt. #	A	B	C = AB
1	1	1	1
2	1	2	2
3	2	1	2
4	2	2	1

Table 4.6 Orthogonal array L8 with three main factors (A/1, B/2, and C/3) and their interactions

Column #	1	2	3	4	5	6	7
Main Factors	A	B		D			
Interactions			C=A*B		E=A*D	F=B*D	G=ABD
Expt. #			3=1*2		5=1*4	6=2*4	7=124
1	1	1	1	1	1	1	1
2	1	1	1	2	2	2	2
3	1	2	2	1	1	2	2
4	1	2	2	2	2	1	1
5	2	1	2	1	2	1	2
6	2	1	2	2	1	2	1
7	2	2	1	1	2	2	1
8	2	2	1	2	1	1	2

assigned to column 15, confounding the four-way interaction ABCD in half fraction mode. Table 4.8 shows the orthogonal array L32, with 32 experiments and 31 columns for 5 main factors and 26 interactions. The decomposition of the interactions from two-way to five-way is listed at the top of Table 4.8, with interactions shown vertically as opposed to horizontally for the other arrays. This will serve as a guide assigning factors to columns in the L32 to allow for analysis of desired interactions.

An example of industrial DoE case study will be given for each type of array L8, L16, and L32 later in this chapter. All case studies will be using the levels 1 and 2 in the OA as opposed to levels −1 and 1 in classical DoE. While the two OA configurations using alternate level symbols might look different in terms of column patterns and interactions, DoE analysis results from either configuration are the same. Usually, a software package for DoE analysis will be used to substitute for hand calculations. Proper labeling of each factor by name and not using software-assigned factor labels will reduce confusion between the two configurations. In

Table 4.7 Orthogonal array L16 with four main factors (A/1, B/2, C/4, D/8) and their 11 interactions

Column #	1	2	3	4	5	6	7	8	9	10	11	12	13	14	15
Factors	A	B		C				D							(E)
Interactions			AB		AC	BC	ABC DE		AD	BD	ABD CE	CD	ACD BE	BCD AE	ABCD
Expt. #			1*2		1*4	2*4	8*15		1*8	2*8	4*15	4*8	2*15	1*15	1234
1	1	1	1	1	1	1	1	1	1	1	1	1	1	1	1
2	1	1	1	1	1	1	1	2	2	2	2	2	2	2	2
3	1	1	1	2	2	2	2	1	1	1	1	2	2	2	2
4	1	1	1	2	2	2	2	2	2	2	2	1	1	1	1
5	1	2	2	1	1	2	2	1	1	2	2	1	1	2	2
6	1	2	2	1	1	2	2	2	2	1	1	2	2	1	1
7	1	2	2	2	2	1	1	1	1	2	2	2	2	1	1
8	1	2	2	2	2	1	1	2	2	1	1	1	1	2	2
9	2	1	2	1	2	1	2	1	2	1	2	1	2	1	2
10	2	1	2	1	2	1	2	2	1	2	1	2	1	2	1
11	2	1	2	2	1	2	1	1	2	1	2	2	1	2	1
12	2	1	2	2	1	2	1	2	1	2	1	1	2	1	2
13	2	2	1	1	2	2	1	1	2	2	1	1	2	2	1
14	2	2	1	1	2	2	1	2	1	1	2	2	1	1	2
15	2	2	1	2	1	1	2	1	2	2	1	2	1	1	2
16	2	2	1	2	1	1	2	2	1	1	2	1	2	2	1

addition, the two configurations have complementary two-way interaction coefficient algebraic signs as mentioned earlier. This can be alleviated by using graphical analysis plots as outcome predictors in analysis software.

4.2.2 Types of Uses for Two-Level OA

There are four uses for each OA in a DoE project. These will be discussed using L8 as an example and summarized in Table 4.9.

1. Use OA as a full factorial array to model main factors and all their interactions. In the case of L8, the model includes three main factors (A, B, and D) at two levels and three two-way and one three-way interactions [C (A*B), E (A*D), and F (B*D) and G (A*B*D)].
2. Use OA as a screening design by assigning main factors to each column and ignoring all interactions. The number of "non-interacting" factors that can be modelled is equal to the number of experiments −1. This technique could be used

Table 4.8 Orthogonal array L32 with 5 main factors (A/1, B/2, C/4, D/8, and E/16) and their 26 interactions

Factor	A	B	A	C	A	B	A	D	A	B	A	C	A	B	A	E	A	B	A	C	A	B	A	D	A	B	A	C	A	B	A
2-way			B		C	C	B		D	D	B	D	C	C	B		E	E	B	E	C	C	B	E	D	D	B	D	C	C	B
3-way							C				D		D	D	C				E		E	E	C		E	E	D	E	D	D	C
4-way															D								E				E		E	E	D
5-way																															E
Expt.#	1	2	3	4	5	6	7	8	9	10	11	12	13	14	15	16	17	18	19	20	21	22	23	24	25	26	27	28	29	30	31
1	1	1	1	1	1	1	1	1	1	1	1	1	1	1	1	1	1	1	1	1	1	1	1	1	1	1	1	1	1	1	1
2	1	1	1	1	1	1	1	1	1	1	1	1	1	1	1	2	2	2	2	2	2	2	2	2	2	2	2	2	2	2	2
3	1	1	1	1	1	1	1	2	2	2	2	2	2	2	2	1	1	1	1	1	1	1	1	2	2	2	2	2	2	2	2
4	1	1	1	1	1	1	1	2	2	2	2	2	2	2	2	2	2	2	2	2	2	2	2	1	1	1	1	1	1	1	1
5	1	1	1	2	2	2	2	1	1	1	1	2	2	2	2	1	1	1	1	2	2	2	2	1	1	1	1	2	2	2	2
6	1	1	1	2	2	2	2	1	1	1	1	2	2	2	2	2	2	2	2	1	1	1	1	2	2	2	2	1	1	1	1
7	1	1	1	2	2	2	2	2	2	2	2	1	1	1	1	1	1	1	1	2	2	2	2	2	2	2	2	1	1	1	1
8	1	1	1	2	2	2	2	2	2	2	2	1	1	1	1	2	2	2	2	1	1	1	1	1	1	1	1	2	2	2	2
9	1	2	2	1	1	2	2	1	1	2	2	1	1	2	2	1	1	2	2	1	1	2	2	1	1	2	2	1	1	2	2
10	1	2	2	1	1	2	2	1	1	2	2	1	1	2	2	2	2	1	1	2	2	1	1	2	2	1	1	2	2	1	1
11	1	2	2	1	1	2	2	2	2	1	1	2	2	1	1	1	1	2	2	1	1	2	2	2	2	1	1	2	2	1	1
12	1	2	2	1	1	2	2	2	2	1	1	2	2	1	1	2	2	1	1	2	2	1	1	1	1	2	2	1	1	2	2
13	1	2	2	2	2	1	1	1	1	2	2	2	2	1	1	1	1	2	2	2	2	1	1	1	1	2	2	2	2	1	1
14	1	2	2	2	2	1	1	1	1	2	2	2	2	1	1	2	2	1	1	1	1	2	2	2	2	1	1	1	1	2	2
15	1	2	2	2	2	1	1	2	2	1	1	1	1	2	2	1	1	2	2	2	2	1	1	2	2	1	1	1	1	2	2
16	1	2	2	2	2	1	1	2	2	1	1	1	1	2	2	2	2	1	1	1	1	2	2	1	1	2	2	2	2	1	1
17	2	1	2	1	2	1	2	1	2	1	2	1	2	1	2	1	2	1	2	1	2	1	2	1	2	1	2	1	2	1	2
18	2	1	2	1	2	1	2	1	2	1	2	1	2	1	2	2	1	2	1	2	1	2	1	2	1	2	1	2	1	2	1
19	2	1	2	1	2	1	2	2	1	2	1	2	1	2	1	1	2	1	2	1	2	1	2	2	1	2	1	2	1	2	1
20	2	1	2	1	2	1	2	2	1	2	1	2	1	2	1	2	1	2	1	2	1	2	1	1	2	1	2	1	2	1	2
21	2	1	2	2	1	2	1	1	2	1	2	2	1	2	1	1	2	1	2	2	1	2	1	1	2	1	2	2	1	2	1
22	2	1	2	2	1	2	1	1	2	1	2	2	1	2	1	2	1	2	1	1	2	1	2	2	1	2	1	1	2	1	2
23	2	1	2	2	1	2	1	2	1	2	1	1	2	1	2	1	2	1	2	2	1	2	1	2	1	2	1	1	2	1	2
24	2	1	2	2	1	2	1	2	1	2	1	1	2	1	2	2	1	2	1	1	2	1	2	1	2	1	2	2	1	2	1
25	2	2	1	1	2	2	1	1	2	2	1	1	2	2	1	1	2	2	1	1	2	2	1	1	2	2	1	1	2	2	1
26	2	2	1	1	2	2	1	1	2	2	1	1	2	2	1	2	1	1	2	2	1	1	2	2	1	1	2	2	1	1	2
27	2	2	1	1	2	2	1	2	1	1	2	2	1	1	2	1	2	2	1	1	2	2	1	2	1	1	2	2	1	1	2
28	2	2	1	1	2	2	1	2	1	1	2	2	1	1	2	2	1	1	2	2	1	1	2	1	2	2	1	1	2	2	1
29	2	2	1	2	1	1	2	1	2	2	1	2	1	1	2	1	2	2	1	2	1	1	2	1	2	2	1	2	1	1	2
30	2	2	1	2	1	1	2	1	2	2	1	2	1	1	2	2	1	1	2	1	2	2	1	2	1	1	2	1	2	2	1
31	2	2	1	2	1	1	2	2	1	1	2	1	2	2	1	1	2	2	1	2	1	1	2	2	1	1	2	1	2	2	1
32	2	2	1	2	1	1	2	2	1	1	2	1	2	2	1	2	1	1	2	1	2	2	1	1	2	2	1	2	1	1	2

to reduce the model to only significant factors. A second DoE could be initiated to study significant factors in the reduced model into a full factorial DoE. This technique reduces the number of total experiments, as in the example of using two L8 experiments to reduce a model of seven factors as opposed to a full factorial $2^7 = L128$.

DoE teams must decide whether to use this sequential DoE scenario of screening DoE followed by full factorial DoE or use classical "foldover" scenarios further outlined in Chap. 8. Foldover begins with a screening L8, and then an

Table 4.9 Two-level OA used in full, half, and partial factorial and screening modes

# of Factors	Full Factorial	# of Experiments Half Fraction	Partial Factorial	Saturated/ Screening
2	4	4	4 (One Interaction)	4
3	8	4	4 (No Interactions)	4
4	16	8	8 (Three Interactions)	8
5	32	16	8 (Two Interactions)	↓
6	64	32	8 (One Interaction)	
7	128	64	8 (No Interactions)	
8	256	128	16 (seven Interactions)	16
↓	↓	↓	↓	↓
15	32,768	16,384	16 (No Interactions)	
16	65,536	32,768	32 (15 Interactions)	32

additional L8 is used for every significant factor, folding one factor at a time to sequentially isolate that factor and its interaction.

For a typical reduced model from 7 to 3 factors, the contrast is between 16 experiments in the sequential DoE scenario and 32 experiments (8 initial screening +3 reduced model factors x 8) in the foldover scenario. The folding of one factor at a time model is much more accurate and avoids the risks of confounding errors.

An alternative to both scenarios is to use non-interacting arrays to be discussed in Chap. 8. A typical non-interacting array is L12 followed by full factorial L8 for a total of 20 experiments in sequential DoE design.

While the assignment of factors to columns in a screening design DoE could theoretically be random, they could be assigned according to perceived value from the highest to the lowest estimated factor coefficients. In the L8 example, up to seven main (primary) factors at two levels with no interactions could be accommodated. Factors should be assigned according to potential significance as follows: the most important three factors should be assigned to the primary columns A, B, and D. The fourth factor should be assigned to column G, which confounds with the three-way interaction A*B*D. The last three least important factors should be assigned to columns C, E, and F which confound with two-way interactions of the primary factors.

3. Use OA as a half fraction factorial resolution IV or V design with primary factors (n) assigned to an OA corresponding to $Lx = 2^{n-1}$. The least important model factor is assigned to the highest (n) interactions and confounds it, and its interactions with other factors are not available for analysis. The least important factor might prove to be not significant and thus can be pooled into the error. If not, then

the complementary half fraction could be performed for accurate model determination.

In the case of four factors in assigned to an L8 in half fraction mode, the first three main (primary) factors could be assigned to columns A, B, and D, and the fourth least important factor can be assigned to column G. There are several confounding and missing interactions in this application of L8: factor G confounds with the three-way interaction ABD and the two-way interaction of the three primary factors A, B, and D with factor G missing and assumed to be very small. Factor G is assumed to be independent of the previous three factors.

In the case of five factors assigned to an L16 in half fraction mode, the first four primary factors could be assigned to columns A, B, C, and D; and the fifth least important factor can be assigned to column E, confounding with interaction ABCD. As shown in Table 4.7, factor E two-way interactions with the other four factors can be analyzed in the L16, but they confound other three-way interactions in resolution V design.

4. Use OA as a partial factorial design by adding more factors to the model that can confound either two-way or three-way interactions, up to using all columns in a screening design. This is the case where the design team can test out their model believing that the additional factors might not be significant and therefore the error due to confounding is negligible. This is a risky plan if the assumption of low-value interactions is not correct. DoE analysis cannot prove or disprove this assumption since only a full factorial DoE can separate the factor from its confounding interactions. In the case of an L8, this method can be used to add additional factors such as a fourth, fifth, or a sixth factor using partial factorial design. The DoE team is assuming that some of the three two-way interactions (C, E, or F) are not significant and therefore can be confounded with small risk.

Using partial factorial allows the opportunity to analyze certain interactions by carefully assigning primary factors to columns. In the voids case study to be discussed later in this chapter, five factors (1/A, 2/B, 4/D, 6/F, and 7/G) were assigned to an L8, leaving two columns (3/C and 5/E) available to study specific interactions of A*B (1*2) and A*D (1*4). Each interaction is confounded by two other two-way interactions:

$$C = A*B \quad \text{or} \quad D*G = D*ABD = A*B \quad \text{or} \quad E*F = AD*BD = A*B$$

$$3 = 1*2 \quad \text{or} \quad 4*7 = 4*124 = 1*2 \quad \text{or} \quad 5*6 = 14*24 = 12$$

$$E = A*D \quad \text{or} \quad B*G = B*ABD = A*D \quad \text{or} \quad C*F = AB*BD = A*D$$

$$5 = 1*4 \quad \text{or} \quad 2*7 = 2*124 = 1*4 \quad \text{or} \quad 3*6 = 12*24 = 1*4$$

The risk of confounding is that one interaction coefficient is assumed large, while the other two confounding interaction coefficients are assumed small. If the coefficient is not significant, then there is no further analysis needed. If it is significant, then additional experiments might be needed to separate the confounding.

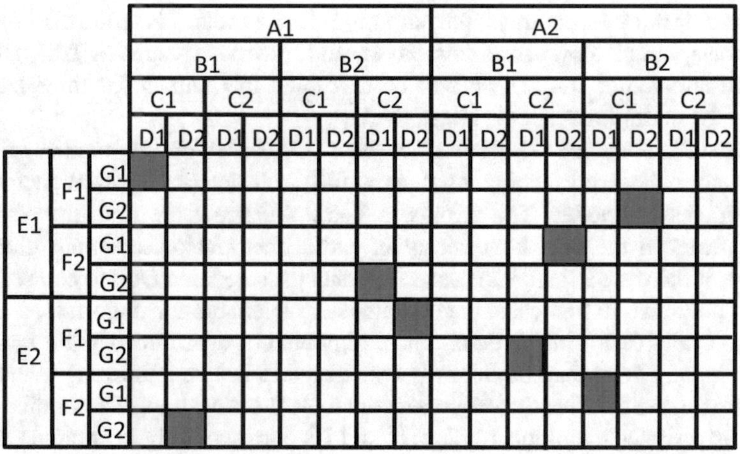

Fig. 4.3 Modelling seven factors (A–G) in screening versus full factorial mode

The use of the L8 as a saturated design with seven independent factors contrasts with their full factorial use of three main factors. A full factorial design with seven factors requires a DoE with L128 (2^7) experiments. What is gained by much less experiments in the screening design (L8) is offset by its inability to calculate interactions as in the full factorial designs (L128). A pictorial presentation of the screening L8 versus full factorial 128 experiments is given in Fig. 4.3, where the 8 experiments are shown as blackened squares in the 128 experiments' matrix.

OA balance can be shown in L8 from Table 4.6. For each level in a column, all of the other levels in the other columns are rotated through their values. Experiments 1–4 have column A with level (1) only, while levels in columns B through G contain both levels 1 and 2, in equal numbers of two each. The balance of the orthogonal arrays allows for a simple solution for the values of the factors in the experiments using Cramer's rule for simultaneous equations. In each array, there are (n) unknown variables that can be solved in ($n - 1$) equations. The L8 can be represented with seven unknown coefficients and eight simultaneous equations, and each factor coefficient can be solved accordingly.

The relationship of successive two-level arrays can be very beneficial. The next higher two-level array is L16, shown in Table 4.7. L16 can be created from two seven-column L8 stacked on top of each other, with additional eight columns used for the fourth factor and its interactions (D, AD, BD, CD, ABD, ACD, BCD, and ABCD). It is not necessary to choose all available columns to be included in a DoE. For example, an L16 DoE can be performed with ten factors in partial factorial design, and the other columns can be left unassigned. This does not jeopardize the utility of the DoE since the analysis of the effect of the ten factors on the outcome is valid. The remaining columns could be used for calculating some of the interactions, according to the assignment of the main factors to columns.

The use of balanced orthogonal arrays allows for multiple arrays to be used with increasing number of factors. For two-level orthogonal arrays, they are L4, L8, L16,

L32, etc. For three-level arrays, they are L9, L27, L81, etc. There are also multiple-level arrays which allows for a mix of two and three levels such as L12, L18, and L36 non-interacting OA. These will be discussed in Chap. 5 for three-level and Chap. 8 for multiple-level OA, respectively.

Table 4.9, shows the selection of two-level OA based on the number of factors and whether the arrays being used as a full, half fraction, partial factorial, or screening design modes. The arrows in Table 4.9 show the transition from one type of array to the next larger array in partial factorial or screening mode. For example, if there are 9 factors in the model, then the screening DoE to be used is L16 with 15 columns. In this case, nine columns are available for main factors and six columns available for interactions. The assignment of a particular main factor to a column in the L16 should be carefully considered to isolate interesting interactions. For example, the first five factors considered to be the most important of nine factors should be assigned to columns 1, 2, 4, 8, and 15, corresponding to columns A, B, C, D, and E. from Table 4.7. The remaining four factors could be assigned to three-way interactions such as ABC, ABD, ACD, and BCD (columns 7, 11, 13, and 14), minimizing the confounding of assigned factors with three-way interactions. The six remaining unassigned columns would allow the analysis of some of two-way interactions AB, AC, BC, AD, BD, and CD to the exclusion of other two-way interactions. The risk of error in the model is multiple layers of confounding as will be discussed later in Chap. 7.

4.2.3 Interaction, Confounding, and Interconnecting Graphs

Interaction occurs when one factor modifies the conditions of another factor. If an interaction is considered to be significant in the model design, the interaction coefficient should be estimated from its own column in the array, and no factor should be assigned to this column. Interactions represent coefficient values of the effect of one factor on others. If an assigned factor confounds an interaction column, then that factor coefficient could be either reduced or amplified by the interaction effect.

In the L8 array shown in Table 4.6 with seven columns, there are four interaction columns (C, E, F, and G) originating from the three main factors (A, B, and D), with the following relationships, using alphabetic and numeric symbols:

$C = A*B$ $3 = 1*2$
$E = A*D$ $5 = 1*4$
$F = B*D$ $6 = 2*4$
$G = A*B*D$ $7 = 1*2*4$

If any factor or interaction of the seven columns is multiplied with another, then the resulting interaction is already in the L8 array. For example:

$A * C = A*A*B = B$

1 * 3   = 1 * 1* 2 = 2
B * E   = B*A*D = A*B*D = G
2 * 5   = 2*1*4 = 1*2*4 = 7

Multiple arrangements of factor and interactions can be combined to form alternative groupings with the same confounding. For example, in the L8, the three main factors could be D/4, F/6, and G/7 and their interactions (B/2, C/3, A/1, and E/5) as follows:

B = D*F = D*B*D = B   2 = 4*6 = 4*2*4 = 2
C = D*G = D*A*B*D = A*B = C   3 = 4*7 = 4*1*2*4 = 1*2 = 3
A = F*G = B*D*A*B*D = A   1 = 6*7 = 2*4*1*2*4 = 1
E = D*F*G = D*B*D*A*B*D = A*D = E   5 = 4*6*7 = 4*2*4*1*2*4 = 1*4 = 5

These arrangements can also be duplicated for half factorial designs. For example, in an L16 resolution V design 2^{5-1}, the main factors could be one of three arrangements with ten interactions each:

1, 2, 4, 8, and 15   or   7, 8, 9, 10, and 12   or   4, 5, 6, 11, and 12

These arrangements of factors and interactions can be grouped together in partial factorial modes for two-level OA. More on these arrangements (called interconnecting graphs) like the discussion above in Chap. 7.

For DoE teams concerned with the complexity of interactions and confounding, non-interacting orthogonal arrays, otherwise known as "Plackett-Burman" design, such as L12, L18, or L36 can be used. In these arrays, any array column does not confound the interaction of any two other columns in the array. L12 is a two-level factor array, while L18 and L36 are a combination of two- and three-level factors. They are discussed further in Chap. 8.

4.3   Two-Level OA Analysis and Model Reductions

Statistical analysis of DoE is based on the model coefficient estimates of each factor as well as the analysis of variance (ANOVA), which is a method of determining the significance of each factor in terms of its effects on DoE outcome(s). These two calculations were introduced in the previous chapter for linear regression, treatments, and full factorial analysis of experiments.

ANOVA apportions the total effect of DoE outcomes' average and variability to each factor in the orthogonal array. The significance test is based on the F distribution, which is a ratio of the degrees of freedom of the factor divided by the degrees of freedom of the error. The least significant factors can be summed together as the error of the experiment since they are not significant in affecting DoE outcomes.

The terms for determining the ANOVA table for (n) DoE outcomes were discussed earlier in Chap. 3 and reviewed again here:

$$\text{Total sum of the squares (SST)} = \sum (y_i - \bar{y})^2 = \sum y_i^2 - (\sum y_i)^2 / n$$

where $(\sum y_i)^2 / n$ is the correction factor (CF)

$$\text{Sum of squares for each factor (SSF)} = \frac{(\sum y_{level1})^2}{n_{level1}} + \frac{(\sum y_{level2})^2}{n_{level2}} - CF$$

Degrees of freedom (DOF) formulas for various terms

DOF orthogonal array	= no. of experiments -1
DOF factor	= no. of levels -1
DOF error	= (Total DOF) – (DOF of significant factors and interactions)
DOF interaction	= product of the DOF of each factor
	= 1, for two-level factor interactions; 2, for three-level factor interactions
MS	= SS/DOF [mean square deviation or variance (V)]
F-ratio for each factor	= MSF/MSE

The F-test critical values are given in Table 2.11 for a confidence level of 95% (one-sided) and the DOF of the factor versus the DOF of the error. The F-test is used to determine the significance of each factor in the DoE. The computed F is a ratio of the factor variance (MSF) over the error variance (MSE). The error of the DoE experiment can be obtained from one of the following options, discussed in Chap. 6:

1. For a single DoE with no repetition, the smallest factor(s) or interactions (with the smallest SS) can be reduced from the model and pooled as the error, especially if it is a higher-order interaction(s).
2. Replicate the whole experiment for one or more times, generating error due to replication(s).
3. Replicate the CenterPoint of the design space/process map of the experiment to generate error based on multiple CenterPoint replications).
4. Replicate some lines of the experiments, such as the corner points of the design space.

For a given confidence level, the F-test determines whether the effect of a factor is due to chance or due to the factor itself (the factor is deemed significant). If a computed factor F-ratio is less than the critical value in the F table given the DOF of factor and error, then it is deemed not significant and can be pooled into the error. The F-ratios are then recalculated, and the F-test is performed on the remaining factors. When a factor is significant to 0.05 (or the confidence is greater than 95%), then the probability of this factor effecting the experiment happens 5% by chance or once every 20 times. Since this is remote in nature, the factor must be significant, and hence it affects DoE outcomes. Pooling should continue until all remaining factors are significant. If no significant factors are left, then the DoE project is considered a failure, and a new DoE will have to be implemented with different factors and levels.

4.4 Two-Level DoE Case Studies

These case studies are drawn from industrial DoE in various disciplines. They demonstrate the use of two-level OA in selecting the best factors and levels to increase the DoE team's knowledge of the factors that influence the design and the levels that best achieve the desired outcome. In addition, they also demonstrate the use of two-level techniques in improving current designs and processes. Each case study is presented both in terms of design and analysis and the reasons behind the team's design decisions. Shortcomings of the design teams' project plans are discussed, results are analyzed, and conclusion is drawn for each case study. Other examples of three- and multiple-level DoE are given in later chapters and summarized in the case study index in Chap. 10.

4.4.1 Full Factorial L8 Case Study, Hipot DoE: Selecting Best Alternative Among Equally Performing Designs

Computer displays for medical products must meet Food and Drug Administration (FDA) standards for protection from discharge of static electricity in case of product failure. This is tested by applying a high-voltage potential probe (Hipot) against the display housing and recording the voltage at which the screen blinks. The display housing design goal is to achieve the maximum Hipot voltage when touched by the high-voltage probe before discharging through the display.

A DoE project team was formed and proceeded with selecting factors and levels. Several options were considered, such as using different types of shielded or grounded cables and connector types and painting the plastic housing with conductive paint to allow static electricity to discharge safely through the product connections to the neutral line in the AC power socket.

The team decided on three factors at two levels in a full factorial L8. The team wanted to study interactions as well as factor coefficients. Factor and level selections were as follows:

A = cable connector type   X or Y
B = contact methods   Spring or screw
D = conductive paint the frame Yes or no

The DoE was designed with a single repetition, and results obtained are shown in Table 4.10. Since this was a full factorial DoE, there is no confounding of the three main factors, and the DoE results can be interpreted directly. The results showed

Table 4.10 L8 Full factorial Hipot DoE design and results

Expt.#	A	B	D	A	B	D	Cable/A	Contact/B	Paint/D	Results (kv)
1	-	-	-	1	1	1	X	Spring	Yes	18.5
2	-	-	+	1	1	2	X	Spring	No	14
3	-	+	-	1	2	1	X	Screw	Yes	18.5
4	-	+	+	1	2	2	X	Screw	No	12.5
5	+	-	-	2	1	1	Y	Spring	Yes	18.5
6	+	-	+	2	1	2	Y	Spring	No	13
7	+	+	-	2	2	1	Y	Screw	Yes	9.5
8	+	+	+	2	2	2	Y	Screw	No	8

Table 4.11 Hipot DoE ANOVA with no pooling

Source	Pool	DOF	SS	MS
Cable	n	1	26.28	26.28
Contact	n	1	30.03	30.03
Cable*Contact	n	1	19.53	19.53
Paint	n	1	38.28	38.28
Cable*Paint	n	1	1.53	1.53
Contact*Paint	n	1	0.78	0.78
A*B*D	n	1	3.78	3.78
Total		7	120.22	17.17

three combinations of the three factors resulting in the same maximum Hipot value of 18.5 Kvolts. The team opted for full statistical analysis to determine the optimum selection of factors and levels.

While only one repetition of the results was conducted, it was necessary to reduce the L8 model of three factors and four interactions to determine the error. The ANOVA table was calculated using Excel®-based DoE analysis developed by the author to perform successive pooling needed for model reduction and shown in Table 4.11, Hipot DoE ANOVA with no pooling. No F-ratio can be calculated since there was no replication, and hence no error can be calculated. One indication is that the interaction "Contact*Paint" has the lowest factor sum of squares (SSF) and therefore can be pooled to create the error, which is shown in Table 4.12, ANOVA after first pooling. The pooled interaction was removed from the model, and its SSF and DOF were transferred to the error, thus allowing for calculating the F-ratio for all remaining factors and interactions in the model. The critical value of $F_{0.05, 1, 1} = 161.4$ shows that there are no significant factors indicating the need for further model reduction. Partial ANOVA calculations in Table 4.11 are shown below:

$\Sigma x = 112.5$, $\Sigma x^2 = 1702.25$, $CF = (\Sigma x)^2/n = 112.5^2/8 = 1582.03$
$SST = \Sigma x^2 - CF = 1702.25 - 1582.03 = 120.22$

Table 4.12 Hipot DoE ANOVA after first pooling

Source	Pool	DoF	SS	MS	F-Ratio	P > F	SS'	p %
Cable	n	1	26.28	26.28	33.64	0.109	25.50	21.2%
Contact	n	1	30.03	30.03	38.44	0.102	29.25	24.3%
Cable*Contact	n	1	19.53	19.53	25.00	0.126	18.75	15.6%
Paint	n	1	38.28	38.28	49.00	0.090	37.50	31.2%
Cable*Paint	n	1	1.53	1.53	1.96	0.395	0.75	0.6%
Contact*Paint	y	0						
A*B*D	n	1	3.78	3.78	4.84	0.272	3.00	2.5%
Replication Error	0	0						
Pooled Error	1	0.78	0.78					
Total Error	1	0.78	0.78				5.47	4.55%
Total	7	120.22	17.17				120.22	100%

Table 4.13 Hipot DoE ANOVA with second pooling

Source	Pool	DoF	SS	MS	F-Ratio	P > F	SS'	p %
Cable	n	1	26.28	26.28	12.94	0.037	24.25	20.2%
Contact	n	1	30.03	30.03	14.78	0.031	28.00	23.3%
Cable*Contact	n	1	19.53	19.53	9.62	0.053	17.50	14.6%
Paint	n	1	38.28	38.28	18.85	0.023	36.25	30.2%
Cable*Paint	y	0						
Contact*Paint	y	0						
A*B*D	y	0						
Replication Error	0							
Pooled Error	3	6.09	2.03					
Total Error	3	6.09	2.03				14.22	11.83%
Total	7	120.22	17.17				120.22	100%

$MST = SST/DOF(T) = 120.22/7 = 17.17 = \sigma^2$ and $\sigma = 4.144$ of the eight results

$SSA = 1/n*[\Sigma x_{level\ 1}{}^2 + \Sigma x_{level\ 2}{}^2] - CF = 1/4$
$[(18.5 + 14 + 18.5 + 12.5)^2 + (18.5 + 13 + 9.5 + 8)^2] - 1582.03 = 26.28$

$MSA = SSA/DOF(A) = 26.28$

F-value $(A) = SSA/SSE = 26.28/0.78 = 33.64$

$SSA' = SSA - DOF(A) * MSE = 26.28 - 0.78 = 25.5$

$p\% = SSA'/SST = 25.5/120.22 = 21.2\%$

Two interactions, Contact*Paint and Cable*Contact*Paint, were then pooled, and the results were shown in Table 4.13. It is recommended that pooling should be performed one factor at a time, but was shortened in this case. Table 4.13 shows that three main factors were all significant with probability P-value > F less than 0.05 and their F-ratios greater than critical value of $F_{0.05,\ 1,\ 3} = 10.13$ from Table 2.11.

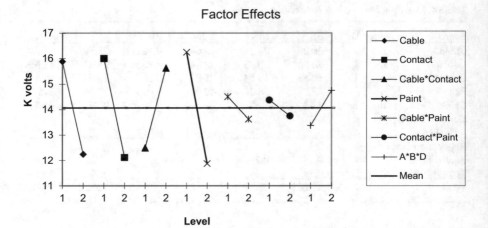

Fig. 4.4 Model coefficient plots for Hipot DoE

Interaction Cable*Contact was close to significance with probability = 0.053 and therefore was not pooled. The model R^2 was calculated as follows:

$R^2 = SSR/SST$
$R^2 = (SST - SSE)/SST$
$R^2 = (120.22 - 6.09)/120.22 = 94.93\%$

Hipot DoE R^2 indicates significance of the reduced model. Once all remaining model factors and interactions are significant, pooling should be stopped and the reduced model verified.

Two observations can be drawn from Table 4.13. One is the contrast between the probability based on the F-test in classical statistics and the $p\%$. They diverge significantly since the interaction Cable*Contact is shown to be close to significance ($P = 0.053$) using the F-test as opposed to $p\%$ high value of 14.6% > 5% indicating significance. If the interaction Cable*Contact is pooled, then the R^2 of the main factor model decreases to 78.68%, indicating non-significance of main factors only model.

Once the model has been reduced, the model coefficient estimates can be calculated. The most direct method for two-level design is to plot the mean of one level to the mean of the second level for factors and interactions. The y-axis span between the two-level means is called the effect. The average of the two-level means should equal the average of the total experiments. The distance between the total experiment means and either of the two levels is equal to the factor coefficient, which is also equal to half the effect. The Hipot DoE factor plots are shown in Fig. 4.4, and model coefficient calculations are shown in Table 4.14. To illustrate some of the calculations, the coefficient values for Cable are shown below:

Cable   level   1   =   Mean   of   results   corresponding   to   Cable   level
     $1 = (18.5 + 14 + 18.5 + 12.5)/4 = 15.875$

Table 4.14 Hipot DoE model coefficient estimates

Source	Level 1	Level 2	Mean	Coefficient	SE	t-value	P > t
Cable	15.88	12.25	14.06	-1.81	0.504	-3.60	0.037
Contact	16.00	12.13	14.06	-1.94	0.504	-3.85	0.031
Cable*Contact	12.50	15.63	14.06	1.56	0.504	3.10	0.053
Paint	16.25	11.88	14.06	-2.19	0.504	-4.34	0.023
Cable*Paint	14.50	13.63	14.06	-0.44	pooled		
Contact*Paint	14.38	13.75	14.06	-0.31	pooled		
A*B*D	13.38	14.75	14.06	0.69	pooled		

Table 4.15 Hipot DoE predicted results based on two models

Significant Factors Only

Expt. #	Cable	Contact	Paint		Result	SE	95% CI
1	L1	L1	L1		20	1.79	(15.03 – 24.97)
3	L1	L2	L1		16.13	1.79	(11.16 – 21.09)
5	L2	L1	L1		16.38	1.79	(11.41 – 21.34)

Significant Factors and One Interaction

Expt. #	Cable	Contact	Paint	Cable*Contact	Results	SE	95% CI
1	L1	L1	L1	L1	18.44	1.13	(14.85 – 22.02)
3	L1	L2	L1	L2	17.69	1.13	(14.10 – 21.27)
5	L2	L1	L1	L2	17.94	1.13	(14.35 – 21.52)

Cable level 2 = (18.5 + 13 + 9.5 + 8)/4 = 12.25
Cable span = Level 2 – Level 1 = 12.25–15.875 = –3.625
Cable coefficient = Cable level 2 – experiment mean = 12.25–14.06 = –1.8125
Cable coefficient = Span/2 = –3.65/2 = – 1.8125

Standard error = $SE_{Coefficient} = \frac{\sigma}{\sqrt{n}} = \sqrt{\frac{MSE}{n}} = \sqrt{\frac{2.03}{8}} = 0.504$

t-value Cable = Coefficient Cable/SE = –1.81/0.504 = – 3.60

Probability of t table value of 3.6 given DOF = 2, two-tailed = Excel® function 2* T.DIST(−3.6, 3, TRUE) = 0.037. This probability is exactly equal to the probability based on the F-test in ANOVA.

The regression equation based on three significant factors and one interaction is as follows:

Result (Kvolts) = 14.0625 − 1.8125 Cable −1.9375 Contact −2.1875 Paint
+1.5625 Contact*Cable

Based on the maximum Kvolt goal for the DoE project, the significant factor levels that provided the larger value are selected from Fig. 4.4 and plugged into the regression equation for maximum predicted (desired) result. Since one interaction is significant and part of the regression equation, it must also be selected corresponding to the chosen main factor levels. In the Hipot example, the interaction will reduce or increase the predicted result, pending the choice of the main factors. Table 4.15 shows the different predicted results with or without the significant interaction for the combination of factors for the highest Kvolt experiments. The three results that produced 18.5 Kvolts (experiments 1, 3, and 5) are recreated using the factor coefficients, with or without the significant interaction. The standard error SE is calculated as follows:

1. For the case of three significant factors and four pooled factors for the error. The SE for each outcome consists of dividing the error into two sets of four experiments for each factor. The MSE from Table 4.13 is recalculated since the interaction Cable*Contact was pooled into the error:
   MSE = (6.09 + 19.53)/4 = 6.41

$$SE_{Predicted\ 1} = \sqrt{MSE/_n} = \sqrt{6.41/_2} = 1.79$$

   CI $= \pm\ t_{0.025,4}$ * SE $= 2.776 * 1.79 = \pm 4.97$ span the predicted values for each outcome

2. For the case of three significant factors and one interaction as well as three pooled factors for the error. The SE for predicted value for each outcome consists of dividing the error into two sets of four experiments for each factor plus one interaction from eight experiment lines. The MSE $= 2.03$ from Table 4.13.

$$SE_{Predicted\ 2} = \sqrt{MSE/_n} = \sqrt{MSE\left(1/_{n_1} + 1/_{n_2}\right)} = \sqrt{2.03(1/_2 + 1/_8)} = 1.13$$

   CI $= \pm\ t_{0.025,3}$ * SE $= 3.182 * 1.13 = \pm 3.59$

All three seemingly equal design alternatives fall within the confidence interval. Several observations can be made from Table 4.15.

1. Significant interactions can increase or decrease predicted DoE result(s).
2. Significant interactions are important in DoE. Predicted result(s) will be different if significant interactions are ignored as in OFAT experiments.
3. When predicting the results based on significant factors, the selected factor levels automatically determine the predicted interaction level. Interaction levels cannot be selected; only main factor levels that constitute the interaction can be selected to determine the interaction level.
4. Another methodology to determine the selection of factor levels is to plot the two-way significant interactions of selected factors. The plot values are obtained from dividing the eight experiments into groups that fit the four values of two-way interactions. For the Hipot L8 example, eight experiments can be divided into four groups and plotted accordingly, as shown in Fig. 4.5:

   • Contact 1/Cable 1 = (18.5 + 14)/2 = 16.25
   • Contact 1/Cable 2 = (18.5 + 13)/2 = 15.75

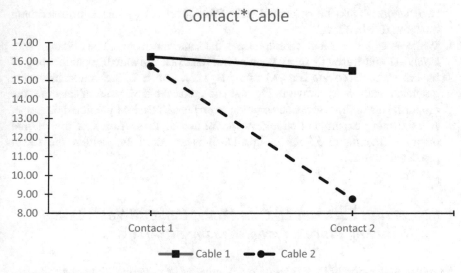

Fig. 4.5 Hipot DoE two-way Contact*Cable interaction plot

- Contact 2/Cable 1 = (18.5 + 12.5)/2 = 15.50
- Contact 2/Cable 2 = (9.5 + 8)/2 = 8.75

Table 4.15 indicates that three 18.5 Kvolts, which are the top 3 outcome combinations above, and Paint level (1) fall within the confidence interval of all three design alternatives. The best level combination should be selected (Contact 1/Cable 1) and added to the Paint level 1 coefficient to obtain the maximum results (16.25 + 2.19) = 18.44 Kvolts shown in Table 4.15 with one interaction. All results in Table 4.15 are derived from the regression equation for three selected factor levels and resulting one interaction.

This selection of maximum Kvolt outcome can be gleaned from several calculations shown above:

1. Figure 4.4, model coefficient plots, shows that the maximum Kvolts can be obtained by selecting level (1) for all three main factors. The interaction Cable*Contact will have level 1 (L1) based on the individual level of the factors Cable (L1) and Contact (L1).
2. The selected levels can be plugged into the regression equation to arrive at the best estimate of the maximum, shown as experiment line (1) in both models of Table 4.15 of Cable (L1), Contact (L1), and Paint (L1) at 20 Kvolts with three factors only and 18.44 Kvolts for three factors and one interaction Cable* Contact (L1).
3. The interaction Cable*Contact counteracts the effect of both factors Cable and Contact and should be included in the model, based on R^2 analysis. It is shown as inverse slope than all factors in Fig. 4.4. Of the four possible combinations, Cable (1)*Contact (1) yields the largest value, shown in Fig. 4.5. This combination of

two factors should be combined with Paint level (1) to select the maximum outcome (Kvolts) level.
4. While using levels 1 and 2 for this DoE, the same results could be obtained with levels −1 and 1 used by many popular DoE analysis software. Proper labeling of factors by the software (instead of A, B, C, . . . or 1, 2, 3,. . .) and using the graphical analysis as shown in Fig. 4.4 can guide the DoE team in selecting the proper levels in the regression equation to arrive at a desired predicted outcome.
5. A confirming experiment should be performed to make sure that the desired outcome (maximum Kvolts in this DoE) is achieved and within the CI of predicted results.

4.4.2 Partial Factorial L8 Case Study, Underfill Voids DoE: Selecting Process Parameters for Zero Defects

A typical application of DoE is using quick projects to solve individual part process quality. An example is the underfilling components in assemblies with epoxy materials in the electronics industry. The underfilling allows for good thermal conductivity as well as mechanical bonding for good vibrations and shock performance of the components. In some cases, the components are placed and refilled on both sides of a printed circuit board (PCB), making the assembly process more vulnerable to voids in the refill due to outgassing of the cured epoxy during soldering. These voids can be detected by X-ray and must be reworked to adhere to industry goals of zero defects. A visual presentation of this process is shown in the YouTube video https://www.youtube.com/watch?v=BntzIAUQItU.

The goal of the DoE project is to select the best materials, methods, machines, and personnel for minimizing voids. A zero-defect process eliminates rework and enhances reliability of the product. Multiple DoE projects can be used to enhance the quality of individual components, rather than attempting to find universal process setting for all possible types of components. Consequently, a focused single component DoE project must be limited in scope and can be accomplished in short time.

The first step of a DoE project implementation is using cause-and-effect diagrams to explore possible sources of quality defects and narrow them down to reduced factors and levels. Figure 4.6 shows a typical team effort for a good cause-and-effect diagram for possible sources of voids in underfill operations. Reducing the causes to be investigated due to time, regulatory, or financial resources results in efficient DoEs. Examples of causes not included in the DoE can be as follows:

1. Material selection (levels), especially in regulated industries such as medical or defense, where a new material might require re-applying for product certification.
2. Factors such as material age in the stockroom or which operator or machine will be used in production should be considered as uncontrolled or noise factors, and relegated to replications, rather than included in the design matrix. Handling of the uncontrolled factors is further discussed in Chap. 9.

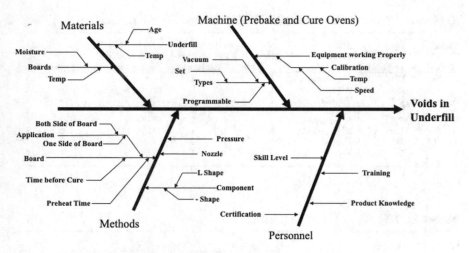

Fig. 4.6 Cause-and-effect diagram for underfill voids DoE

3. Factors related to maintaining a quality operation such as skill levels of operators, training, certifications, or calibrations should be assumed to be in control and not tested in the DoE project.

Five factors were selected, in order of importance that the team considered: Preheat Temperature, Cure Oven Type, Pressure applied to the nozzle, dispensing (fill) Application method, and PreCure Time. Three alternatives to conducting this project can be considered using Table 4.9: L32 in full factorial mode, L16 in half fraction, and L8 for partial factorial with interactions down to L8 screening design mode with no interactions considered. It was decided to select L8 partial factorial for the five factors at two levels, leaving two columns to investigate two interactions. The team decided on Preheat*Oven Type and Preheat*Time before Cure.

To investigate these two interactions, the assignment of the selected factors to L8 columns has to be very specific. The first three important factors were assigned to the main orthogonal columns A, B, and D in the L8, while the fourth factor was assigned to column G and the fifth factor assigned to column E in the L8 shown in Table 4.6. The five factors and levels chosen and columns that were assigned to were:

Column 1/A. Preheat Temperature, °F = 195, 210
Column 2/B. Cure Oven Type = Pr°ogrammable, Vacuum
Column 4/D. PreCure Time (minutes) = 5, 15
Column 6/F. Application (Method) = L Shape (L), Single Side (S)
Column 7/G. Pressure (nozzle), PSI = 30, 40

The corresponding DoE design is highlighted in Table 4.16 with two replication results for number of voids, showing the factor to column assignment to select the two interactions of interest at columns 3 (1*2 or A*B) and 5 (1*4 or A*D).

Table 4.16 L8 partial factorial underfill voids DoE factor and level assignments

Factor	1 (A) Preheat	2 (B) Cure	3 (C) (A*B)	4 (D) PreCure (A*D)	5 (E)	6 (F) (B*D)	7 (G) (A*B*D)	Results.	
Expt. #	Temp	Oven Type	Interaction	Time	Interaction	Pressure	Application		
1	195	Programmable	1	5	1	30	L	4	5
2	195	Programmable	1	15	2	40	S	4	3
3	195	Vacuum	2	5	1	40	S	3	4
4	195	Vacuum	2	15	2	30	L	2	2
5	210	Programmable	2	5	2	30	S	2	1
6	210	Programmable	2	15	1	40	L	0	1
7	210	Vacuum	1	5	2	40	L	1	1
8	210	Vacuum	1	15	1	30	S	0	0

Table 4.17 Underfill voids DoE confounding

Source	L8 Column	Confounding	
Preheat	A/1	F*G	6*7
Oven Type	B/2	D*F	4*6
Interaction	C/3	A*B, D*G	1*2, 4*7
Pre Cure Time	D/4	B*F	2*6
Interaction	E/5	A*D, B*G	1*4, 2*7
Pressure	F/6	B*D, A*G	2*4, 1*7
Application	G/7	A*F	1*6

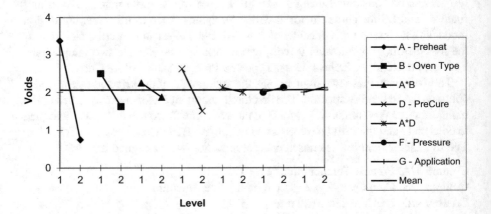

Fig. 4.7 Coefficient plots for underfill voids DoE

A partial factorial experiment with the symbol 2^{5-2} design will generate multiple confounding of main factors with interactions (resolution III), as well as two-way interactions confounding other two-way interactions (resolution IV) as shown in Table 4.17. The confounding is given for only two-way interactions in both numeric

Table 4.18 Voids DoE statistical analysis

Factor	Effect	Coef	SE Coef .	t-value	P > t
DoE Mean		2.063	0.130	15.85	0.000
Preheat	-2.625	-1.313	0.130	-10.09	0.000
Oven Type	-0.875	-0.438	0.130	-3.36	0.006
PreCure	-1.125	-0.563	0.130	-4.32	0.001

Source	Pooled	DoF	SS	MS	F-Value	Prob > F
A - Preheat	n	1	27.56	27.56	101.77	0.000
B - Oven Type	n	1	3.06	3.06	11.31	0.006
A*B	y	1	0.56	0.56	1.80	0.217
D - PreCure	n	1	5.06	5.06	18.69	0.001
A*D	y	1	0.06	0.06	0.20	0.667
F - Pressure	y	1	0.06	0.06	0.20	0.667
G - Application	y	1	0.06	0.06	0.20	0.667
Replication Error		8	2.50	0.31		
Pooled Error		4	0.75	0.19		
Total Error		12	3.25	0.27		
Total		15	38.94	2.60		

and alphabetic assignments of factors. It was hoped that statistical analysis of the two least important factors (Pressure and Application) that confounded with two-way interactions will show non-significance to reduce their effect on the analysis.

The graphical analysis for voids DoE is shown in Fig. 4.7, and the ANOVA is shown Table 4.18, using an Excel®-based DoE analysis software developed by the author. ANOVA should be performed first to reduce the model to significant factors only, and then model coefficient estimates should be performed to determine the model regression equation and predicted value for minimum voids.

Table 4.18 shows that only the main factors Preheat (A), Oven Type (B), and PreCure (D) should be considered in the regression equation. The remaining two factors Pressure and Application and two interactions A*B and A*D are not significant, shown as (pooled = y) values in Table 4.18, and their levels should be selected for lowest operation time or cost. This is based on the critical value for $F_{0.05,1,8} = 5.32$ from Table 2.11. The probabilities for factor coefficients and ANOVA are the same, using independent t- and F-tests. The predicted value based on selecting the levels for minimum voids of significant factors (A2, B2, and D2) result in predicted value of -0.25 with a good $R^2 > 0.9$ as follows:

Regression equation = 2.0625 − 1.3125 Preheat − 0.4375 Oven Type − 0.5625 PreCure.

Voids model regression equation = 16 experiments' average − factor coefficients (A + B + D).

Predicted voids = 2.0625 − 1.3125 − 0.4375 − 0.5625 = −0.25.

Voids   95%   CI $= \pm SE * t_{\alpha=0.025,12} = \pm \frac{s}{\sqrt{n}} * t_{\alpha=0.025,12} = \pm \sqrt{\frac{MS}{n}} * 2.179 =$ $\pm \sqrt{\frac{0.27}{4}} * 2.179 = \pm 0.567.$

Predicted voids 95% CI $= -0.25 \pm 0.567 = (-0.817, 0.317)$

$R^2 = 1 - (SST - SSE)/SST = 1 - (38.94 - 3.25)/38.94 = 91.65\%$

The predicted voids model CI values include the zero voids goal. The recommended levels for significant factors (A2, B2, and D2) based on Fig. 4.7 for least voids, match the levels in experiment 8 with two void-free results indicating low variability. The voids DoE project was successful in reaching the goal of zero defects, starting with the 16 experiments' average of 2 defects per component. Further enhancements to the project include selecting the vacuum oven and narrowing the process map using treatment analysis for Preheat Temperature and PreCure Time in search of better process map performance.

4.4.3 Partial Factorial L16 Example Case Study, Rivet Design DoE: Selecting Part Dimension Design for Best Product Performance

One of the issues of DoE is whether it is an alternative to traditional design and analysis techniques for developing new products. DoE is not a design system since it can only analyze a design once it is conceptualized. It can be used to enhance a design, but not to create it. DoE does not replace the mathematical simulation of design, such as finite element analysis (FEA), but helps in understanding the relative importance of design elements and their interactions with each other. A good combination of classical design and DoE might be using simulations with OA as inputs and analyzing simulation outputs as DoE outcomes, such as the half adder chip case study in the next chapter.

In this case study, a DoE project was undertaken to enhance an existing shear ring bulb blind rivet design by increasing the rivet strength. A typical aerospace rivet type is military pin M7885 made by Allfast Fastening Systems Inc., with drawing in http://trsaero.com/allfastinc/wp-content/uploads/technical-drawings/allmax/AF3243-REV-T.pdf.

There are three types of loads in rivet installations:

1. Donut load which indicated the maximum sheet take-up and stays constant during hole fill.
2. Shear load where the rivet shearing ring comes off the rivet shank.
3. Break load where the pin breaks off the rivet assembly. This is the specification most important to the customer.

The balance of these loads determines the quality of rivet installation. The break load was selected as the quality characteristic, and a riveting gun was modified with a

Table 4.19 Rivet DoE factor and level assignments

Factor	Operating Range	L16 Column #		Level 1	Level 2
Head Style	-	1	A	Flush	Dome
Grip Length	01 - 12	2	B	1	8
Shear Ring Width	0.025 - 0.030	4	C	0.025	0.028
Pin Diameter	0.120 - 0.130	8	D	0.124	0.127
Sleeve Diameter	0.133 - 0.140	13	G	0.134	0.137
Radius Shear Ring	0.001 - 0.004	14	R	0.002	0.004
Shear Ring Diameter	0.170 - 0.180	15	E	0.173	0.177

Table 4.20 Rivet DoE factor confounding

L16 Columns		2-Way Confounding		3-Way	Column Description
1	A	E*R	14*15		Head Type
2	B	E*G	13*15		Grip Length
3	A*B	G*R	13*14		Interaction of A*B
4	C				Shear Ring Width
5	A*C	D*G	8*13		Interaction of A*C
6	B*C	D*R	8*14		Interaction of B* C
7	D*E			A*B*C	Interaction of D*E
8	D				Pin Diameter
9	A*D	C*G	4*13		Interaction of A*D
10	B*D				Interaction of B*D
11	C*E			A*B*D	Interaction of C*E
12	C*D				Interaction of C*D
13	G	B*E	2*13	A*C*D	Sleeve Diameter
14	R	A*E	1*15	B*C*D	Radius shear Ring
15	E	A*R – B*G	1*14 - 2*13	ABCD	Shear Ring Diameter

load cell to measure the break load. The DoE goal was to maximize the break load within the selected DoE design space and to study the effects of critical rivet assembly dimensions and their interactions.

The rivet assembly consists of the sleeve (body) and the mandrel (pin). The project team used the techniques of cause-and-effect diagrams to narrow down the selection of factors and levels. Five independent dimensions of the pin were selected as factors. Other dimensions are either dependent or directly proportional to the five pin factors. The levels selected were within the operating range of the pin dimensions. Two other factors were added, including rivet head type and grip size for a total of seven factors.

A screening L8 design was considered, but the team opted for a partial factorial L16 from Table 4.9, which allowed for studying interactions between factor levels.

Table 4.21 Rivet DoE design and results

Column# 1	2	4	8	13	14	15				
		Shear			Radius	Shear				
	Head	Grip	Ring	Pin	Sleeve	Shear	Ring				
Expt. #	Style	Length	Width	Diam.	Diam.	Ring	Diam.	Test 1	Test 2	Test 3	Test 4
1	Flush	1	0.025	0.124	0.134	0.002	0.173	1506	1591	1547	1560
2	Flush	1	0.025	0.127	0.137	0.004	0.177	1523	1519	1528	1521
3	Flush	1	0.028	0.124	0.137	0.004	0.177	1598	1606	1588	1600
4	Flush	1	0.028	0.127	0.134	0.002	0.173	1606	1532	1581	1589
5	Flush	8	0.025	0.124	0.134	0.004	0.177	1573	1571	1577	1572
6	Flush	8	0.025	0.127	0.137	0.002	0.173	1499	1574	1592	1528
7	Flush	8	0.028	0.124	0.137	0.002	0.173	1666	1601	1546	1592
8	Flush	8	0.028	0.127	0.134	0.004	0.177	1575	1568	1582	1574
9	Dome	1	0.025	0.124	0.137	0.002	0.177	1689	1647	1598	1652
10	Dome	1	0.025	0.127	0.134	0.004	0.173	1501	1493	1511	1498
11	Dome	1	0.028	0.124	0.134	0.004	0.173	1626	1618	1632	1630
12	Dome	1	0.028	0.127	0.137	0.002	0.177	1514	1575	1595	1615
13	Dome	8	0.025	0.124	0.137	0.004	0.173	1603	1602	1603	1608
14	Dome	8	0.025	0.127	0.134	0.002	0.177	1523	1615	1606	1592
15	Dome	8	0.028	0.124	0.134	0.002	0.177	1782	1711	1749	1773
16	Dome	8	0.028	0.127	0.137	0.004	0.173	1652	1660	1652	1652

The L16 includes 4 main factors and 11 interactions. Four of the most important factors were assigned to the main columns in the L16 (A/1, B/2, C/3, and D/4). The other three factors were assigned to three- and four-way interaction columns (13, 14, and 15). Table 4.19 illustrated the Rivet DoE factor and level assignments.

- Using a partial factorial L16 $= 2^{7-3}$ reduces the numbers of experiments but increases confounding of two- and three-way interactions as shown in Table 4.20. Ignoring three-way interactions, only one factor (E) is confounded with two two-way interactions, and four factors (A, B, G and R) are confounded with one two-way interaction, resulting in a resolution III design. Two factors are not confounded (C and D). Some of the two-way interactions are confounded with other two-way interactions. If the DoE project team has chosen a different three-way interaction column such as column 7 (A*B*C) or column 11 (A*B*D) for the factor Shear Ring Diameter, then the design becomes a resolution IV design where no factor is confounded by a two-way interaction; only two-way interactions are confounded by other two-way interactions. The next L32 case study in this chapter avoided main factor confounding with interactions by assigning non-main factors to three-way interactions only. However, two-way interactions could not be analyzed since they are confounded by other two-way interactions. More details about resolution design and interactions will be shown in Chap. 7.

The DoE sequence and the results of four replications of break load testing are shown in Table 4.21. The experiments' patterns are obtained by assigning the seven factors to the columns shown in Table 4.19, with the chosen levels reflected in the

Table 4.22   Rivet DoE coefficient analysis

Source	Effect	Coef .	SE Coef	T-Value	P > t
Constant		1593.16	3.53	450.97	0.000
Head Style	49.75	24.87	3.53	7.04	0.000
Grip Length	37.00	18.50	3.53	5.24	0.000
Shear Ring Width	53.69	26.84	3.53	7.60	0.000
Pin Diam	-52.25	-26.13	3.53	-7.40	0.000
Sleeve Diameter	1.06	0.53	3.53	0.15	0.881*
Radius Shear Ring	-22.81	-11.41	3.53	-3.23	0.002
Shear Ring Diam	20.63	10.31	3.53	2.92	0.005
Head Style*Grip Length	24.81	12.41	3.53	3.51	0.001
Head Style*Shear Ring Width	14.75	7.37	3.53	2.09	0.042
Head Style*Pin Diam	-27.06	-13.53	3.53	-3.83	0.000
Grip Length*Shear Ring Width	14.87	7.44	3.53	2.11	0.041
Grip Length*Pin Diam	9.44	4.72	3.53	1.34	0.188*
Shear Ring Width*Pin Diam	2.50	1.25	3.53	0.35	0.725*
Shear Ring Width*Shear Ring D.	-10.00	-5.00	3.53	-1.42	0.163*
Pin Diam*Shear Ring Diam	-26.56	-13.28	3.53	-3.76	0.000

Table 4.23   Rivet DoE ANOVA after pooling

Source	DOF	SS	MS	F-Value	P > F
Head Style	1	39601	39601.0	49.58	0.000
Grip Length	1	21904	21904.0	27.42	0.000
Shear Ring Width	1	46118	46117.6	57.74	0.000
Pin Diam	1	43681	43681.0	54.69	0.000
Shear Ring Diam	1	6806	6806.2	8.52	0.005
Sleeve Diameter	1	18	18.1	0.02	0.881*
Radius Shear Ring	1	8327	8326.6	10.42	0.002
Head Style*Grip Length	1	9851	9850.6	12.33	0.001
Head Style*Shear Ring Width	1	3481	3481.0	4.36	0.042
Head Style*Pin Diam	1	11718	11718.1	14.67	0.000
Grip Length*Shear Ring Width	1	3540	3540.3	4.43	0.041
Grip Length*Pin Diam	1	1425	1425.1	1.78	0.188*
Shear Ring Width*Pin Diam	1	100	100.0	0.13	0.725*
Shear Ring Width*Shear Ring D.	1	1600	1600.0	2.00	0.163*
Pin Diam*Shear Ring Diam	1	11289	11289.1	14.13	0.000
Pooled Error	4	3143	785.8		
Replication Error	48	38340	798.8		
Total Error	52	41483	797.8		
Total	63	247798			

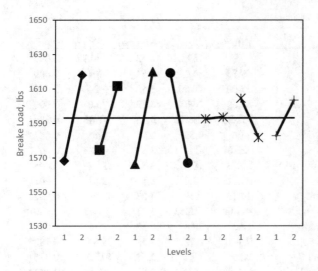

Fig. 4.8 Model coefficient plots for Rivet DoE

pattern of each experiment. A total of 64 experiments were conducted with 4 replications of L16 patterns and shown in Table 4.21.

The Rivet DoE coefficient analysis is shown is Table 4.22, and the rivet and
ANOVA after pooling is shown in Table 4.23. There were one factor (Sleeve
Diameter) and three interactions which were not significant, shown with an asterisk
(*). The reduced model R^2 was closed to desired 90%:

$$R^2 = 1 - (SST - SSE)/SST = (247{,}798 - 41{,}483)/247{,}798 = 83.26\%$$

The reduced model was significant with six factors and five interactions. The rivet
coefficient plots are shown in Fig. 4.8, and the levels can be selected for the
maximum outcome and plugged into the regression equation. The recommended
factors and levels were as follows:

Head Style	Level 2—Dome
Grip Length	Level 2—8
Shear Ring Width	Level 2—0.028
Pin Diameter	Level 1—0.124
Sleeve Diameter	Not significant
Radius Shear Ring	Level 1—0.002
Shear Ring Diameter	Level 2—0.177

Regression equation for break load = 1593.16 + 24.87 Head Style + 18.5 Grip Length + 26.84 Shear Ring Width − 26.13 Pin Diam. + 10.31 Shear Ring Diam. − 11.41 Radius Shear Ring +12.41 Head Style*Grip Length + 7.37 Head Style*Shear Ring Width − 13.53 Head Style*Pin Diam. + 7.44 Grip Length*Shear Ring Width + 13.28 Pin Diam.*Shear Ring Diam.

The predicted value for the maximum break load is 1765.25 lbs., based on the regression equation, which is higher than the mean of 64 experiments of 1593.16 lbs., an 11% increase. The top 3 factors that were deemed most important were Shear Ring Width, Pin Diameter, and Head Style, all with coefficients greater than 24, as shown in Fig. 4.8 and Table 4.22. The interactions were not as important as the team expected with most having coefficients that were less than those of the top 3 factors as shown in the regression eq. A confirming experiment was conducted which matched the predicted values.

4.4.4 Partial Factorial L32 Case Study, APOS for Robotics DoE: Selecting Process Parameters for Multiple Adjustment Production Machines

A good use of DoE projects is the selection of process parameters for automatic machines with multiple adjustments on the manufacturing floor. Many of these machines do not come with specific operating instructions for adjusting the machine for each part. Production engineers constantly fiddle with these adjustments trying to reduce defects or improve performance. DoE offers a **structured** methodology to "tune" these machines for selected parts.

A good case study is a DoE project to improve the performance of a robotic line with multiple part positioning (APOS) stations for part assembly into a complex electrotechnical product. Each APOS station allowed for a hopper containing parts to deposit them into preformed trays (pallets) by using circular vibration sources in two axes with controlled amplitude (width), frequency, and durations. Excess parts not properly positioned are air blown away from the pallets. The parts are thus positioned correctly in the pallet for a robotic arm to pick and then assemble them into the product. These techniques are used in mass consumer products such as ball point pens and smartphones assembly. They work very well for simple geometry parts but are difficult to implement for complex parts. Some of the positioning stations could contain tens of adjustments (factors), making a simulation of how best to adjust each part difficult.

An example of a robotic line that was discussed in the author's book *Successful Implementation of Concurrent Engineering Products and Processes* is the SMART® robotic line concept shown in Fig. 4.9. Elements of the line include APOS stations, parts' hoppers, tray, and pallet transport systems to convey parts from APOS stations or suppliers and present them to robotic arms.

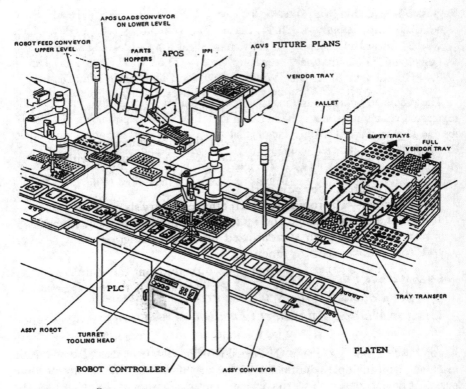

Fig. 4.9 SMART® robotic line concept

Table 4.24 APOS DoE factors and interactions

#	Factors	Interactions
A	Vibration Frequency of XY Motor	A*B
B	Vibration Width of XY Brake	F*H
C	Slant Angle during Loading	A*D
D	Vibration Frequency of XZ Motor	B*E
E	Vibration Width of XZ Brake	L*N
F	Frequency Air Blow during Vibration	D*E
G	Vibration Frequency of Hopper	G*J
H	Time Air Blow during Vibration	A*M
J	Number of Parts in Hopper	G*L
L	Time Final Air	G*M
K	Time Gate Open	G*N
M	Time Vibration.	
N	Gate Open Interval.	

Table 4.25 APOS DoE factor assignment and experiment results

Col. # 1	2	4	7	8	11	13	14	16	19	21	22	25		
Confounding			A		A	A	B	E	A	A	B	A		
			B		B	C	C		B	C	C	D		
			C			D	D	D		E	E	E	E	Results
Exp. # A	B	C	D	E	F	G	H	J	K	L	M	N	Spring	Beam
1	1	1	1	1	1	1	1	1	1	1	1	1	9	23
2	1	1	1	1	1	1	1	2	2	2	2	2	21	21
3	1	1	1	1	2	2	2	2	1	1	1	1	37	21
4	1	1	1	1	2	2	2	2	2	2	2	1	58	29
5	1	1	2	2	1	1	2	2	1	1	2	2	1	63	14
6	1	1	2	2	1	1	2	2	2	2	1	1	2	33	9
7	1	1	2	2	2	2	1	1	1	1	2	2	2	21	13
8	1	1	2	2	2	2	1	1	2	2	1	1	1	5	10
9	1	2	1	2	1	2	1	2	1	2	1	2	1	8	25
10	1	2	1	2	1	2	1	2	2	1	2	1	2	2	18
11	1	2	1	2	2	1	2	1	1	2	1	2	5	54	18
12	1	2	1	2	2	1	2	1	2	1	2	1	1	38	27
13	1	2	2	1	1	2	2	1	1	2	2	1	1	60	33
14	1	2	2	1	1	2	2	1	2	1	1	2	2	64	19
15	1	2	2	1	2	1	1	2	1	2	2	1	2	15	6
16	1	2	2	1	2	1	1	2	2	1	1	2	1	13	7
17	2	1	1	2	1	2	2	1	1	2	2	1	2	62	29
18	2	1	1	2	1	2	2	1	2	1	1	2	1	54	28
19	2	1	1	2	2	1	1	2	1	2	2	1	1	14	11
20	2	1	1	2	2	1	1	2	2	1	1	2	2	10	0
21	2	1	2	1	1	2	1	2	1	2	1	2	2	7	16
22	2	1	2	1	1	2	1	2	2	1	2	1	1	32	17
23	2	1	2	1	2	1	2	1	1	2	1	2	1	68	23
24	2	1	2	1	2	1	2	1	2	1	2	1	2	66	15
25	2	2	1	1	1	1	2	2	1	1	2	2	2	62	28
26	2	2	1	1	1	1	2	2	2	2	1	1	1	26	14
27	2	2	1	1	2	2	1	1	1	1	2	2	1	25	22
28	2	2	1	1	2	2	1	1	2	2	1	1	2	11	13
29	2	2	2	2	1	1	1	1	1	1	1	1	2	7	12
30	2	2	2	2	1	1	1	1	2	2	2	2	1	36	13
31	2	2	2	2	2	2	2	2	1	1	1	1	1	35	7
32	2	2	2	2	2	2	2	2	2	2	2	2	2	45	19

The goal for the robotic project is enhancing palletizing efficiency, which is the percentage of parts that are correctly positioned in the pallet after each part filling operation for presentation to the robotic pickup arms. A DoE approach for maximizing APOS performance would provide for increased knowledge of the effect of each factor selected, interactions of different factors, and ranking factors by importance.

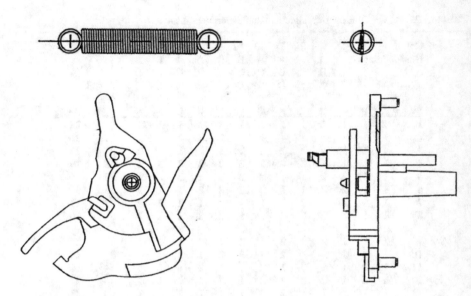

Fig. 4.10 Spring and walking beam APOS parts

Table 4.26 APOS DoE ANOVA of two parts

Spring	DOF	SS	MS	F-Value	P > F
C	1	195	195	3.43	0.076
D	1	236.5	236.5	4.16	0.052
G	1	10841.3	10841.3	190.53	0.000
H	1	621.3	621.3	10.92	0.003
K	1	1001.3	1001.3	17.60	0.000
L	1	770.3	770.3	13.54	0.001
Error	25	1422.5	56.9		
Total	31	15088.2			

Beam	DOF	SS	MS	F-Value	P > F
C	1	276.1	276.12	11.8	0.002
E	1	190.1	190.12	8.12	0.009
F	1	190.1	190.13	8.12	0.009
G	1	351.1	351.12	15	0.001
H	1	190.1	190.12	8.12	0.009
K	1	153.1	153.13	6.54	0.017
Error	25	585.3	23.41		
Total	31	1936			

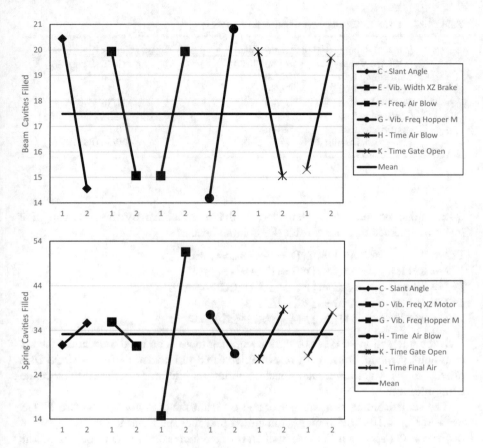

Fig. 4.11 APOS DoE coefficient plots for two parts

After brainstorming and cause-and-effect sessions, it was decided to select 13 factors and 11 interactions shown in Table 4.24. An L32 OA was selected as a partial factorial from Table 4.9 and shown in Table 4.25 with results of testing the palletizing of two parts of simple and complex geometry (Spring or Walking Beam). The two parts are shown in Fig. 4.10, with the pallet for springs having 80 cavities to be filled, while the pallet for walking beams had 36 cavities. The L32 design in Table 4.25 was created by using the full factorial L32 in Table 4.8 and assigning the 13 factors sequentially to the 5 main columns 1, 2, 4, 8, and 16 and 8 successive three-way interaction columns 7(ABC), 11(ABD), 13(ACD), 14(BCD), 19(ABE), 21(ACE), 22(BCE), and 25(ADE). Factor assignments to four-way interaction columns were avoided to prevent main factors confounding two-way interactions, as was discussed in the L16 Rivet DoE case study (Table 4.20).

The L32 configuration in Table 4.25 is a resolution IV design where two-way interactions confound other two-way interactions. This prevents a direct study of

Table 4.27 APOS DoE outcome validation for two parts

Run #	Spring Cavities filled	Beam Cavities filled
Predicted Value	72/80 (90%)	33/36 (92%)
Confirming Run 1	72/80	28/36
Confirming Run 2	75/80	31/36
Confirming Run 3	79/80	31/36
Confirming Run 4	74/80	30/36
Confirming Run 5	77/80	28/36
Mean 5 confirming Runs	75/80 (94%)	30/36 (83%)

interactions. For example, interaction A*B is confounded by the following five two-way interactions, using Table 4.8 column headings:

$$1*2 = 4*7 \quad \text{or} \quad A*B = C*D = C*ABC = A*B$$
$$1*2 = 8*11 \qquad A*B = D*F = D*ABD = A*B$$
$$1*2 = 13*14 \qquad A*B = G*H = ACD*BCD = A*B$$
$$1*2 = 16*19 \qquad A*B = J*K = E*ABE = A*B$$
$$1*2 = 21*22 \qquad A*B = L*M = ACE*BCE = A*B$$

If the interaction A*B is not significant, then there is no need to separate the six confounding interactions. If any one of the selected 11 interactions in the APOS DoE proves to be significant, additional experiments might be needed to separate confounding.

The statistical analysis was performed on main factors only for the two APOS parts in Fig. 4.10. The Spring symmetrical part indicated that the most important factor was G (Vibration Frequency of Hopper) in increasing the DoE outcome. The Beam, which was a complex non-symmetrical part with many features in different axis, had different but equally significant factors.

APOS DoE analysis for the two parts was compared using the top 6 factors (in terms of their SS values) using ANOVA pooling for insight into their effects. Both model coefficients for the six factors in Spring and Beam were plotted in Fig. 4.11, showing that different factors affect each part outcome differently, with factor G (Hopper Vibrations) being the most important in Spring but less so in Beam. It was determined that one universal set of factors and levels for APOS is not possible and each part to be assembled in an APOS station should have its own DoE to tune it. The ANOVA analyses for both spring and beam parts are shown in Table 4.24. Both models are shown to be significant with R^2 of 91% for spring and 70% for beam. All six factor coefficients are significant for beam parts, and four factors are significant for the spring, while two other factors (C and D) are borderline significant.

The regression equation for both parts is given below, with the predicted value based on selecting the proper level for maximum parts positioned in the pallets:

Spring = 33.15625 − 2.46875 C (Slant Angle) + 2.71875 E (Vibration Frequency
XZ Motor) − 18.40625 G (Vibration Frequency Hopper) + 4.40625 H (Time Air
Blow) − 5.59375 K (Time Gate Open) − 4.90625 L (Time Vibration).
Beam = 17.5 + 2.9375 C (Slant Angle) − 2.4375 E (Vibration Width XZ
Brake) − 3.3125 F (Frequency Air Blow) − 3.3125 G (Vibration Frequency
Hopper) + 11.17 H (Time Air Blow) + 2.4375 K (Time Gate Open).

The predicted values for spring are 71.7 parts filled in pallets and 95% CI
(64.3901, 78.9224), and the predicted values for beam are 33.25 parts filled in
pallets and 95% CI (28.5894, 37.9106). A set of five confirming experiments were
conducted using the proposed factors and levels for the experiment and shown in
Table 4.27. All five confirming experiment runs showed that the DoE was successful
in filling 94% of the cavities in the spring pallet and 83% for the beam and all were
within the confidence interval of the prediction. An 80% filling rate was the required
minimum since it would keep the robotic line from shutting down due to lack of
parts.

The APOS DoE demonstrates the benefits and pitfalls in conducting large
experiments with many factors such as an L32, which can be summarized as follows:

1. The DoE was successful in achieving the goal of increased palletizing efficiency
   for different geometry parts.
2. It was shown that different parts' positioning needs to be optimized by an
   individual DoE for each part.
3. The choice of large number of factors (13) with eight factors confounding three-
   way interactions resulted in compromising the study of factor interactions.
4. Since factor G (Hopper Vibrations) level (2) was the most important for both
   parts, subsequent part DoE could run with the top 5 factors in a half fraction L16
   or partial factorial L8 to study each part factor and interaction for optimal
   palletizing.

4.5   Conclusions

This chapter focused on the merits of using two-level OA to conduct successful DoE.
Different methodologies for factor and level selections, analyzing interactions and
confounding, and model reduction through pooling were discussed. In addition, DoE
statistical analysis of model coefficient and ANOVA were demonstrated with formulas and example calculations. The use of regression equations to predict model
outcome(s) and confirming experiments was illustrated.

Several case studies in this chapter demonstrated that DoE is an excellent tool to
investigate, evaluate, improve, or optimize new and existing designs and processes.
There are several techniques in DoE that should be thought out well in advance: the
definition of outcomes to be optimized, the selection of factors and levels, the
treatment of factor interactions, the selection of experiment arrays, and how to
simulate and measure variability and error.

The use of two-level DoE offers interesting opportunities to reduce DoE project scope by reducing the number of experiments in an organized approach, with focus on selecting factors and levels, as well as managing factor interactions. The types of two-level OA were defined with their special use as full, half fraction, or partial factorial as well as screening design. Careful attention was given to different strategies of handling interactions and confounding.

Each case study in this chapter was discussed in depth and its design critiqued and summarized with results obtained. The constraints of each DoE were presented with the conflicts of meeting initial goals and determining project scope versus time and resources needed to successfully complete DoE projects. Decisions made by DoE project teams were analyzed and suggestions made for improvement to increase design resolution.

Additional Reading Material

Anderson, M. and Whitcomb, P., *DOE Simplified: Practical Tools for Effective Experimentation, 3rd Edition*, Boca Raton, FL: CRC press, 2015

Anderson, V. and McLean, R., *Design of Experiments: A Realistic Approach*, New York: Marcel Dekker, 1974

Antony, J., *Design of Experiments for Engineers and Scientists,* 2nd Edition, Elsevier Insights: London 2014

Box, G., Hunter, J. and Hunter, W., *Statistics for Experimenters: Design, Innovation, and Discovery*, 2nd Edition. New York: Wiley-Interscience, 2005.

Box, G., Bisgaard S. and Fung C., *An explanation and Critique of Taguchi's Contribution to Quality Improvement*, University of Wisconsin. 1987.

Camasso, C., "The effects of APOS Operating Parameters on Palletizing Efficiencies of Two Specific Dissimilar Parts", Master of Management Science Thesis, UMass Lowell, May 1989.

Cochran W. and Cox G. *Experimental Designs,* 2nd Edition, New York: Wiley, 1981.

Diamond W., *Practical Experiment Design*. Van Nostrand Reinhold, New York, 1981.

Ealy, L., "Taguchi Basics." *Quality Journal*, November 1988, 26–30.

Field, A. and Hole G., *How to Design and Report Experiments*, London: Sage Publishing, 2003.

Hicks, C., *Fundamental Concepts in the Design of Experiments*, New York: McGraw Hill, 1964.

Hinkelmann, K. and Kempthorne O., *Design and Analysis of Experiments, Volume 1: Introduction to Experimental Design*, New York: Wiley-Interscience, 2007.

John, P., *Statistical Analysis and Design of Experiments*, New York: Macmillan, 1971.

Johnson, R., *Probability and Statistics for Engineers*, 5th Edition, Englewood Cliffs, NJ: Prentice-Hall, 1994.

Mason, R., Gubst R. and Hess, J., *Statistical Design and Analysis of Experiments, with Applications to Engineering and Science 2nd Edition*, Hoboken, NJ: Wiley 2003.

Ram, K., "Design of an Aircraft Industry Riveting System", Master of Mechanical Engineering Thesis, UMass Lowell, May 1992.

Montgomery, D., *Design and Analysis of Experiments,* 3rd Edition, New York: Wiley, 1991.

Ross, P., *Taguchi Techniques for Quality Engineering*, New York: McGraw Hill, 1987.

Roy, R., *A Primer on the Taguchi Method*, New York: Van Nostrand Reinhold, 1990.

Phadke, M., *Quality Engineering Using Robust design*, Englewood Cliffs, NJ: Prentice-Hall. 1989.

Chapter 5
Three-Level Factorial Design and Analysis Techniques

The use of three-level orthogonal arrays (OAs) is well suited for optimizing current designs and process parameters. These parameters can be improved with DoE using current design or process value as the middle level of selected factors. The other two levels can be extensions of this middle level into other alternatives, including materials, geometry, equipment, or process control such as speed, time, and environmental conditions.

Three-level OAs follow the same principles outlined in the previous chapters for two levels. The evolving definitions of quality, DoE arrangement of elements, and their objectives shown in Figs. 4.1, 4.2 and 4.3 are also applicable for three levels. Three-level DoEs are conducted in the same sequence as Table 4.1. They provide additional information than one-factor-at-a-time (OFAT) experiments due to their ability to measure interactions between main (primary) factors.

Three-level DoEs allow for design space or process map to be extended to offer outcomes for each factor using three levels. They can show any curvature or non-linear relationship between factor levels and outcomes as opposed to the linear relationships of two-level DoE. A DoE team using three levels can focus on certain segments of the relationship by performing additional DoE. The next DoE can clarify an outcome segment of interest, either by choosing closer level values or by extending the range of levels in either direction of interest.

There are several sections in this chapter to illustrate design, analysis, and case studies of three-level DoE:

5.1 Three-Level Factorial Design

DoEs using predetermined OA with three-level factors are presented with emphasis on factor interactions and confounding. The first three-level OA, L9 and L27, will be discussed in detail with case studies. They represent two and three factors in full factorial designs and can be used in additional modes such as screening or partial factorial design. The next higher three-level OA for four factors (L81) is impractical to use since the sheer number of experiments can result in costly and time-consuming efforts, with a high probability of error in

© The Author(s), under exclusive license to Springer Nature Switzerland AG 2022
S. Shina, *Industrial Design of Experiments*, https://doi.org/10.1007/978-3-030-86267-1_5

conducting 81 experiments. Alternative DoE methods for multiple three-level factors can use sequential DoE by performing a screening followed by full factorial design. Another method of handling three-level factors is using partial factorial designs with confounding to reduce the number of experiments.

5.2 Three-Level DoE Analysis and Model Reductions

The analysis approach of two-level DoE from the last chapter is extended to include three-level factors and their interactions. Pooling to reduce the model is explored with regression equation, model prediction, and outcome optimization. Factor interactions and the role they play in determining factor significance and interdependence will be explored.

Different techniques for creating model error in ANOVA are presented with formulas, sample calculations, and verification using software analysis packages. They include pooling of the smallest model coefficient or replicating existing or additional experiment lines.

5.3 Three-Level DoE Case Studies

These case studies are drawn from various uses of three-level DoE in design and manufacturing in different industries. They illustrate the use of the different types of three-level arrays in screening and partial factorial modes. In each case study, DoE team strategy is discussed, experiment layout is shown, and results are analyzed. Conclusion of each case study is given with suggestions for future DoE planning. Other three-level case studies will be presented in later chapters, including follow-on (sequential) experiments in Chap. 8 and variability reduction in Chap. 9.

5.4 Conclusions

5.1 Three-Level Factorial Design

The DoE definitions, expectations, and process discussed in Chap. 4 for two levels are applicable to three-level design as well as multiple-level design discussed in Chap. 8. The model reduction and error handling techniques for three-level DoE are the same as those for two levels.

Three-level OAs have different interaction attributes than two levels. Each three-level factor has two degrees of freedom (DOF), and their two-way interactions have $2 * 2 = 4$ DOF, expressed in two columns of three levels each. Three-way interactions are expressed as $2 * 2 * 2 = 8$ DOF or four three-level columns. For example, if two factors are considered (A, B) for an L9 (3^2) experiments, then the interaction A*B consists of two columns A*B and A*B$_2$. If three factors (A, B, C) are considered for an L27 (3^3) experiments, then the interaction ABC consists of four columns A*B*C, A*B$_2$*C$_2$, A*B$_2$*C, and A*B*C$_2$. Tables 5.1 and 5.2 show compositions of L9 and L27 OA for factor and interaction columns, respectively.

The pattern of level sequences for the main factors and interactions allows for solving model coefficient estimates for the first three-level arrays L9, L27, and L81 as shown in Table 5.3. For four factors at full factorial (L81), the number of factors

Table 5.1 L9 with two main factors (1/A and 2/B) and interactions (3/C/AB and 4/D/AB$_2$)

		Columns		
Number	1	2	3	4
Factors	A	B	C	D
Expt. #			AB	AB$_2$
1	1	1	1	1
2	1	2	2	2
3	1	3	3	3
4	2	1	2	3
5	2	2	3	1
6	2	3	1	2
7	3	1	3	2
8	3	2	1	3
9	3	3	2	1

and their interactions is 40, and the number of coefficients to be solved is 40 * 3 = 120, making it difficult to interpret the results and provide good conclusions.

Three-level orthogonal arrays can measure interactions in three methods, similar to two-level arrays, and are summarized in Table 5.4:

1. Full factorial DoEs are used to evaluate the effects of all factors and their interactions. As shown in Table 5.3, for every number of factors (*n*), there are multiple two- to n-way interactions to be considered. The interaction column levels are derived from sequential arrangements of the originating factor levels in the following sequences: 1 2 3, 2 3 1, and 3 2 1. These three sequences can be arranged in different orders down the interaction column, as can be seen in Table 5.2 for L27. In the example of L9, the pattern of the levels in the main factors and interaction columns allows for solving 12 coefficients (4 factors with 3 levels) in 9 equations, as will be demonstrated in the following sections.
2. Partial factorial DoEs provide a cost-effective way of determining significance of selected factors and interactions while increasing chances for DoE analysis error. They can be used with additional factors confounding two-, three-, or four-way interactions, similar to methods discussed in two-level OA, assuming some interactions are not significant. By selectively choosing where to confound, a fractional factorial DoE could be used to study more factors and some of their interactions with fewer experiments. The adverse effect of this strategy is that confounded interaction might be significant, resulting in incorrect DoE analysis.
3. Screening design DoE allows all columns in OA to be assigned to different factors. All factors are confounded by two-way interactions. An example in Table 5.4 is using four factors in a screening L9 as opposed to full factorial L81. They represent a minimum set of experiments for the number of factors considered. They are used to whittle down the number of factors quickly through

Table 5.2 L27 with three main factors (1/A, 2/B and 5/C) and ten interactions

Number 1	2	3	4	5	6	7	8	9	10	11	12	13
Factors A	B			C								
Expt. #		AB	AB2		AC	AC2	BC	ABC	AB2C2	BC2	AB2C	ABC2
1	1	1	1	1	1	1	1	1	1	1	1	1
2	1	1	1	2	2	2	2	2	2	2	2	2
3	1	1	1	3	3	3	3	3	3	3	3	3
4	1	2	2	2	1	1	1	2	2	2	3	3	3
5	1	2	2	2	2	2	2	3	3	3	1	1	1
6	1	2	2	2	3	3	3	1	1	1	2	2	2
7	1	3	3	3	1	1	1	3	3	3	2	2	2
8	1	3	3	3	2	2	2	1	1	1	3	3	3
9	1	3	3	3	3	3	3	2	2	2	1	1	1
10	2	1	2	3	1	2	3	1	2	3	1	2	3
11	2	1	2	3	2	3	1	2	3	1	2	3	1
12	2	1	2	3	3	1	2	3	1	2	3	1	2
13	2	2	3	1	1	2	3	2	3	1	3	1	2
14	2	2	3	1	2	3	1	3	1	2	1	2	3
15	2	2	3	1	3	1	2	1	2	3	2	3	1
16	2	3	1	2	1	2	3	3	1	2	2	3	1
17	2	3	1	2	2	3	1	1	2	3	3	1	2
18	2	3	1	2	3	1	2	2	3	1	1	2	3
19	3	1	3	2	1	3	2	1	3	2	1	3	2
20	3	1	3	2	2	1	3	2	1	3	2	1	3
21	3	1	3	2	3	2	1	3	2	1	3	2	1
22	3	2	1	3	1	3	2	2	1	3	3	2	1
23	3	2	1	3	2	1	3	3	2	1	1	3	2
24	3	2	1	3	3	2	1	1	3	2	2	1	3
25	3	3	2	1	1	3	2	3	2	1	2	1	3
26	3	3	2	1	2	1	3	1	3	2	3	2	1
27	3	3	2	1	3	2	1	2	1	3	1	3	2

Table 5.3 Column assignments for three-leve cl OA

Array #	Main Factors	Two-Way	Interactions Columns Three-Way	Four-Way	Total Columns	Total Coefficients
L9 ($3^{2)}$)	2	2	-	-	4	12
L27 (3^3)	3	6	4	-	13	39
L81 (3^4)	4	12	16	8	40	120

Table 5.4 Three-level OA use in full and partial factorial and screening modes

# of Factors	# of Experiments Full Factorial	# of Experiments Partial Factorial	# of Experiments Screening (Saturated)
2	9	9 (Two Interactions)	9
3	27	9 (One Interaction)	9
4	81	No Interactions	9
5	243	27 (Eight Interactions)	27
↓	↓	↓	↓
13	$3^{13} = 1{,}594{,}323$	No Interactions	27
14	81	81(26 Interactions)	81

smaller DoEs, followed by full factorial DoEs (L9 or L27) that can be performed on the top 2 or 3 significant factors. The assumption in screening designs is that all interactions are small and can be ignored compared to the main factor effects.

5.1.1   Commonly Used Three-Level Orthogonal Arrays

The most used three-level OAs are full factorial implementations of $n = 2, 3$, and 4 factors, commonly known as L9, L27, and L81 in full factorial mode. The Lx nomenclature is the (x) number of experiments corresponding to $x = 3^n$. The first two arrays are listed in Tables 5.1 and 5.2, using labels 1, 2, and 3 to identify the levels for each column. In some DoE analysis software, the three levels are labeled -1, 0, and 1. The analysis is the same using either level labels, but interactions' algebraic sign might be different for the two-level types. Plotting factor coefficients reduces the error when making DoE outcome predictions, since it is very apparent as to the proper level selection for minimum, maximum, or target outcome.

In Table 5.1 (L9) and 5.2 (L27), the top rows show the factors and interactions with their alphabetical and numerical labels. The interaction columns are labeled as two or three ways, considering that two columns are required for each two-way interaction and four columns for three-way interactions in Table 5.2.

The use of L81 in industry for four-factor full factorial is rare due to the effort and time required to perform the experiments and with higher error probability in collecting data. Usually, it is substituted with a screening L9 DoE, followed by a full factorial L9 or L27 for the significant factors, or the use of partial factorial L27, with the fourth factor confounding one of the four three-way interaction columns (9, 10, 12, or 13).

Use of Three-Level OA: Full and Partial Factorial and Screening Modes

There are three uses for each three-level OA in a DoE project. These will be discussed using L9 and L27 as examples and summarized in Table 5.4.

1. OA use as a full factorial array to model main factors and all their interactions. In the case of L9, the model includes two main factors (A and B) at three levels and one two-way interaction in two columns [C (AB) and D (AB$_2$)]. For L27, there are three factors (A, B, and C) with three two-way interactions with two columns each (AB, AB$_2$, AC, AC$_2$, BC, BC$_2$) and one three-way interaction ABC with four columns (ABC, AB$_2$C$_2$, AB$_2$C, and ABC$_2$).
2. OA use as a screening design by assigning main factors to each column and ignoring all interactions. The number of "non-interacting" factors that can be modelled is equal to the number of columns: 4 for L9 and 13 for L27. This technique can be used to reduce the model to only significant factors. A second DoE could be initiated to study significant factors in the reduced model in a full factorial DoE mode. A reasonable alternative to the two DoE scenarios is to use a non-interacting array to be discussed in Chap. 8. A typical three-level non-interacting array is L18 and could be followed a full factorial L9 or L27.

   While the assignment of factors to columns in a screening design could theoretically be random, they could be assigned according to perceived value from the highest to the lowest estimated factor coefficients. In the L9 example, four factors should be assigned according to potential significance as follows: the two most important factors should be assigned to the main columns A and B, and the next two factors should be assigned to columns C and D which confound with two-way interactions AB and AB$_2$ of the main factors.
3. OA use as a partial fraction factorial by adding more factors to the model that confound either two-way or three-way interactions, up to using all columns in screening design. This can be used when model testing while presuming that additional factors might not be significant. This is a risky plan if the assumption of low-value interaction confounding is not correct. DoE analysis cannot prove or disprove this assumption since only a full factorial DoE can determine the significance of the interaction that is being confounded. If the confounded factor is significant, then a full factorial DoE might be required to separate the confounding of factor assigned to an interaction.

   An L9 can be used in partial factorial mode with three factors. The fourth column can be considered as an interaction of the first two factors. The significance of that interaction is relevant since it can increase or decrease the predicted value of the DoE outcome.

While there is no half fraction DoE design in three-level OAs, each array can be used as a cornerstone of a third of the next array if desired. If the L9 shown in Table 5.1 is compared against the L27 in Table 5.2, nine lines of L9 are equivalent to nine lines 1, 4, 7, 10, 13, 16, 19, 22, and 25 of the L27 for the first four columns corresponding to factor C level 1. An L9 with three-factor partial factorial design can be converted to a full factorial L27 with the addition of 18 experiments for factor C levels 2 and 3.

Table 5.4 can be used as a guide for which three-level array mode is best suited for the DoE goals, balancing the project effort versus results expected, within the constraint of time and cost. The number of experiments increases exponentially as the number of factors increases. This is the reason that most industrial three-level DoEs are partial factorial or screening design.

5.1.2   Use of Three- Versus Two-Level DoE

A common question is which DoE level basis to use and what are the advantages or disadvantages of each level system. The following can be used as a guideline for selecting the most appropriate two- or three-level OA according to DoE goals:

1. Exploring new design or process alternatives or improving current ones. Two-level arrays have the capability of examining more factors than comparative three-level arrays. For example, a screening L8 can investigate design or process alternatives using seven factors, while an L9 screening design can only use four factors. This is helpful when investigating a new design or a process by using as many factors as possible in DoE to examine the effects of each factor.

   Using three levels is more appropriate for improving current design and processes, where DoE teams can collectively pool their knowledge to reduce the number of factors to be investigated. The current parameters can be used as the mid-level, while the other two levels can investigate a possible optimal outcome in the design space or process map.

2. Relationship of the factor and level inputs to the DoE outcome(s). Two-level DoEs provide a linear relationship between input variables (factor and levels) and DoE outcomes. Three levels can provide the curvature in the relationship if any, or a polynomial fit. This is important for follow-on DoE in two ways:

   (i) Guiding the DoE team on where to investigate additional outcome improvements in selecting factor levels for the next DoE. The levels could be made more divergent or closer together depending on the curvature of the input levels to DoE outcomes.

   (ii) Selecting the direction of steepest accent. This concept was discussed in Fig. 1.14, which can extend the design space accordingly to find factor levels for optimal results. This is accomplished by taking factors with increasing or decreasing outcome and extending their level further in the following DoE.

3. Interaction handling. Two-level DoEs are easier to handle interactions in one column as opposed to two-column interactions of three-level factors. This is important in partial factorial design as more factors are added to full factorial DoE, causing confounding. Adding more factors assumes interactions are not significant, which might result in faulty design and process decisions. The remedy

could be creating additional experiments to restore full factorial analysis, which is easier to perform in two- rather than three-level DoE. In some cases, DoE team can select non-interacting arrays discussed in Chap. 8 to eliminate the problems of interactions and confounding.

5.2   Three-Level DoE Analysis and Model Reductions

Statistical analysis of three-level DoEs is based on model coefficient estimates of each factor as well as ANOVA in determining significance of each factor in terms of its effects on DoE outcomes. These two methodologies were introduced in previous chapters with formulas and example analyses.

The model coefficient estimates for three-level DoE take advantage of the level patterns in OAs for solving each value of factor levels. It is very similar to the two-level analysis discussed in Sect. 4.3. For example, in the L9 shown in Table 5.1 with nine outcomes Y1–Y9, it takes nine experiments to perform a coefficient solution of four factors at three levels. The mean of the first three experiments' results, Y1, Y2, and Y3, subtracted by the total nine experiments' mean, is the coefficient estimate due to selecting level (1) of factor A, while the other factors negate themselves by averaging out their three levels. The mean of Y2, Y5, and Y8 subtracted by the nine experiments' mean is the coefficient for level (2) of factor B. In this manner, the coefficient estimates of all 12 possible combinations (factors A, B, C, and D and their levels 1, 2, and 3) are examined in terms of attaining the desired result for the product or process design.

For L9 with (n) repetitions and nine results Y1–Y9, the level coefficient estimates for factor A can be calculated as follows:

$$A1 = \frac{\sum_n Y1 + Y2 + Y3}{n*3}; A2 = \frac{\sum_n Y4 + Y5 + Y6}{n*3}; A3 = \frac{\sum_n Y7 + Y8 + Y9}{n*3}$$

$$B1 = \frac{\sum_n Y1 + Y4 + Y7}{n*3}; B2 = \frac{\sum_n Y2 + Y5 + Y8}{n*3}; B3 = \frac{\sum_n Y3 + Y6 + Y9}{n*3}$$

$$C1 = \frac{\sum_n Y1 + Y6 + Y8}{n*3}; C2 = \frac{\sum_n Y2 + Y4 + Y9}{n*3}; C3 = \frac{\sum_n Y3 + Y5 + Y7}{n*3}$$

$$D1 = \frac{\sum_n Y1 + Y5 + Y9}{n*3}; D2 = \frac{\sum_n Y2 + Y6 + Y7}{n*3}; D3 = \frac{\sum_n Y3 + Y4 + Y8}{n*3}$$

The factor and level coefficients can be plotted for significant factors after ANOVA is completed. The appropriate levels can then be selected to meet the desired outcome of maximum, minimum, or targeted values.

The predicted value of DoE output is the result of applying all the recommended levels based on desired outcome, using the factor plots as a guide in the regression equation. This is constructed from the overall experiment mean and then adding the coefficient contribution of each recommended level of significant factors to the prediction, with a confidence interval based on coefficient standard error.

The predicted value is calculated for significant factors and interactions in the reduced model regression equation. The significant factors are determined by performing ANOVA F-tests or coefficient t-tests. The contribution of non-significant factors is within the error of the experiment (the confidence interval of prediction). Interaction levels cannot be selected for predicted value in the regression equation; only the levels from the originating factors that form interactions can be considered in the predicted value for significant interactions, even if one of the originating factors is not significant.

ANOVA for three-level DoE is similar to the previous chapter, with the difference of degrees of freedom (DOF) of three-level factors equal to $3 - 1 = 2$. When not using screening design, two-way interactions for two factors consisting of two columns should be treated as one interaction sum of the squares with four degrees of freedom. An example of that is shown later in this chapter for the green soldering case study. A similar scheme should be used for three-way interaction with eight degrees of freedom.

5.3 Three-Level DoE Case Studies

These case studies are drawn from industrial DoE in various disciplines. They demonstrate the use of three-level DoE in selecting the best factors and levels to increase the DoE teams' knowledge of factors that influence the design and levels that best achieve desired outcomes. In addition, they also demonstrate the use of three-level techniques in improving current designs and processes. Each case study is presented both in terms of design and analysis and the reasons behind the design decisions. Shortcomings of the design teams' project plans are discussed, results are analyzed, and conclusion is drawn for each case study. Other examples of three-level DoE are given in later chapters and summarized in the case study index in Chap. 10.

5.3.1 Screening Design L9 Case Study: Bonding I DoE

The problem in this case study was medical transducers comprising of two halves glued together that were separating at the customer site (ICU beds at hospitals). The DoE goal was to maximize bond strength by measuring the peel force required for transducer separation. The DoE was constrained due to the desire not to seek FDA recertification; therefore, no changes in materials used in the transducer can be part of the DoE project.

The process consisted of using an RTV adhesive as the bonding agent. Parts were cleaned prior to bonding in an ultrasonic bath filled with a cleaning chemical, and a measured volume of glue was applied to both halves of the parts. Parts were then cured in an oven after bonding. The levels for cleaning time and chemical as well as curing temperature were arbitrarily selected in the design stage of the product. The product was not performing adequately in the field, as several parts separated during customer use. A DoE was launched to increase the bonding process maximum peel strength. The peel force was measured by a special spring force tool, which is commonly used to determine the maximum outside force necessary to cause the parts to separate. This case study is labeled Bonding I DoE to distinguish it from another bonding case study in Chap. 9.

The DoE team decided on a three-level L9 screening design. Four three-level factors were selected with the current process parameters set to the mid-level for each factor as follows:

A. Cure temperature, °C
B. Cleaning time, minutes
C. RTV volume, cc
D. Soak chemical

The first two factors were considered the most important, and the third and fourth factors confounded the two-column interactions AB and AB_2.

The three levels of the four factors were selected based on the current process parameters. Cure temperature levels of 30 °C (elevated room temperature), 50 °C (current level of oven bake), or 70 °C (higher level of oven bake). Ultrasonic cleaning time for parts' immersion in a chemical for 1, 3, or 5 min were chosen, with 3 min being the current time. RTV dispensed volumes, using three different dispensing heads, were selected around the current volume of 1.7 cc. The levels were 1.2 cc, 1.7 cc, and 2.5 cc, respectively, representing commercially available dispensing heads. The ultrasonic bath soak chemical was varied from the current methylene (MET) to other cleaners which either were harsher or were environmentally safe such as methyl ethyl ketone (MEK) or plain water (H_2O).

The team did not consider other possible factors in the bonding process DoE such as different bonding materials or manufacturing process humidity because of cost of resources necessary to change the production environment or approval required by the FDA for all medical device material changes. Table 5.5 shows the L9 screening bonding process DoE design and peel force results.

Each experiment of the L9 was a unique combination of factor levels selected prior to running the DoE. For example, in experiment 3, 30 °C was the cure temperature in the oven, the RTV volume was 2.5 cc, the ultrasonic soak chemical was MEK, and the cleaning time in the ultrasonic bath was 5 minutes. The experiment resulted in an average of 24.12 pounds of pressure applied to test nine transducers made by each experiment line before peeling (separating) the assembly into two halves.

The initial ANOVA of the bonding DoE ANOVA is shown in Table 5.6, using DoE analysis software developed by the author. No significance testing with F-test can be performed without having an error in the experiments. A typical source of error for DoE is the replication of all or some of the experiment lines including the CenterPoint, which was not done in this case study. To provide for error, the soak chemical with the lowest sum of squares (SS) was deemed not significant and pooled

Table 5.5   Bonding I DoE design and outcomes

Factors Selected	Levels of Factors			
A = Cure Temperature	30	50	70	°C
B = Ultrasonic Cleaning	1	3	5	Minutes
C = RTV Volume	1.2	1.7	2.5	cc
D = Soak Chemical	H2O	MET	MEK	

L9 (3 ^ 4) Orthogonal Array Screening Design.

Exp. #	A	B	C	D	A	B	C	D	Peel Force (lbs.)
1	1	1	1	1	30	1	1.2	H2O	11.5
2	1	2	2	2	30	3	1.7	MET	22.7
3	1	3	3	3	30	5	2.5	MEK	22.6
4	2	1	2	3	50	1	1.7	MEK	19.0
5	2	2	3	1	50	3	2.5	H2O	28.5
6	2	3	1	2	50	5	1.2	MET	24.0
7	3	1	3	2	70	1	2.5	MET	25.2
8	3	2	1	3	70	3	1.2	MEK	30.3
9	3	3	2	1	70	5	1.7	H2O	33.3

Table 5.6   Bonding I DoE ANOVA before pooling

Factor	Pool	DOF	SS
Cure Temperature	n	2	171.04
Ultrasonic Cleaning	n	2	139.32
RTV Volume	n	2	21.84
Soak Chemical	n	2	0.436
Total		8	332.64

into the error. While this decision appears to be correct in this case owing to the small SS of the pooled factor, it could result in faulty conclusions if the SS was large.

Pooling of the soak chemical transferred its DOF and SS into the error, making it possible to calculate the F-value for the remaining three factors, as shown in Table 5.7. The remaining three factors are all significant, since their calculated F-value is greater than the critical $f_{05, 2, 2} = 19$ from Table 2.11 and their probability is less than 0.05 using Excel© function F.DIST.RT(F-Value,2,2). Table 5.7 shows the p% contribution value for the first three factors being greater than 5%, indicating its significance with the same conclusion as the F-test.

Table 5.8 is the Bonding I DoE coefficient estimates, with 12 values for all 4 three-level factors, taken from summing of all experiments with the same factor levels and dividing by three. Selected levels are those that contribute to increasing the peel force, with each contribution noted:

Table 5.7   Bonding I DoE ANOVA after pooling

Factor	Pool	DOF	SS	MS	F-Value	P > F	SS'	p %
Cure Temperature	n	2	171.04	85.52	392.70	0.0025	170.61	51%
Ultrasonic Cleaning	n	2	139.32	69.66	319.86	0.003	138.88	42%
RTV Volume	n	2	21.84	10.92	50.15	0.020	21.41	6%
Soak Chemical	y							
Replication Error		0	0					
Pooled Error		2	0.436	0.218				
Total Error		2	0.436	0.218			1.74	1%
Total		8	332.64				332.64	100%

Table 5.8   Bonding I DoE coefficient estimates

Factor/Level	Coefficient	SE Coefficient	t-Value	P > t
DoE Mean	24.122	0.156	155.07	0.000
Cure Temperature				
30	-5.189	0.220	-23.59	0.002
50	-0.289	0.220	-1.31	0.320
70* (selected)	5.478	0.220	24.90	0.002
Ultrasonic Cleaning				
1	-5.556	0.220	-25.25	0.002
3* (selected)	3.044	0.220	13.84	0.005
5	2.511	0.220	11.41	0.008
RTV Volume				
1.2	-2.189	0.220	-9.95	0.010
1.7	0.878	0.220	3.99	0.057
2.5* (selected)	1.311	0.220	5.96	0.027
Soak Chemical (Error)				
H_2O	0.311			
MET	-0.156			
MEK	-0.156			

Factor	Level 1	Level 2	Level 3	Mean	Selected Level	Contribution
Cure Temperature	18.93	23.83	29.60	24.12	3	5.48
Ultrasonic Cleaning	18.57	27.17	26.63	24.12	2	3.04
RTV Volume	21.93	25.00	25.43	24.12	3	1.33
Soak Chemical	24.43	23.97	23.97	24.12	1	0.31

Each factor and their three-level coefficients are shown in Table 5.8 and plotted in Fig. 5.1, with each factor importance rank corresponding to F-value or p% in the ANOVA of Table 5.7. The selected levels, noted by an (*) in Table 5.8, should be the highest value, and its additive contribution is equal to the level coefficient.

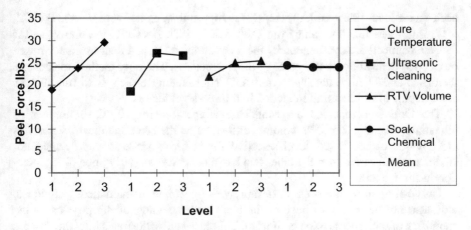

Fig. 5.1 Bonding I DoE coefficient plots

For example, factor A contribution is level A1 – DoE mean $= 29.60 - 24.12 = 5.48$ or A1 coefficient.

The DoE mean (24.12) is also the mean of each factor level. Each level coefficient shares the same standard error (SE) using $n = 9/2$, with two DOFs for each factor. The DoE mean standard error (SE) with $n = 9$ and predicted SE with $n = 9/7$ with 7 representing the total DOF (6) in the model +1 due to the mean. They have the following formulas:

$$SE_{\text{Coefficient}} = \sqrt{\frac{MSE}{n/DOF_{\text{Factor}}}} = \sqrt{\frac{MSE}{n/2}} = \sqrt{\frac{0.218}{9/2}} = 0.220$$

$$SE_{\text{DoE Mean}} = \sqrt{\frac{MSE}{n}} = \sqrt{\frac{0.218}{9}} = 1.56$$

$$SE_{\text{Predicted}} = \sqrt{\frac{MSE}{n/DOF_{\text{Model+Mean}}}} = \sqrt{\frac{MSE}{n/6+1}} = \sqrt{\frac{0.218}{9/7}} = 0.412;$$

$$CI = \pm\, SE_{\text{Predicted}} * t_{0.025,2} = \pm 0.412 * 4.303 = \pm 1.77$$

The t-values are the division of each factor coefficient over SE. For level A3, the t-value $= 5.378/0.220 = 24.90$ with corresponding probability of 0.002, indicating significance.

The regression equation for the bonding DoE contains the DoE mean with nine terms corresponding to three levels for each of the significant factors. There are no interaction terms in the regression equation since they were not included in the model. The predicted maximum peel force is the addition of all significant factor contributions to the experiments' mean, which is equal to $24.12 + 5.48$ (A3) $+ 3.04$ (B2) $+ 1.31$ (C3) $= 33.96$. This prediction does not include the contributions of factor D (0.31) since it was not significant and pooled into the error. The CI for prediction is ± 1.77 or (32.19, 35.73). The predicted maximum peel force based on each factor level selection (A3, B2, C3) is higher than experiment 9 (A3, B3, C2, D1) resulting in 33.3 lbs. value from Table 5.5. Both factors C (RTV Volume) and D (Soak Chemical) are confounded by the interactions A*B, and their estimates might be in error depending on the value of the interactions. The only method to separate factors C and D from the interaction A*B confounding effects is to run all L81 experiments in full factorial mode for four three-level factors.

The DoE team decided on using cure temperature of 70 °C, 3 minutes of ultrasonic cleaning, 2.5 RTV volume, and water as the soak chemical. Water was the obvious choice in the level selection for factor D since it was deemed not significantly different than the other two harsh chemicals requiring operator protection in fume hoods.

DoE can optimize a process or design by using simple mathematical techniques. It is also not necessary to have an in-depth understanding of the physics or the chemistry of the process to be optimized. This case study illustrates how the process mean can be shifted to the desired level, in this case to the maximum possible. A similar method can be applied to reduce the variability, with several replications for each experiment line. Four replications are preferable $4 \times 9 = 36$ to realize more than 30 data points, approximating the population distribution of bonding. A mathematical transformation can convert the four numbers for each experiment line into a single number indicating variability. The same coefficient estimates and graphical analysis for variability can be performed on the transformed numbers for each experiment. The two analyses for average and variability can be contrasted and factor level selected for the most efficient process improvement, through trade-offs of average and variability, if any. Further examples of these techniques will be provided in Chap. 9 on variability reduction.

There are two important terms used in DoE. One is the design space or process map, which is the limit of the investigation of the factors, as bounded by the selection of the levels for each factor. The other is the "direction of steepest ascent." This is direction of increasing or decreasing the amount of factor level values when expanding the current DoE analysis results into follow-on DoE.

In the process map for the bonding DoE, the selection of the levels for factor B, the time for ultrasonic cleaning, was optimal as shown by Fig. 5.1. The best level position was in the middle of the three levels. The maximum point can be calculated by drawing a best fit curve through the three points and can be determined accurately, rather than declaring that level 2 (3 min) is better than 1 or 5 min of cleaning. A second-order equation can be fitted through the three points, and the maximum point can be determined by setting the derivative to zero.

For factor A oven temperature, the process map is not optimal. It can be seen from Fig. 5.1 that level 3 temperature (70 °C) has the highest peel force. The current process map does not allow for any conclusions regarding higher temperatures than 70 °C. If more information is desired regarding the bonding process, then a second DoE could be performed. Some factors could be expanded in the direction of steepest ascent such as having higher temperature levels, while other factors could be removed from the experiment (such as soak chemical) in favor of partial or full factorial analysis.

A possible follow-on DoE could be a full factorial DoE with two factors, cure temperature and ultrasonic cleaning time. The levels could be 75, 85, and 95 °C and 3.5, 4, and 4.5 min, respectively. Four repetitions of the full factorial L9 should be completed in order to provide 36 data points for DoE significance and investigation of the interaction of the 2 factors, if any.

5.3.2 Screening Design L9 Case Study: Zero-Defect Mixed Soldering DoE

In this case study, a quick DoE project can resolve specific quality problems when using automatic machines in product assembly. This case is like the robotic assembly machine discussed in the previous chapter, where an L32 two-level array was used to investigate the setting of multiple controls for the APOS assembly machine. In this case, an L9 three-level array was used to solve a quality problem in the wave soldering of a specific mixed technology printed circuit board (PCB) assembly produced with special requirements by an electronics manufacturing supplier. The PCB was double sided with mixed technology of through-holes (TH) and surface-mount technology (SMT) components on the bottom side. The current soldering process resulted in a mean of 13 defects per PCB, requiring extensive rework, which was unacceptable to the company's customers.

A DoE was initiated, and a cause-and-effect analysis was performed to evaluate the causes of defects. It was concluded that SMT bottom side components were the mostly likely reason for the defects by creating a shadowing effect on the rest of the PCB components, resulting in solder defects as defined by industry standards. It was decided to perform a DoE on the process parameters for this PCB to see if it required a different operational setup with the solder wave machine than the rest of the PCB population.

Four factors were selected (Preheat Temperature, Belt Speed, Wave Height, and Solder Pot Temperature), and an L9 with three levels was chosen for the DoE. The smaller is better defect outcome was selected with a target of zero defects. Five production PCBs with SMT components were used for each experiment to generate enough defect opportunities for statistical analysis of results, and the defects' mean was recorded for each experiment line. The DoE and its results are shown in Table 5.9.

The factors and levels selected for this DoE were very easy to manipulate, since they were generated by turning knobs on the machine control panel. The second level for each factor was the current process operational settings. Preheat temperature could be set automatically using a machine control knob. The belt speed in feet

Table 5.9 Mixed soldering DoE and results

Factors Selected			Levels of Factors			
A = Preheat Temperature			400	425	450	°F
B = Belt Speed			2.5	3.0	3.5	feet/minute
C = Wave Height			4	5	6	Knob setting
D = Solder Pot Temperature			470	480	490	°F

L9 (3 ^ 4) Orthogonal Array Screening Design.

Exp. #	A	B	C	D	A	B	C	D	Mean Defects / PCB.
1	1	1	1	1	400	2.5	4	470	7.6
2	1	2	2	2	400	3.0	5	480	11.8
3	1	3	3	3	400	3.5	6	490	2.6
4	2	1	2	3	425	2.5	5	490	3.8
5	2	2	3	1	425	3.0	6	470	4.4
6	2	3	1	2	425	3.5	4	480	15.2
7	3	1	3	2	450	2.5	6	480	0.6
8	3	2	1	3	450	3.0	4	490	6.0
9	3	3	2	1	450	3.5	5	470	12.6

per minute was adjusted by using a potentiometer setting in the machine. The solder pot re-circulating pump was adjusted with a potentiometer setting of 4, 5, or 6 to create the desired level of solder wave height. The solder pot temperature was varied in increments of 10 °F. Because of the thermal mass of the solder pot, changing the pot temperature operation took a long time, and the experiment lines sharing the same solder pot temperature were run in sequence. For example, when the solder pot temperature was set at 470 °F, then experiments 1, 5, and 9 were run sequentially, although DoE practitioners recommend a random order when running experiments. In addition, the choice of levels for this experiment has to be within the operating range (process map) of the machine. If the solder pot temperature is too high and the conveyor speed is set too slow, the components could sustain thermal damage.

The selected four factors were assigned to columns based on the prevailing knowledge of the process, with Preheat Temperature followed by Belt Speed placed in primary (main) columns A and B. The other two considered less important factors, Wave Height and Solder Pot Temperature, were placed in columns C and D, confounding the two interactions of A*B and A*B$_2$. Special wax temperature indicators were placed on the PCB which would melt at the specified temperature to indicate the proper preheat levels just before reaching the soldering wave. This was done rather than just setting the preheat knob on the machine in order to reduce the interaction of the Preheat Temperature*Belt Speed (A*B and AB$_2$) which confound factors C and D in L9 screening design.

ANOVA is shown in Table 5.10 after pooling the Preheat Temperature. Unfortunately, the results' column in Table 5.9 shows the mean defects per PCB, rather than showing the actual five repetitions, which could be used to generate the error in

Table 5.10 Mixed soldering DoE ANOVA after pooling

Factor	Pool	DOF	SS	MS	F-Value P > F		SS'	p%
Preheat Temperature	y	2	3.05	1.52	--			
Belt Speed	n	2	56.65	28.32	18.58	0.051	53.60	27%
Wave Height	n	2	97.13	48.56	31.86	0.030	94.08	47%
Pot Temperature	n	2	43.21	21.60	14.17	0.066	40.16	20%
Replication Error		0						
Pooled Error		2	3.049	1.524				
Total Error		2	3.049	1.524			12.20	6%
Total		8	200.036				200.04	100%

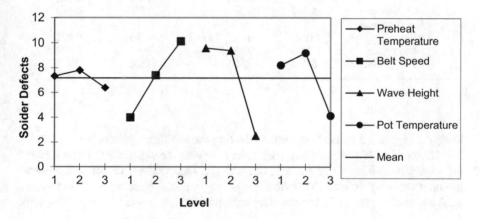

Fig. 5.2 Mixed soldering DoE coefficient plots

ANOVA. It is recommended not to use the replication mean in DoE, since it tends to smooth out the outcomes and will result in pooling one of the factors to generate experiment error, rather than using replication error. After pooling factor A, Table 5.10 shows two factors B ($P = 0.051$) and D ($P = 0.066$) close to significance, and their F-values are also smaller than the critical value of $F_{05,2,2} = 19$, while factor C is significant ($P = 0.030$). However, the reduced L9 R^2 is 98.48% indicating good model significance. If factor D is pooled, then the model of the remaining two factors (B and C) is not significant at $R^2 = 77\%$. The p % shows that factor D could be considered significant since its contribution is 20% > 5%, showing the disparity of the two DoE analysis tools of F-value and p%.

Table 5.11 shows the 12 model coefficient estimates derived from the 9 experiments with 4 three-level factors, using the methodology outlined in the previous section for the Bonding I DoE case study. The model coefficient plots for L9 screening DoE for mixed soldering machine are shown in Fig. 5.2, which correspond to the ranking of factors in the ANOVA of Table 5.10. Factor A, the Preheat Temperature, is least important and therefore used as the error for the F-value

Table 5.11 Mixed soldering DoE coefficient estimates

Factor/Level	Coefficient	SE Coefficient	t-Value	$P > t$
DoE Mean	7.178	0.412	17.44	0.003
Preheat Temperature (Error)				
400	0.156			
425	0.622			
490	-0.778			
Belt Speed				
2.5* (selected)	-3.178	0.582	-5.46	0.032
3.0	0.222	0.582	0.38	0.739
3.5	2.956	0.582	5.08	0.037
Wave Height				
4	2.422	0.582	4.16	0.053
5	2.222	0.582	3.82	0.062
6* (selected)	-4.644	0.582	-7.98	0.015
Pot Temperature				
470	1.022	0.582	1.76	0.221
480	2.022	0.582	3.47	0.074
490 * (selected)	-3.044	0.582	-5.23	0.035

analysis. Figure 5.2 points to the level selection for minimum defects, which are A3, B1, C3, and D3. These recommended levels are close to experiment 7 result of 0.6 defects/PCB with levels of A3, B1, C3, and D2. During the conduct of the experiment, it was very difficult to convince the production managers not to forgo the DoE mathematical analysis and immediately switch to the levels used in experiment line 7.

The DoE mean (7.178) is also the mean of each factor level. Each level coefficient shares the same standard error (SE) using $n = 9/2$, with two DOFs for each factor. The DoE mean SE with $n = 9$ and predicted SE with $n = 9/7$ with 7 representing the total DOF (6) in the model +1 due to the mean. They have the following formulas:

$$SE_{Coefficient} = \sqrt{\frac{MSE}{n/DOF_{Factor}}} = \sqrt{\frac{MSE}{n/2}} = \sqrt{\frac{1.524}{9/2}} = 0.582$$

$$SE_{DoE\ Mean} = \sqrt{\frac{MSE}{n}} = \sqrt{\frac{1.524}{9}} = 0.412$$

$$SE_{Predicted} = \sqrt{\frac{MSE}{n/DOF_{Model+Mean}}} = \sqrt{\frac{MSE}{n/6+1}} = \sqrt{\frac{1.524}{9/7}} = 1.089;$$

$$CI = \pm SE_{Predicted} * t_{0.025,2} = \pm 1.089 * 4.303 = \pm 4.68$$

The regression equation for the mixed soldering DoE contains the DoE mean with nine terms corresponding to three levels for each of the significant factors. There are no interaction terms in the regression equation since they were not included in the model. The predicted defect rate from selecting the minimum value levels for the three significant factors B1, C3, and D3 and subtracting each level contribution from the experiment mean is 7.18 − 3.18 (B1) − 4.64 (C3) − 3.04 (D3) = −3.69, with CI = ± 4.68 or (−8.37, 1) that is much lower than the lowest defect average obtained in experiment line 7, which was 0.6 per PCB.

None of the recommended levels matched the current process at A2, B2, C2, and D2. The negative value of the EV is obviously within the confidence limits, since there are no negative defects. The s_{error} can be quickly calculated from the square root of the variance error or MS_{error}. The t-value of each selected coefficient (factor contribution) is based on the coefficient divided by standard error (SE), which leads to the P-value of each coefficient. The coefficient t-value of level B1 (the selected level of belt speed) from Table 5.11 is equal to the factor contribution divided by the standard error or − 3.178/0.582 = −5.46 which is significant ($P = 0.032$). The prediction 95% CI of (−8.37, 1) includes the zero-defect goal.

The mixed soldering machine was changed for this one PCB type to the levels recommended, and it resulted in zero defects per PCB in the short term. This is much lower than the pre-DoE 13 defects per PCB. For the medium term of up to 6 months after the process change, histograms or normal plots should be kept of the process before and after the DoE.

5.3.3 Partial Factorial DoE L27 Case Study: Green Electronics Phase I DoE

This case study was the first attempt for the New England LeadFree Consortium in investigating the quality and reliability of green electronics by selecting lead-free solder materials and associated processing parameters. The goal was achieving zero-defect quality and with reliability at least equal to legacy lead solder materials. This phase I was followed by several phases including phase II and for selecting materials and process parameters for the chosen green solder material from phase I in Chap. 3.

A test vehicle PCB for phase I was designed with various electronic devices, as shown in Fig. 5.3. Quality was measured by the number of solder defects on all components based on industry standards, and reliability was measured by thermal cycling (aging) of the PCB and then pulling off the corner leads for various components. The pull force for both green and legacy test vehicles was compared using t-tests.

A partial factorial L27 was selected since the consortium desired to test three leading candidates for green soldering material, with four associated processing parameters for each candidate. The first three factors deemed important were assigned to the primary (main) columns in the L27 (1, 2, and 5) from Table 5.2. The remaining two factors were assigned to two out of four three-way interaction columns (9 and 10), thus confounding them. Some of the processing factors were two levels, with one level, deemed important by the consortium, assigned to both levels 1 and 3.

Fig. 5.3 Green Electronics
Phase I DoE test vehicle

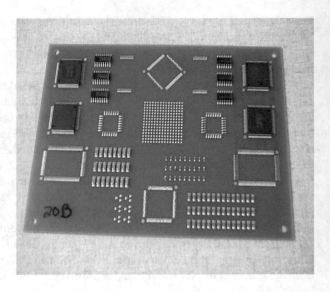

An L27 was selected by the consortium based on their collective experience and the available resources and materials. The factors and levels selected were as follows:

1. Solder Pastes: these were selected based on published performance and actual use in consumer products.

   - Level 1: 95.5/3.8/0.7 Tin/Silver/Copper (SnAgCu) or SAC.
   - Level 2: 57/43 Tin/Bismuth (Sn/Bi).
   - Level 3: 96.5/3.8 Tin/Silver (SnAg).

2. PCB Surface Finishes: These were selected based on low price and wide use.

   - Levels 1 and 3: Organic Solderability Preservative (OSP).
   - Level 2: Electroless Nickel/Immersion Gold (ENIG).

3. and 4. Reflow Process: Two factors were selected to fully understand the impact of the solder reflow process and reduce interactions between two factors: Time Above Liquidus (TAL) and the Reflow Profile. Levels were selected to examine the impact of lessening the thermal shock to the electronic components by trading off the lengthening of the reflow time (TAL) versus a lower peak temperature or applying a longer preheat exposure time to the reflow process (a linear or a cash register reflow profile).

   - TAL: 60, 90, or 120 seconds for levels 1, 2, and 3.
   - Profile: Linear (soak) for level 1 or (cash) register profile for levels 2 and 3.

5. Reflow Atmospheres: Nitrogen was selected as well as air to investigate its possible effect on the process.

Table 5.12 Green Electronics Phase I DoE design and results

Col. #	A/1	B/2	C/5	D/9	E/10		
	Solder	Surface	TAL	Reflow		Results	
Expt.#	Paste	Finish	(secs)	Profile	Atmosphere	Def. 1	Def. 2
1	Sn/Ag/Cu	OSP	60	Linear	Nitrogen	797	944
2	Sn/Ag/Cu	OSP	90	Register	Air	1213	1146
3	Sn/Ag/Cu	OSP	120	Register	Nitrogen	874	890
4	Sn/Ag/Cu	ENIG	60	Register	Air	544	584
5	Sn/Ag/Cu	ENIG	90	Register	Nitrogen	0	0
6	Sn/Ag/Cu	ENIG	120	Linear	Nitrogen	0	0
7	Sn/Ag/Cu	OSP	60	Register	Nitrogen	828	819
8	Sn/Ag/Cu	OSP	90	Linear	Nitrogen	902	960
9	Sn/Ag/Cu	OSP	120	Register	Air	1182	1164
10	Sn/Bi	OSP	60	Register	Nitrogen	1134	963
11	Sn/Bi	OSP	90	Register	Nitrogen	875	1136
12	Sn/Bi	OSP	120	Linear	Air	967	1146
13	Sn/Bi	ENIG	60	Register	Nitrogen	1024	960
14	Sn/Bi	ENIG	90	Linear	Air	1016	1002
15	Sn/Bi	ENIG	120	Register	Nitrogen	843	560
16	Sn/Bi	OSP	60	Linear	Air	1148	1067
17	Sn/Bi	OSP	90	Register	Nitrogen	781	606
18	Sn/Bi	OSP	120	Register	Nitrogen	765	882
19	Sn/Ag	OSP	60	Register	Air	1212	1279
20	Sn/Ag	OSP	90	Linear	Nitrogen	1131	988
21	Sn/Ag	OSP	120	Register	Nitrogen	1027	933
22	Sn/Ag	ENIG	60	Linear	Nitrogen	0	0
23	Sn/Ag	ENIG	90	Register	Nitrogen	0	0
24	Sn/Ag	ENIG	120	Register	Air	180	240
25	Sn/Ag	OSP	60	Register	Nitrogen	796	829
26	Sn/Ag	OSP	90	Register	Air	1205	1146
27	Sn/Ag	OSP	120	Linear	Nitrogen	868	935

- Levels 1 and 3: Nitrogen (20 PPM oxygen).
- Level 2: Air.

The factors, levels, and defects (def.) for two replications are shown in Table 5.12. Three two-level factors, Surface Finish, Reflow Profile, and Atmosphere, had the most important level assigned to levels 1 and 3 (reflow register profile was assigned to levels 2 and 3). Two factors, Reflow Profile and Atmosphere, were confounded with two out of four three-way interactions (9/ABC and 10/AB_2C_2), assuming that three-way interactions are small, and confounding is minimum. The other two unassigned three-way interactions (12/AB_2C and 13/ABC_2) were pooled into the first ANOVA in Table 5.13.

Table 5.12 provided the first affirmation that green materials can provide high-quality electronic products with zero solder defects. Four experiment lines showed that zero defects can be achieved with the following green materials and processes:

Table 5.13 Green Electronics Phase I DoE first ANOVA

Col.#	Source	DOF	SS	Contribution	MS	F-Value	P > F	p%
A/1	Solder	2	611320	7.33%	305660	42.66	0.000	7.17
B/2	Finish	2	4412023	52.88%	2206012	307.89	0.000	52.73
C/5	TAL	2	61309	0.73%	30655	4.28	0.023	0.58
D/9	Profile	2	7301	0.09%	3651	0.51	0.606	0
E/10	Atmosphere	2	909057	10.90%	454529	63.44	0.000	10.74
3-4	Solder*Finish	4	1887270	22.62%	471818	65.85	0.000	22.32
6-7	Solder*TAL	4	83840	1.00%	20960	2.93	0.037	0.70
8-11	Finish*TAL	4	148895	1.78%	37224	5.20	0.003	1.48
12-13	Pooled interct.	4	50255	0.60%	12564	1.97	0.127	0.30
	Repetition	27	171858	2.06%	6365			4.00
	Total Error	31	222112	2.66%	7165			4.30
	Total	53	8343129	100.00%	1574175.53			100%

Expt.		Surface		Reflow	
#	Solder	Finish	TAL	Profile	Atmosphere
5	Sn/Ag/Cu	ENIG	90	Register	Nitrogen
6	Sn/Ag/Cu	ENIG	120	Linear	Nitrogen
22	Sn/Ag	ENIG	60	Linear	Nitrogen
23	Sn/Ag	ENIG	90	Register	Nitrogen

The four zero-defect experiment lines indicate that ENIG Surface Finish and Nitrogen will be significant in achieving zero defects, while TAL and Reflow Profile would not. ANOVA affirmed the significance based on probabilities in Table 5.13.

The L27 ANOVA with two replications resulted in $54-1 = 53$ total DOF. Each factor has two DOFs, and each two-way interaction has four DOFs, based on two three-level columns. The total error consisted of repetition error DOF = 27 and pooled error due to pooling of two out of four three-way interaction columns with DOF = 4. A new ANOVA column was added to indicate the % contribution, which is the ratio of SS of each factor or interaction divided by SST. This column contrasts with the percent contribution ($p\%$) column discussed in Chap. 3. The results are similar due to the small total error.

Five factors and three interactions were analyzed using the first ANOVA in Table 5.13. Two factors (TAL, Reflow Profile) and two interactions Solder *TAL = A*C and Finish*TAL = B*C were pooled as shown in Table 5.14. The factor C = TAL and interactions A*C and B*C were pooled even though ANOVA in Table 5.13 indicates significance resulting in probabilities of 0.023, 0.037, and 0.003, respectively. This is based on other indicators such as the $p\%$ that were less than 2% for all three factors/interactions, and the coefficient plots in Fig. 5.4 indicate a lower span for TAL than the other three main factors.

Table 5.14 Green Electronics Phase I DoE final pooled ANOVA

Col. #	Source	DOF	SS	MS	F-Value	P > F
A/1	Solder	2	611320	305660	25.11	0
B/2	Finish	2	4412023	2206012	181.22	0
E/10	Atmosphere	2	909057	454529	37.34	0
3-4	Solder*Finish	4	1887270	471818	38.76	0
Pooled Factors		16	351600	21975	3.45	0.002
Repetition		27	171858	6365		
Total Error		43	523457	12173		
Total		53	8343129			

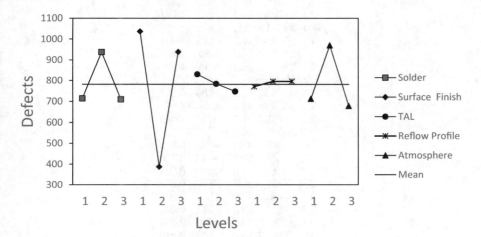

Fig. 5.4 Green Electronics Phase I DoE coefficient plots

The reduced model of three main factors (Solder, Finish, and Atmosphere) and one interaction (Solder *Finish) was significant at $R^2 = 93.73\%$. The regression equation based on the reduced model has 18 coefficients: 3 each for 3 significant factors (9) and 9 for the 1 interaction of 3 * 3 levels (11, 12, 13, 21, 22, 23, 31, 32, 33). All coefficients for the reduced model of A, B, E, and A*B are shown in Table 5.15.

The 25 coefficients for DoE mean (1), 5 factors with 3 levels (15), and 1 interaction (9) are given in Table 5.15, and their factor and level plots are given in Fig. 5.4. The standard error (SE) and t- and P-values shown in Table 5.15 for the mean and the factors and interactions are given below:

Table 5.15 Green Electronics Phase I DoE coefficient analysis

Factor	Coef.	SE	t-Value	P > t
Mean	787.1	15.0	52.42	0.000
A (solder)				
1	-72.8	21.2	-3.43	0.001
2	150.4	21.2	7.09	0.000
3	-77.7	21.2	-3.66	0.001
B (Surface Finish)				
1	249.3	21.2	11.74	0.000
2	-400.2	21.2	-18.85	0.000
3	150.9	21.2	7.11	0.000
C (TAL)				
1	42.8	16.3	2.63	0.013
2	-3.3	16.3	-0.20	0.839
3	-39.5	16.3	-2.42	0.021
D (Reflow Profile)				
1	-16.4	16.3	-1.01	0.321
2	8.2	16.3	0.50	0.617
3	8.2	16.3	0.50	0.617
E (Atmosphere)				
1	-74.3	21.2	-3.50	0.001
2	182.4	21.2	8.59	0.000
3	-108.2	21.2	-5.09	0.000
A*B (Solder*Finish)				
1 1	13.7	30.0	0.46	0.650
1 2	-124.4	30.0	-4.14	0.000
1 3	110.7	30.0	3.69	0.001
2 1	-150.0	30.0	-5.00	0.000
2 2	363.6	30.0	12.11	0.000
2 3	-213.6	30.0	-7.11	0.000
3 1	136.3	30.0	4.54	0.000
3 2	-239.2	30.0	-7.96	0.000
3 3	102.9	30.0	3.43	0.001

$$SE_{DoE} = \sqrt{\frac{MSE}{n}} = \sqrt{\frac{12,173}{9}} = 15.01$$

$$SE_{Factors} = \sqrt{\frac{MSE}{n/DOF_{Factor}}} = \sqrt{\frac{MSE}{n/2}} = \sqrt{\frac{12,173}{54/2}} = 21.2$$

$$SE_{Interactions} = \sqrt{\frac{MSE}{n/DOF_{Interactions}}} = \sqrt{\frac{MSE}{n/4}} = \sqrt{\frac{12,173}{54/4}} = 30.0$$

$$SE_{Pre-Pooled\ Factors} = \sqrt{\frac{MSE}{n/DOF_{Factor}}} = \sqrt{\frac{MSE}{n/2}} = \sqrt{\frac{7165}{54/2}} = 16.3$$

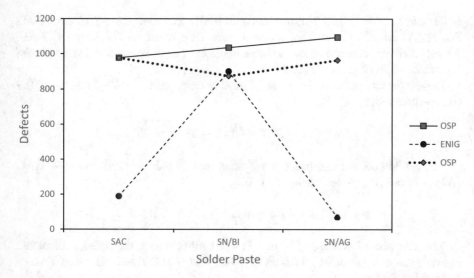

Fig. 5.5 Green Electronics Phase I DoE interaction A*B

The coefficients of factors TAL and Profile have different SE, since they are derived from the first (pre-pooled) ANOVA of Table 5.13. The remaining three factors (Solder, Finish, and Atmosphere) and interaction Solder*Finish are from the pooled ANOVA of Table 5.14. All levels of significant factors and one interaction are significant except for A1*B1 (Sn/Ag/Cu*OSP).

The model factor coefficients are plotted in Fig. 5.4. Solder materials Sn/Ag/Cu and Sn/Ag (A1 and A3) perform the same, ENIG B2 is superior to OSP (B1 and B3), and nitrogen (E1 and E3) is superior to air. Two factors, TAL and Reflow Profile, are not significant based on p%. Figure 5.5 is a plot of the interaction Solder*Finish (A*B) derived from the bottom nine combinations of A*B coefficients in Table 5.15. All nine combination coefficients are significant except for A1*B1.

Each coefficient of nine interaction points of A*B plot in Fig. 5.5 is the mean of six defects values from two replications*three experiment lines for each of nine interactions. For example, the coefficient A2*B2 (Sn/Bi*ENIG) is equal to 900.83 and is the mean of all defects in experiments 13, 14, and 15. The top 2 plots in Fig. 5.5 represent the interaction of OSP (levels 1 and 3 of Surface Finish) with all three levels of Solder. The lower plot is the interaction of the three levels of Solder Paste with ENIG (B2) Surface Finish, which is the same shape of the Solder levels in Fig. 5.4. The defect rate of the two lowest points in Fig. 5.5 corresponds to selecting the best levels of Solder (Sn/Ag/Cu or Sn/Ag) with the best Surface Finish (ENIG).

The prediction for the L27 was not calculated since it was shown that four experiment lines achieved the goal of zero defects in Table 5.12. All four experiments included ENIG (B2), Nitrogen (E1 or E3), and either Sn/Ag/Cu (SAC) or Sn/Ag (A1 or A3). The total contribution % of the top 2 factors A and B (Solder and Finish) and their interactions A*B is 82.83%, and their combined p% is 82.22% from the first ANOVA in Table 5.13. The focus on the nine interaction levels of A*B as demonstrated by Fig. 5.5 shows that with ENIG (B2) and either SAC or SN/AG Solder (A1 or A3), defects are reduced from DoE mean of 787.1 to 189.7 and 70, respectively, because of the additive benefits of each selected A and B levels, regardless of all other factor levels:

Lowest defects based on Solder and Finish = DoE mean − SN/AG (A3) − ENIG (B2) − Interaction (A3*B2)

$$= 787.1 - 77.7 - 400.2 - 239.2 = 70.$$

Second lowest defects based on Solder and Finish = DoE mean − SAC (A1) − ENIG (B2) − Interaction (A1*B2)

$$= 787.1 - 72.8 - 400.2 - 124.4 = 189.7.$$

The selection of Nitrogen (E1 or E3) will further reduce the defects calculated above by an average of 91.25 (mean of −74.3 and − 108.2), regardless of TAL or Reflow Profile levels.

The consortium's first green design DoE proved that zero-defect quality can be achieved. The reliability of green materials was demonstrated as equal to or better than legacy lead materials by comparing pull forces after aging using t-tests. The consortium decided to perform a follow-on phase II full factorial DoE with more levels of materials and processes to demonstrate that green electronics can be achieved with multiple materials in air without expensive use of nitrogen or ENIG surface finish, which proved successful. This phase II DoE was previewed in Chap. 3.

5.3.4   Screening Design Software L27 Case Study: Minimizing Half-Adder Chip Delay Time DoE

Most DoEs are hardware based, where a model of a design is developed and factors with levels are selected for experiments to test the model using physical parts. Each experiment line specifies how parts are made of designs, materials, or processing parameters. Parts are then measured or tested for certain attributes and results analyzed through statistical methods of model reduction, coefficient estimates, and sources of variation.

The development of software tools for simulation, modelling, and analysis resulted in the ability to investigate the performance of designs and processes through different conditions of material properties, geometrical dimensions, and operating conditions. Using these software tools is uniquely advantageous in DoE projects: higher-order experiments can be performed and results obtained quickly and without investing in the expense of preparing samples and testing of results. The

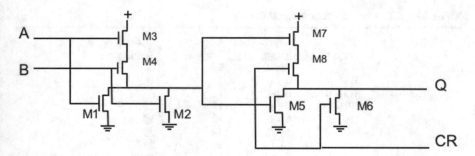

Fig. 5.6 Half-adder chip design

effective use of software tools for DoE implementation falls narrowly within a constrained space of design complexity and maturity of the software tools.

A case study of using software tools for DoE implementation is the semiconductor integrated circuit (IC) industry specializing in digital circuits and devices. Software design and manufacturing process tools are mature in this industry, driven by the high cost of developing new IC devices and their potential impact for use in commodity products such as computers and phones. The use of DoE in this or other industries can be initiated by studying whether legacy designs can be improved using DoE techniques through software design and analysis tools.

A DoE pilot project was formed to investigate improved performance for a half-adder IC, commonly called half-adder chip. The circuit diagram is shown in Fig. 5.6, where A and B are logic inputs, with outputs Q and carry bit CR. The circuit consists of 14 MOS transistors, M1–M14, each with physical channel gate length and width. Some of these transistors are shown in Fig. 5.6. Chip performance is quantified by the SPICE (Simulation Program with Integrated Circuit Emphasis) used in integrated circuit and board-level design to check the integrity of circuit designs and to predict behavior.

An important feature of chip output performance is the chip delay time between input and output signal changes. Total chip delay is calculated by summing up four representative delay times within the chip. The DoE goal was to investigate whether the chip delay could be shortened using SPICE simulator.

Each of the 14 transistors in the chip was originally designed specifically with a predetermined gate length and width. The delay time was decreased by setting each transistor length to 2 micrometers. The DoE project goal was to investigate the effect of varying transistor widths within the design operating range. Three transistor pairs W1–W2, W5–W6, and W11–W12 were design constrained to have identical widths, leaving 11 independent transistor widths that can be varied to achieve the goal of minimizing delay time. The chip performance was to be studied within its operating range of temperature (0–70 °C) and supply voltage (4.5–5.5 v) variations.

The DoE design team decided to use three-level OA, starting with current width of the 14 transistors as the middle level and then ranging 50% above and below that for the other two levels.

Table 5.16 L27 half-adder chip design DoE results

L27 Factors	Experimental Range	Original Settings	Optimal Settings
W14	12-36	24	36
W13	6-18	12	18
W11-W12	6-18	12	6
W10	6-18	12	18
W9	6-18	12	18
W8	24-72	48	24
W7	24-72	48	24
W5-W6	6-18	12	18
W4	12-36	24	12
W3	12-36	24	12
W1-W2	4-8	6	8

The total factor and level combinations in a full factorial design with 11 three-level factors will amount to 3^{11} or 177,147 experiments which is impractical. A partial factorial design would require ranking the importance of each factor so that the design team selects some factors with interactions while ignoring interactions for other factors. The design team selected an L27 array to be used in screening mode with 4 repetitions for a total of 108 experiments. The repetitions represented the four operating range limits (0 °C, 4.5 v; 0 °C, 5.5 v; 70 °C, 4.5 v; and 70 °C, 5.5 v). It was decided not to consider interactions to initiate this pilot project, and the 11 transistor widths were assigned sequentially to an L27 with 13 columns, leaving 2 three-way interaction columns with no factor assignments.

The experiment result summary is given in Table 5.16, with 11 transistor widths as three-level factors. Each original transistor width setting was selected for level 2, and the other levels (L1 and L3) were labeled experimental range of ±50%. The experiment was conducted using the SPICE software, and results were analyzed using the methodologies shown in this and previous chapters. Most of the factors were shown to be significant due to the large error caused by three repetitions (DOF = 80) and two unassigned columns (DOF = 6).

The DoE-recommended (optimal) levels resulting from the L27 experiment to minimize chip delay are shown in Table 5.16. None of the optimal settings matched the original settings. The optimal level selections resulted in the reduction of delay time from the current 6.825 s to 5.5 microseconds, a reduction of 19.4% as was shown in Fig. 1.14. Repeated attempts at reducing or extending the level ranges using principles of steps in the direction of steepest ascent did not significantly decrease chip delay time.

5.4   Conclusions

The use of three-level DoE offers opportunities to investigate new as well as improve current designs or processes. Three-level DoE differs from two-level designs in allowing for more in-depth analysis of the curvature of DoE outcomes and for exploring the design space around current designs and processes. In addition, a mix of two- and three-level factors can be analyzed as was shown in green soldering case study. Three-level DoE causes proportionally more experiments to be performed for the same number of factors in two levels, with more complex factor interactions. Handling of interactions and confounding with three levels is complex with two interaction columns for each two-way factor interactions. Typically, more screening or partial factorial three-level DoEs are performed rather than full factorial designs, especially for those projects where the DoE team has in-depth experience in the product or process to be optimized.

Four case studies were discussed in this chapter, with in-depth analysis of results obtained and conclusions drawn. Three were screening and one was partial factorial designs. Each case study was discussed in depth and its design justified and critiqued. Its results were carefully analyzed with reference to significance and confounding. The constraints of each DoE were presented with the conflicts of meeting project goals while reducing DoE execution time and effort.

Additional Reading Material

Eaton W., et al., "Using circuit simulation with Taguchi Design of Experiment Techniques to Optimize the Performance of a Digital Half-adder Integrated Circuit", Quality Progress, Journal of Quality Engineering, 1993, Vol. 5, Number 4, pp. 589–600.

Ross, P., Taguchi Techniques for Quality Engineering, New York: McGraw Hill, 1987.

Roy, R., A Primer on the Taguchi Method, New York: Van Nostrand Reinhold, 1990.

Phadke, M., Quality Engineering Using Robust design, Englewood Cliffs, NJ: Prentice-Hall. 1989.

Shina, S., "Taguchi Experiments for Improving Solder Quality", Journal of Surface Mount Technology, July 1992, pp. 4–13.

Shina, S., "The use of the Taguchi Method to Optimize Manufacturing", Technologia en Marsha, published by the Instituto Technologico de Costa Rica, Volume 10, Number 2, 1990, pp. 3–7.

Shina, S., et al., "A Comparative Analysis of Lead Free Materials And Processes Using Design Of Experiments Techniques", SMTI International, Chicago, IL, September 2003.

Taguchi, G., Introduction to Quality Engineering, Tokyo: Asian Productivity institute. 1986.

Taguchi G., El Sayed A. and Hsiang T., Quality Engineering in Production Systems, New York: McGraw Hill, 1988.

Taguchi, G., System of Experimental Design, White Plains, NY: UNIPUB- Kraus, 1976.

Chapter 6
DoE Error Handling, Significance, and Goal Setting

One of the most important elements of industrial DoE is the ability to conduct a project for design feasibility or improvements within reasonable constraints of time, budget, and resources. These constraints might lead to shortened project scope by making too many assumptions to reduce the number of experiments. Previous chapters and their case studies showed the balance of using shortened alternatives to full factorial DoE versus its consequences of not meeting DoE goals, handling of interactions, confounding, and analysis of error.

In this chapter, DoE design and analysis will be further explored in terms of different techniques for resolving DoE planning, execution, and sources of error. The focus is on using alternative approaches within the project steps as outlined in previous chapters. The professional experience of DoE team members can be leveraged to minimize the model of design or process to be investigated. This can be accomplished by proper selection of input factors and levels and fitting them into the most efficient orthogonal array (OA) that matches project needs. DoE results can then be used to analyze model significance and sources of variations due to the error in making samples for measurements.

Initial decisions made in defining the model to be used for DoE projects can greatly influence subsequent steps to be taken. The need for follow-on experiments is directly influenced by the modelling method used. The design team can either select a step-by-step approach in incrementally using sequential experiments to achieve ultimate results or take some calculated risks in attempting to reach desired goals quickly and efficiently in one comprehensive DoE, using their collective experience. This chapter is aimed in helping DoE teams strike a reasonable balance between these two options. The adverse consequences of making too many assumptions or ignoring statistical methodologies developed in previous chapters will also be shown.

A case study will be presented with alternatives for developing good error analysis through partial or full replications. Alternative DoE planning and execution will also be presented under the most desirable conditions of experiment dividing.

Topics to be discussed in this chapter are as follows:

6.1 DoE Error Handling Techniques for Significance Testing

In previous chapters, only two methods for handling error in DoE analysis were discussed: pooling of the lowest effect factor and replicating the entire set of experiments to generate error in ANOVA tables. The first method might result in ignoring the significance of a pooled factor, and the second might result in extensive testing and data collection and possible error caused by many replicates. This section will discuss alternate methods of generating error including using multiple CenrterPoints as an independent set of additional experiments to generate error, as well as repeating some experiment lines and pooling of three-way interactions or higher as sources of error. A case study DoE will be discussed with all the mentioned error handling techniques to contrast their use and relative benefits. All error and significance calculations will be shown with visual techniques to highlight concepts used.

6.2 DoE Project Goal Setting and Design Space

These techniques are critical to proper DoE execution but sometimes ignored in the rush to use DoE as a design problem-solving tool. DoE project team might be distracted by their focus on other issues of DoE design such as the selection of factors and levels, their interactions or confounding, and the proper OA to use. The issues discussed in this section provide a good foundation to begin proper project planning and execution. While some topics such as project goals and design space were discussed in earlier chapters, a review is made in this section with visual tools to make best level section. All principles for design space are also applicable to process mapping equivalent for improving manufacturing processes.

6.3 Experiment Dividing (Blocking)

Techniques for conducting experiments over a prolonged period might require experiment dividing, sometimes called blocking. They are presented to reduce errors caused by environmental conditions as well as materials, operators, and equipment variation in conducting different parts of planned experiments. Proper sequencing of experiments will minimize the effects of uncontrolled factors, making better statistical results due to reduced variation error.

6.4 Conclusions

6.1 DoE Error Handling Techniques for Significance Testing

DoE error handling methodology is dependent on project goals. They range from using experiments as initial design alternative investigations, improving current designs, or solving a design performance problem. These goals should be within the budgetary, schedule, and staffing constraints of the project.

Once DoE model factors and levels are identified and an OA is selected, a predetermined set of experiments can be performed as was demonstrated in earlier chapters. To test out the model, two types of statistical analysis are conducted using DoE outcomes:

1. Model coefficient estimates, which is a measure of the effect of each factor on DoE outcomes. Using the standard error formulas, which are the same for each factor, a 95% confidence interval (CI) of each value can be calculated. If a factor level ± CI crosses the total experiment mean, then the factor is not significant, since the CI range exceeds the factor coefficient. This indicates that error effect is greater than factor effect and the factor should be pooled into the error.

   Figure 6.1 is a plot of model coefficients for a Plastics DoE case study to be discussed later in this chapter. A visual tool is to plot the estimated results versus each factor level, with the 95% CI at each level shown as a vertical line in Fig. 6.1, for factors Pressure and Time. If the CI crosses the mean of all experiments, then it is not significant. In Fig. 6.1 example, factor Pressure is significant, while factor Time is not.

   A t-value can be calculated for each factor based on the coefficient estimate and standard error. It can be compared to the critical t-value using error degrees of freedom (DOF). A probability can be calculated for each factor coefficient from the t distribution when compared to the critical value. A visual tool to demonstrate this model reduction is the use of a Pareto diagram for all factor t-values as shown in Fig. 6.2 as a horizontal bar graph after pooling three factors. The critical t-value is the dashed line corresponding to the desired % confidence level (usually at 95% two-sided). Its value is 3.182 corresponding to error DOF = 3 and 97.5% confidence or 0.025 significance from Table 2.6. Factors to the right of dashed line are significant (shown in black), and those to the left (shown in gray) are not, including Pressure*Time which is equal to 0. In Fig. 6.2, Pressure is significant, while Time is not.

2. ANOVA which uses the sum of the squares (SS) method to allocate the source of variation from each factor in the model. Error sum of squares (SSE) is calculated from subtracting the sum of the squares for all model factors (SSR or sum of regression) from total sum of squares (SST). Each SS is converted into a

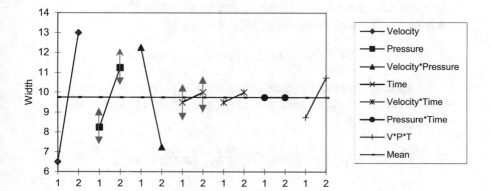

Fig. 6.1 Plastics DoE model coefficient plots

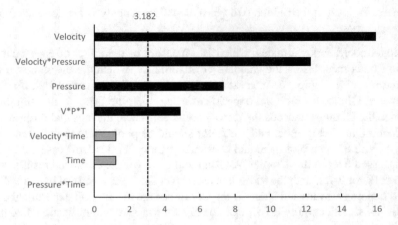

Fig. 6.2 Ranked Pareto diagram of coefficient t-values for Plastics Single DoE after pooling

variance or mean square (MS) through dividing by DOF. An F-value can be calculated by dividing the variance of each factor by the variance of the error and compared to the critical F-value for significance testing. A probability can also be assigned to each factor F-test, which is equivalent to the probability for the same factor coefficient t-test.

Both statistical analysis methods above can be achieved by identifying the error caused when redundancy is introduced into the model. This can be achieved by the following methods:

1. Reducing a single replication model by declaring one or more factors or interactions as part of the error, chosen either by having the smallest coefficient estimates or the smallest factor SS. Their combined sum of the squares can be divided by their total DOF to create the error variance or the standard error of the experiments, resulting in factor coefficient t-values or F-values mentioned above.
2. Replicating a single or multiple experiment line(s) and using the variation from the repetition to generate error. Each replication adds more DOF to the error, and the SSE can be calculated from the ANOVA table:

$$\text{SSE} = \text{SST} - \text{SSR} = \text{SST} - \text{SS}_{\text{Factor 1}} - \text{SS}_{\text{Factor 2}} - \text{SS}_{\text{Factor 3}} - \cdots \qquad (6.1)$$

An alternate method for determining mean square error (MSE) or error variance can be calculated using pooled variance of each replicated experiment line $V_{Expt\ n}$ as follows:

$$MSE = \frac{DOF_{Expt\ 1} * V_{Expt\ 1} + DOF_{Expt\ 2} * V_{Expt\ 2} + DOF_{Expt\ 3} * V_{Expt\ 3} + \ldots + DOF_{Expt\ n} * V_{Expt\ n}}{DOF_{Expt\ 1} + DOF_{Expt\ 2} + DOF_{Expt\ 3} + \ldots + DOF_{Expt\ n}}$$

$$(6.2)$$

Once the error is quantified, each factor can be tested for significance. The model can be reduced by pooling non-significant factors until only significant factors remain in the model or no significance is shown. In the latter case, the DoE project is assumed to be a failure, and a new project can be launched using a different model of factors and levels or a different design space.

6.1.1   Regression Equation and Predicted Outcome with Interaction

A regression equation can be developed using model coefficients from significant factors and interaction estimates. Levels of each significant factor can be selected depending on the DoE project goals (largest, smallest, or target value). A prediction can be made based on the level selection, with a CI based on the standard error.

Interaction levels cannot be selected in the regression equation; only the source (constituent) factor levels can be. If a pooled factor has significant interaction(s) with other factors, then a corresponding level should be selected for this factor in the regression equation to account for its interaction, even if the factor is not in the regression equation. Careful selection of non-significant factor levels should result in the correct coefficient sign for the interaction, as interactions can increase or decrease predicted outcomes by virtue of source factor level selection. Interaction Table 4.4 provides a method of selecting the proper interaction level in the regression equation.

An example of an interaction counteracting the effects of its source factors is the Hipot DoE discussed in Chap. 4. Figure 4.4 shows that selecting level (1) for factors Cable and Contact results in maximum desired outcome (together with Paint level 1). Interaction Cable*Contact has the reverse slope, and its resulting level (1) from interaction Table 4.4 reduces the predicted outcome. An alternative is to plot the four outcomes of the interaction and select the largest value of the four interaction combinations as shown in Fig. 4.5. The regression equation for Hipot DoE is:

Outcome (Kvolts) = 14.0625 − 1.8125 Cable −1.9375 Contact −2.1875 Paint +1.5625 Contact*Cable.

Choosing level (1) for factors Cable and Contact results in the following prediction (without including factor Paint in the model):

Outcome (Kvolts) = 14.0625 + 1.8125 Cable (level 1) + 1.9375 Contact (level 1) − 1.5625 Contact*Cable (level 1) based on interaction Table 4.4.
Outcome (Kvolts) = 16.25 which is also the top left point in interaction Fig. 4.6 corresponding to Cable (1) and Contact (1).

6.1.2   DoE Error Handling Types

DoE error can be generated through different variants of methodologies discussed earlier as follows:

1. Conducting a single OA experiment. This is the fastest method to conduct a DoE project with the anticipation of quick results and corresponding analysis. The error can be generated in two alternatives:

   (a) Designate the factor with the smallest coefficient or SS as not significant and pool it into the error. Adverse consequences of this method could be that the pooled factor(s) are significant which can only be shown if total experiment replication is performed.
   (b) Ignore three-way or higher interactions and pool them as error in a single replication experiment if they are not confounded by additional factors. This technique can be used for higher-order OA such as L8, L9, L16, L27, and L32. However, these pooled factors can possibly be significant, resulting in reduced regression equation and prediction accuracy.

2. Conducting at least one full replication of an OA, allowing for both mean and variability analyses (to be discussed in Chap. 9) and providing error DOF equal to number of repeated experiments. Repeating an OA multiple times will result in statistically significant 30 outcomes. Examples would be repeating an L8 four times, L9 three times, and L16 two times.

3. Repeating the CenterPoint of the design space. For two levels, a middle value (CenterPoint) of the levels is chosen for each factor. Additional experiments are performed using these CenterPoint levels which form the basis of experiment error. The standard error (SE) of each coefficient in the original experiment is calculated from previous chapters as follows:

$$SE_{\text{coefficient}} = s_{\text{Error}} / \sqrt{n_{Expts}} = \sqrt{MSE / n_{Expts}} = \sqrt{\frac{SSE}{DOF_{\text{Error}}} / n_{Expts}} \qquad (6.3)$$

$$SE_{\text{CenterPoint}} = s_{\text{Error}} \sqrt{\frac{1}{n_{\text{CenterPoint}}} + \frac{1}{n_{Expts}}} \qquad (6.4)$$

   The model coefficients as well as the SS for each factor remain the same since the CenterPoint is not part of the original factor level calculations. When plotting the coefficients, the CenterPoint mean should be noted for each factor as well as the total experiment mean, to observe any significant mean difference.

4. Replicating some OA experiment lines. This method is easier to execute than the CenterPoint replications, since no setup is required for additional experiments. The error is generated by the additional replicated experiments, one error DOF for each replication when using two-level OA. The model coefficients and ANOVA must be recalculated since the replication and error for each factor is changed from the original number of experiments before the replications, unlike the

CenterPoint method for error. The error variance or MSE can be calculated by using either the ANOVA Eq. (6.1) or pooled variance methods Eq. (6.2) outlined above.

6.1.3   Error Handling and Significance Technique L8 Case Study: Plastics Injection Molding DoE

In this section, techniques for handling error will be discussed using a two-level full factorial L8 with three factors. All error handling methods and conclusion made for this sample L8 are applicable to all OA of two and three levels, as well as partial and screening designs. This sample L8 array was chosen from an industrial case study, and some results were altered to highlight the different techniques of error handling.

The experiments were conducted using a screw-driven injection molding machine to fabricate Lexan 5000 Polycarbonate tensile strength test specimen, normally used to measure the material response under tensile stress. The goal was to optimize the specimen features by maximizing the width of the narrow part of the specimen and minimizing the weight. For this example, only increasing the width will be analyzed, and the experimental results were modified to illustrate the error handling methods discussed in the earlier section.

The DoE team brainstormed the injection molding process and considered the three phases of the process: injection of the plastic charge, packing of material in the mold, and holding parts in the mold before ejecting them. It was decided to maintain some of the process parameters in place such as the feed temperature, melt temperature, mold temperature, and mold cooling time. The team focused on relevant factors in each process phase:

1. Injection phase: Velocity, Pressure, and Runner Size.
2. Packing phase: Pressure, Time, and Gate Seal.
3. Holding phase: Pressure, Time, and Gate Seal.

The team decided on the following three two-level factors:

1. Injection Velocity (in/sec), with levels 1 and 4.5.
2. Pack Pressure (psi), with levels 2000 and 4000.
3. Pack Time (sec), with levels 0.2 and 0.5.

Four sets of experiments were conducted, corresponding to the following:

1. A single L8 orthogonal array.
2. Adding 2 replications to the original L8 for a total of 24 experiments.
3. Adding 4 experiments with CenterPoint to the original L8, for a total of 12 experiments.
4. Repeating 2 lines in the original L8 (experiment lines 2 and 6), with 2 repeats each for a total of 12 experiments.

Table 6.1 Plastics DoE error handling case study

Expt. #	Velocity	Pressure	Pack Time	Outcomes	Error Method
1	1	2000	0.2	6	Single Expt.
2	1	2000	0.5	9	Single Expt.
3	1	4000	0.2	6	Single Expt.
4	1	4000	0.5	5	Single Expt.
5	4.5	2000	0.2	10	Single Expt.
6	4.5	2000	0.5	8	Single Expt.
7	4.5	4000	0.2	16	Single Expt.
8	4.5	4000	0.5	18	Single Expt.
9	1	2000	0.2	6	Replication 1
10	1	2000	0.5	8	Replication 1
11	1	4000	0.2	5	Replication 1
12	1	4000	0.5	4	Replication 1
13	4.5	2000	0.2	9	Replication 1
14	4.5	2000	0.5	8	Replication 1
15	4.5	4000	0.2	16	Replication 1
16	4.5	4000	0.5	14	Replication 1
17	1	2000	0.2	9	Replication 2
18	1	2000	0.5	10	Replication 2
19	1	4000	0.2	8	Replication 2
20	1	4000	0.5	4	Replication 2
21	4.5	3000	0.2	8	Replication 2
22	4.5	3000	0.5	7	Replication 2
23	4.5	4000	0.2	15	Replication 2
24	4.5	4000	0.5	16	Replication 2
C9	2.75	3000	0.35	10	CenterPoint
C10	2.75	3000	0.35	11	CenterPoint
C11	2.75	3000	0.35	10	CenterPoint
C12	2.75	3000	0.35	9	CenterPoint
R9	1	2000	0.5	9	Replicate Expt. 2
R10	1	2000	0.5	8	Replicate Expt. 2
R11	4.5	2000	0.5	9	Replicate Expt. 6
R12	4.5	2000	0.5	8	Replicate Expt. 6

The results were modified in terms of $0.001''$ increments beyond the minimal width. The Plastics DoE and its outcomes are shown in Table 6.1.

6.1.4 Error Handling and Significance for Single Repetition DoE Analysis

For the single L8 experiments, there is no experiment outcome replication. The error must be generated by reducing the model through pooling of smallest effect or smallest sum of square factor(s). Two types of statistical analysis, model coefficient and ANOVA, were performed on the single L8 experiments before pooling, as shown in Table 6.2, and after pooling as shown in Table 6.3. The steps for reducing the model and finding the maximum prediction are as follows:

Table 6.2 Plastics Single DoE analysis before pooling

Model Coefficients

Source	Effect	Coefficient
DoE Mean		9.75
Velocity	6.50	3.25
Pressure	3.00	1.50
Time	0.50	0.25
Velocity*Pressure	5.00	2.50
Velocity*Time	-0.50	-0.25
Pressure*Time	0	0
Velocity*Pressure*Time	2.00	1.00

ANOVA Before Pooling

Source	DOF	SS	Contribution	MS
Velocity	1	84.50	52.32%	84.50
Pressure	1	18.00	11.15%	18.00
Time	1	0.50	0.31%	0.50
Velocity*Pressure	1	50.00	30.96%	50.00
Velocity*Time	1	0.50	0.31%	0.50
Pressure*Time	1	0	0	0
Velocity*Pressure*Time	1	8.00	4.95%	8.00
Error	0			
Total	7	161.50	100%	

Table 6.3 Plastics Single DoE analysis after pooling

Model Coefficients

Source	Effect	Coef.	SE Coef.	95% CI	t-Value	$P > t$
DoE Mean		9.75	0.204	(9.10, 10.40)	47.79	0
Velocity	6.50	3.25	0.204	(2.60, 3.90)	15.93	0.001
Pressure	3.00	1.50	0.204	(0.85, 2.15)	7.35	0.005
Velocity*Pressure	5.00	2.50	0.204	(1.85, 3.15)	12.25	0.001
Velocity*Pressure*Time	2.00	1.00	0.204	(0.35, 1.65)	4.90	0.016

ANOVA After Pooling

Source	DOF	SS	Contribution	MS	F-Value	$P > F$	$p\%$
Velocity	1	84.50	52.32%	84.50	253.50	0.001	52.12%
Pressure	1	18.00	11.15%	18.00	54.00	0.005	10.94%
Velocity*Pressure	1	50.00	30.96%	50.00	150.00	0.001	30.75%
Velocity*Pressure*Time	1	8.00	4.95%	8.00	24.00	0.016	4.75%
Regression Model Total	4	160.50	99.38%	40.13	120.37	0.001	98.56%
Pooled Error	3	1.00	0.62%	0.33	-	-	1.44%
Total	7	161.50	100%				100%

1. Model coefficient estimates, calculated for the total experiment mean and for each factor. The effect (span) of each factor is calculated from the difference of the means of all four sets of levels (1) and (2). The coefficient of each factor is 1/2 the effect or the span of that factor. No other terms such as standard error (SE), confidence interval (CI), t-value, or P-value can be calculated since there is no mean square error (MSE) to quantify s_{error} before pooling.

   For example, the effect of factor Velocity is $[(10 + 8 + 16 + 18)-(6 + 9 + 6 + 5)]/4 = (52-26)/4 = 6.5$ and the coefficient $= 3.25$. For estimating interactions, the interaction column must be generated from the original factors with two levels, and then the effect can be calculated like the Velocity factor above. The OA preset values of factor and interaction levels are given in Chap. 4 for two levels and Chap. 5 for three levels.

   Factor coefficient (Pack) Time and interactions Velocity*Time and Pressure*Time are either zero or small and could be pooled to reduce the model and generate error. Fig. 6.1 shows the model coefficient plots for the Plastics DoE, which visually demonstrate the least effect factors and interactions.

2. Analysis of variance (ANOVA) before pooling. Table 6.2 shows the ANOVA for each factor before pooling. The factor SS calculations were shown in Chap. 3 but are repeated here for factor Velocity as an example:

$$SS(\text{Velocity}) = \left[(10 + 8 + 16 + 18)^2 + \left[(6 + 9 + 6 + 5)^2\right]\right]/4 - \text{Correction Factor (CF)}$$

$$= (52^2 + 26^2)/4 - (6 + 9 + 6 + 5 + 10 + 8 + 16 + 18)^2/8 = 845 - 760.50$$
$$= 84.50.$$

   A new column (Contribution %) was added to the ANOVA in Table 6.2. It is different than the $p\%$ contribution discussed in Chap. 3, since it is a simple percentage of the total sum of squares (SST). Once the error is identified by pooling, this column could be used for determining the R^2 of the L8 DoE.

3. After pooling one factor (Time) and two interactions (Velocity*Time and Pressure*Time), the model coefficient estimates and ANOVA can be recalculated using the reduced model, shown in Table 6.3. The three pooled factors are summed up into SSE with $DOF_{error} = 3$. MSE is equal to the SSE divided by DOF_{error}, and it is also equal to error variance. Some of the calculations in Table 6.3 are shown below:

   (i) $MSE = SSE/DOF$ of pooled factors $= (0.5 + 0 + 0.5)/3 = 1/3 = 0.33$
   (ii) $s_{error} = \sqrt{MSE} = \sqrt{0.333} = 0.577; SE_{Coef.} = s_{error}/\sqrt{n_{Expts.}} = 0.577/\sqrt{8} = 0.204$
   (iii) The values for the model coefficient table are:
         CI is equal to the coefficient $\pm SE_{Coefficient} * t_{0.025,n(error)} = $ coefficient $\pm$ SE
         $* t_{0.025, 3 = 3.182}$
         For Velocity 95% CI $= 3.25 \pm 0.204 * 3.182 = 3.25 \pm 0.65 = (2.6, 3.9)$.

None of the CI for the remaining factors cross the zero horizontal axis point, indicating that all factors in Table 6.3 are significant as well as the $P > t$-value <0.025.

t-value for each factor $=$ coefficient/$SE_{coefficient}$. For Velocity, t-value $= 3.25/0.204 = 15.93$.

P-value $(P > t) =$ Excel$^{\circledR}$ T.DIST.2 T $(4.9, 3) = 0.00054$ or 0.001

(iv) The values for the ANOVA table are:

F-value of each factor $=$ MS factor/MSE.

For Velocity F-value $= 84.50/0.33 = 253.50$. Each F-value can be compared to the critical $F_{0.05,\ 1,\ 3} = 10.13$ from Table 2.11. If the factor F-value >10.13, then the factor is significant.

A P-value $(P > F)$ from the Excel$^{\circledR}$ F function F.DIST.RT $(253.5,1,3) = 0.00054$ for Velocity shows the probability of that factor, which is the same probability from the model coefficient $P > t$. Any probability <0.05 indicates significance. Table 6.3 ANOVA shows that the remaining factors and interactions are all significant and the model reduction is complete. No further pooling is necessary.

(v) The $p\%$ column indicates the percentage value discussed in Chap. 3. A $p\% > 5\%$ is considered significant. This is true for the first three values in the ANOVA that are significant using the F-test. It is not true for the three-way interaction since its $P >$ F-value is 0.016 and its $p\% = 4.75\%$. The DoE team must adopt one of these two methods of significance to make sure it is consistent with the selected software analysis results.

(vi) It is recommended that pooling occurs one factor at a time for accuracy of model reduction. In this case study, the smallest factor SS was zero, and the next two factors had identical SS. This is the reason for pooling all three factors together.

Another indication of significance is the ranked Pareto diagram of model coefficient t-values, shown in Fig. 6.2. The critical $t_{0.025,\ n\ =\ 3(error)} = 3.182$ is plotted as dashed line. Any factor or interaction with a t-value <3.182 is shown in gray, indicating non-significance. In addition, Fig. 6.2 shows the ranking of the factors in terms of importance.

The contribution column in Table 6.3 shows the $\%$ of significant factors and interactions in the regression equation, adding up to $R^2 = 99.38\% > 90\%$, which indicates model significance. All remaining factors and interactions are significant, using five separate methods indicated above in Table 6.3 and Fig. 6.2:

1. CI range of coefficients not containing value $= 0$.
2. $P > t$-values from the model coefficients.
3. $P > $ F-values in ANOVA.
4. Ranked Pareto of coefficient t-values.
5. $p\%$ contribution, which might conflict with probability from either t- or F-values depending on analysis software used and for high error $p\%$ contribution.

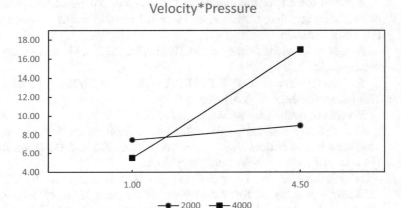

Fig. 6.3 Velocity*Pressure interaction plot for Plastics Single L8 DoE

The factor plots in Fig. 6.1 remain the same after pooling. The CI ($\pm$ 0.65) from previous section is universal for each significant factor since it is derived from coefficient SE. When CI is superimposed on each factor level shown in Fig. 6.1, it indicates significance if the CI does not cross the mean, shown for Pressure. Inversely, if CI for a factor or interaction crosses the mean, it is not significant. This is true for the pooled factor Time and interactions Velocity*Time and Pressure*Time.

4. There is only one significant interaction, Velocity*Pressure, which is shown as the third line from the Y-axis in Fig. 6.1. The DoE team cannot set the level of this interaction since it is controlled by factor levels of Velocity and Pressure independently. It is important to carefully choose the levels of these two factors, especially since the interaction coefficient > Pressure coefficient. Fig. 6.3 is a plot of the interaction, showing four combinations of two-level factors. They can be estimated by averaging four sets of two results each:

(a) Experiments 1 and 2 (Velocity 1 and Pressure 2000 = 7.5).
(b) Experiments 3 and 4 (Velocity 1 and Pressure 4000 = 5.5).
(c) Experiments 5 and 6 (Velocity 4.5 and Pressure 2000 = 9).
(d) Experiments 7 and 8 (Velocity 4.5 and Pressure 4000 = 17).

5. For the goal of maximum width, Velocity level 4.5 (2) and Pressure level 4000 (2) result in a value = 17 for best selection for both factors. The two source factors for the interaction can be treated as one factor with four levels, and the best of the four levels will satisfy the desired outcome. The DoE team can select the best of four combinations (for two-level arrays) or nine combinations for three levels from each significant interaction plot.

   An alternative method is selecting the interaction level according to the two source (constituent) factor levels of the interaction based on interaction Table 4.4. In this case, selecting interaction level 1 based on combined levels (2) of source factors for both or Velocity (2) * Pressure (2) = Velocity*Pressure (1). In this Plastics DoE, interaction Velocity*Pressure coefficient is additive to outcome, which is the reverse of the Hipot DoE in Chap. 4 mentioned earlier in Sect. 6.1.1.

6. The three-way interaction Velocity*Pressure*Time is shown to be significant. Even if one of the factors in the interaction (Time) is not significant and does not affect outcome prediction, this interaction should be included in the prediction. Some DoE practitioners recommend pooling three-way or higher interaction factor(s) as the error in single experiments, considering that their effect might be small. There is one three-way interaction in L8 and four in each of L16 and L27. While this might seem as a reasonable plan, there could be instances where a three-way interaction is significant as in this Plastics DoE. The recommended method of using the least important factor or interaction as the initial pooling in single experiments demonstrated in this section is a more reasonable approach to handling the error.

7. The regression equation can be generated from the Plastics DoE outcome mean and the coefficients of each significant factor and interaction. The sign of the coefficient points to the proper level to be selected for the project goal. These signs might be different depending on whether the two levels are (1 or 2) or (-1 and + 1) using different analysis software. A two-way interaction of source factor levels (2) and (2) results in reverse interaction level (1), while a two-way interaction of levels (+1) and (+1) results in similar interaction level (+1) as was shown in interaction Table 4.1. Three-way interactions have the same sign in either level scheme.

   This discrepancy of level schemes is not critical since DoE teams can examine factor plots such as Fig. 6.1 to select the appropriate level for the project goal. The slopes of Velocity and Pressure and the three-way interaction Velocity*Pressure*Time are positive, and the slope for the two-way interaction Velocity*Pressure is negative. This is reflected in the regression equation coefficient signs for Plastics Single DoE as follows:

   Regression equation (predicted result) = 9.750 + 3.250 Velocity + 1.500 Pressure − 2.500 Velocity*Pressure + 1.000 Velocity*Pressure*Time.

   With the selected levels of Velocity level 2 (4.5), Pressure level 2 (4000), Velocity*Pressure level 1, and Velocity*Pressure*Time level 2, the predicted outcome is 18. The contributions of three non-significant factors cancel each other out: Time level 2 (+0.25) and Velocity*Time level 1 (−0.25) and Pressure*Time (contribution = 0). The total prediction using model coefficient for maximum width outcome is 18, which happens to be experiment 8 results. This is always true for full factorial OAs, since all combinations of factors and interactions are available. This is not true for partial factorial or screening DoE since they are reduced number of experiments.

   For a project goal of specific target value outcome, there might be many combinations of factor levels that yield desired target value. In this case, it is

recommended that variability analysis (to be discussed in Chap. 9) be used to select best alternative combinations with reduced variability as well as achieving the target goal.
8. At the conclusion of the analysis phase of DoE projects, a confirming experiment with the recommended factor levels should be undertaken to verify the statistical analysis. Any decision to change the design or process to new levels should be undertaken with careful observations of new design or process for an extended period until the team is confident that the new design is successful. Usually, data from a 6-month period should be collected before and after the DoE project as shown in Fig. 1.15.

6.1.5   Error Handling and Significance for Multiple Replication DoE Analysis

Replicating the total experiment one or multiple times provides a source for error due to duplication of results. Pooling an initial factor or interaction that might be significant in single experiments is not required, rendering the analysis more accurate. An additional benefit is that an incorrect or damaged specimen outcome from one experiment line can be excluded from the analysis without having to repeat the experiment line to produce an alternate specimen. Having an outcome for each experiment line will keep the orthogonality of the data and integrity of the analysis.

Some DoE projects require summing multiple results, especially in the case of attribute counts. This is the case of using DoE to reduce defects or solve intermittent problems. If a design or process produces low defects, it would take many repetitions to produce reliable outcomes. A DoE with multiple zero outcomes produced in certain experiment line(s) cannot be analyzed. In most cases, the team should make specimen within the extreme limits of the specification to purposely generate the maximum number of defects to be reduced.

The DoE team should count the sum of defects directly if they are within a reasonable range such as singles to tens of defects per specimen. If the range of defects per specimen is large, the use of transformed results is recommended. This can be done by multiplying each result with a fraction to reduce the result range or take the natural logarithm of the results to reduce data dispersion. An alternative method would be ranking the results in equal ranges or bins and using the ranges in sequential numbers for recording as results (from 1 to 5 or 1 to 10).

While it might be easy to produce multiple results from a single experiment line, DoE project constraints such as cost of manufacturing or time required could limit DoE execution. In this case, actual production line(s) or machine(s) might be taken out of service to conduct experiments. Many DoE projects using production equipment to produce results are performed on the weekend with production staff to reduce the impact on manufacturing schedules but increasing the project cost due to overtime.

Table 6.4 Plastics three-repetition DoE analysis before pooling

Model Coefficients

Source	Effect	Coef.	SE Coef.	95% CI	t-Value	$P > t$
DoE Mean		9.375	0.25	(8.84, 9.91)	36.99	0
Velocity	5.42	2.71	0.25	(2.17, 3.25)	10.69	0
Pressure	2.42	1.21	0.25	(0.67, 1.75)	4.77	0
Time	-0.25	-0.13	0.25	(-0.66, 0.41)	-0.49	0.629
Velocity*Pressure	5.08	2.54	0.25	(2.00, 3.08)	10.03	0
Velocity*Time	-0.25	-0.13	0.25	(-0.66, 0.41)	-0.49	0.629
Pressure*Time	-0.58	-0.29	0.25	(-0.83, 0.25)	-1.15	0.267
Velocity*Pressure*Time	1.42	0.71	0.25	(0.17, 1.25)	2.79	0.013

ANOVA Before Pooling

Source	DOF	SS	Contribution	MS	F-Value	$P > F$
Velocity	1	176.04	43.40%	176.04	114.19	0
Pressure	1	35.04	8.64%	35.04	22.73	0
Time	1	0.38	0.09%	0.38	0.24	0.629
Velocity*Pressure	1	155.04	38.22%	155.04	100.57	0
Velocity*Time	1	0.38	0.09%	0.38	0.24	0.629
Pressure*Time	1	2.04	0.5%	2.04	1.32	0.267
Velocity*Pressure*Time	1	12.04	2.97%	12.04	7.81	0.013
Regression Model Total	7	380.96	93.92%	54.42	35.30	0
Replication Error	16	24.67	6.08%	1.54		
Total	23	405.63	100%			

The DoE team should consider all the above constraints, benefits, and shortcomings of error handling techniques and make a balanced decision on how many repetitions of experiments, if any, for project planning purposes.

For the Plastics DoE, the outcomes of the original L8 experiment and two replications are shown in Table 6.1. The statistical analysis before pooling is shown in Table 6.4. All calculations for Table 6.4 are based on the top 24 outcomes from Table 6.1. They were discussed in previous examples of DoE in this and other chapters.

Two replications of L8 produce 16 error DOFs. All factors' and interactions' attributes from the L8 can be calculated using the error from the two replications. All model coefficients and ANOVA can be completed without initial pooling of any factor or interaction. The coefficient standard error (SD) can be calculated as follows:

$$SE_{\text{Coeff.}} = s_{\text{error}} / \sqrt{n_{Expts.}} = \sqrt{MSE/n} = \sqrt{1.54/24} = 0.25$$

The model for three-repetition Plastics DoE before pooling shows good significance with R^2 from the regression model total of 93.92 > 90%. The same factor Time and interactions Velocity*Time and Pressure*Time that were pooled in the single L8 experiment show lack of significance and can be pooled into the error. Pooling should be performed one factor or interaction at a time since the F-value changes with each round of pooling.

There are many measures of significance for factors and/or interactions mentioned above. Some are qualitative, while others are quantitative, using the tools discussed in the previous section. The following is a review of these methods based on Table 6.4 for three-repetition DoE and Figs. 6.1, 6.2 and 6.3 for single-repetition DoE:

1. A plot of factors and interactions, as shown in Fig. 6.1 for single-repetition DoE, will show the same observations and suggest which factor(s) to be pooled first.
2. Factor/interaction coefficients vary in value, and some may have smaller effect than others and could be pooled.
3. The CI value in the model coefficient tables is another indication of significance. If the CI range crosses zero, then the factor or interaction is not significant. Alternatively, if the coefficient $\pm$ CI crosses the experiment mean as shown in Fig. 6.1, then the factor is not significant, as in the examples of factor Time and interactions Velocity*Time and Pressure*Time.
4. The contribution % is a good indicator of importance ranking of factors and interactions. It is different than the $p\%$ discussed in Chap. 3. The use of either method for significance will vary depending on the software analysis used and the error $p\%$.
5. The computed t-value in the model coefficient tables can be compared to the critical t-value ($t_{0.025, \text{DOFerror}}$) for desired significance (± 0.025) and error DOF, which implies the use of the absolute value of t. The probability based on each t-value will indicate which factor to pool if >0.05.
6. The Pareto diagram shown in Fig. 6.2 for single-repetition DoE is a good method to rank the effect of each factor to initiate the first or subsequent pooling processes. It is plotted after pooling three non-significant factors, indicating the remaining four factors are all significant.
7. The computed F-value (ratio) and P > F in ANOVA, shown in Table 6.4, are used to indicate significance. The computed F-value can be compared to the critical F-value with the DOF of the factor and the error ($F_{0.05, \text{DOF factor, DOF error}}$). The probability based on each F-value will indicate which factor to pool if >0.05.
8. The P-value obtained in either coefficient estimates (P > t) or ANOVA (P > F) tables is equal and compared to the critical probability of 95% confidence or $\alpha = 0.05$ significance. A factor or interaction with a P-value >0.05 should be considered not significant.

Once the initial pooling is selected, it should be continued until all remaining factors and interactions are significant. The factor and interaction significance were the same for one or three repetitions. Table 6.5 shows the results of three-repetition Plastics DoE after pooling is completed with the following observations:

1. The SST remains the same as pre-pooled DoE ANOVA from Table 6.4, but the % contribution of the regression and error SS change slightly. The R^2 is slightly lower for three repetitions after pooling (93.23% versus 93.92%) due to increased

Table 6.5   Plastics three-repetition DoE analysis after pooling

Model Coefficients

Source	Effect	Coef.	SE Coef	95% CI	t-Value	P $> t$
DoE Mean		9.375	0.245	(8.86, 9.89)	38.20	0
Velocity	5.42	2.71	0.245	(2.20, 3.22)	11.04	0
Pressure	2.42	1.21	0.245	(0.70, 1.72)	4.92	0
Velocity*Pressure	5.08	2.54	0.245	(2.03, 3.10)	10.36	0
Velocity*Pressure*Time	1.42	0.71	0.245	(0.20, 1.22)	2.89	0.009

ANOVA After Pooling

Source	DOF	SS	Contribution	MS	F-Value	P $>$ F
Velocity	1	176.04	43.40%	176.04	121.81	0
Pressure	1	35.04	8.64%	35.04	24.25	0
Velocity*Pressure	1	155.04	38.22%	155.04	107.28	0
Velocity*Pressure*Time	1	12.04	2.97%	12.04	8.33	0.009
Regression Model Total	4	378.17	93.23%	94.54	65.42	0
Pooled Error	3	2.792	0.69%	0.93	0.6	0.622
Replication Error	16	24.67	6.08%	1.54		
Error Total	19	27.46	6.77%	1.45		
Total	23	405.625	100%			

total error from pooling. The regression model and pooled error probabilities P > F are 0 and 0.622, respectively, indicating significance of the model and non-significance of pooled factor and interactions.

2. The total SS error (27.46) includes both adding replication SS and pooling SS. It is divided by the total error DOF (19) to calculate the total MSE (1.45).
3. The two replications resulted in the same three significant factors and one interaction as the single L8 DoE, but with more significance as indicated by either the P > t or P > F probabilities. This is an indication of the robustness of the model for three replications.

After pooling, model reduction is complete, and a prediction of design or process performance can be calculated from the regression eq. A confirming experiment should be undertaken to verify the efficacy of the new design or process. These steps were discussed in the previous section on single experiment outcome.

6.1.6   Error Handling and Significance for Multiple CenterPoint Replications

The use of additional experiments to resolve DoE error was demonstrated with replicating all L8 experiments. A shortened version of that methodology is to repeat the design space CenterPoint.

An advantage of CenterPoint replications is observing whether input variable effect on DoE outcomes is linear or not since there is a midpoint between two-level

Table 6.6 Plastics DoE CenterPoint replication analysis before pooling

Model Coefficients

Source	Effect	Coef.	SE Coef.	95% CI	t-Value	$P > t$
DoE Mean		9.75	0.289	(8.83, 10.70)	33.77	0.000
Velocity	6.50	3.25	0.289	(2.33, 4.17)	11.26	0.002
Pressure	3.00	1.50	0.289	(0.58, 2.42)	5.20	0.014
Time	0.50	0.25	0.289	(-0.67, 1.17)	0.87	0.450
Velocity*Pressure	5.00	2.50	0.289	(1.58, 3.42)	8.66	0.003
Velocity*Time	-0.50	-0.25	0.289	(-1.17, 0.70)	-0.87	0.450
Pressure*Time	0	0	0.289	(-0.20, 0.20)	0	1.000
Velocity*Pressure*Time	2.00	1.00	0.289	(0.08, 1.92)	3.46	0.041
CenterPoint		0.25	0.500	(-1.34, 1.84)	0.50	0.651

ANOVA Before Pooling

Source	DOF	SS	MS	F-Value	$P > F$
Velocity	1	84.50	84.50	126.75	0.002
Pressure	1	18.00	18.00	27.00	0.014
Time	1	0.50	0.50	0.75	0.450
Velocity*Pressure	1	50.00	50.00	75.00	0.003
Velocity*Time	1	0.500	0.50	0.75	0.450
Pressure*Time	1	0	0	0	1
Velocity*Pressure*Time	1	8.00	8.00	12.00	0.041
CenterPoint (Curvature)	1	0.17	0.17	0.25	0.651
Regression Model Total	8	161.67	20.21	30.31	0.009
CenterPoint Error	3	2.00	0.67		
Total	11	163.67			

factors. CenterPoint replication adds three-level benefits of observing possible outcome curvature to two-level advantage of more factors and columns with less experiments.

CenterPoint replication should be treated as a different sample than the original OA experiments. The two samples can be pooled as demonstrated in Chap. 2 for t-tests, using Eq. 2.8. The CenterPoints should be added to the outcome analysis, both in model coefficients and ANOVA. If the CenterPoint is significant, then its bias to the original experiment mean should be added into the regression equation. CenterPoint replications also provide a source of variation (usually called curvature) to the DoE analysis as well as providing error (SSE).

An example of using CenterPoint for error handling was given in Table 6.1. The original L8 experiment is recorded in the first eight lines, and the CenterPoint is repeated four times. The CenterPoint replications are shown as line C9–C12 outcomes of 10, 11, 10, and 9 in Table 6.1.

Table 6.6 shows the statistical analysis of the original L8 and the CenterPoints before pooling. Several observations can be made from the data analysis:

1. Model coefficients and SS of all factors and interactions remain the same as the original L8 since the CenterPoint levels are different. The mean of four CenterPoint replications is 10, which represents a bias of 0.25 to the 9.75 L8 experiment mean. The 0.25 mean bias is the CenterPoint coefficient.

2. CenterPoint replications provide ANOVA SSE and are included in the SST for the 12 experiments (L8 original and 4 CenterPoints). The original L8 SSR remains the same from Table 6.2 = 161.50. The calculations for Table 6.6 are as follows:

$$\text{Centerpoint } (CP) = [\,10 \ \ 11 \ \ 10 \ \ 9\,]; \ \sum(CP) = 40; \ \sum(CP)^2 = 402; \ CF$$

$$= \frac{\left(\sum CP\right)^2}{N} = \frac{40^2}{4} = 400$$

$$SSE = \sum(CP)^2 - CF = 2; \ MSE = \frac{SSE}{n_{CP} - 1} = \frac{SSE}{3} = 0.67; \ s_{\text{error}} = \sqrt{MSE}$$

$$= 0.8165$$

$$\sum(12\text{Exps}) = 118; \ \sum(12\text{Exps})^2 = 1324; \ CF = \frac{(12\text{Exps})^2}{n} = \frac{118^2}{12}$$

$$= 1160.33; \ SST = 163.67$$

$$SS_{\text{Centerpoint}} = SST - SSR - SSE = 163.67 - 161.50 - 2 = 0.17$$

3. The Standard error (SE) of the coefficients for the factors and CenterPoint can then be calculated:

$$SE_{\text{Coeff.}} = s_{\text{error}} \Big/ \sqrt{n_{\text{Expts.}}} = \frac{0.8165}{\sqrt{8}} = 0.289; \ SE_{CP} = s\sqrt{\frac{1}{n_1} + \frac{1}{n_2}} = s\sqrt{\frac{1}{8} + \frac{1}{4}} = 0.50$$

4. While the coefficients of the L8 factors and interactions are the same, their SE is different from Tables 6.2 and 6.4 due to CenterPoint replications. The CenterPoint coefficient 95% CI $(-1.34, 1.84)$ and t-value of $(0.25/0.50) = 0.50$ are not significant since the CI range crosses the zero axis and the $P > t = 0.651 > 0.05$. The CenterPoint mean is not significantly different than the experiment mean, indicating non-significant curvature due to the CenterPoint. This also indicates that linear response is a good approximation for all factors and interactions in the design space.
5. CenterPoint curvature source of variation adds one more DOF to ANOVA, increasing model DOF from the original seven to eight. The four CenterPoint replications produce SSE = 2 with DOF = 3. The 12 experiments' ANOVA SST was calculated as 163.67, and the model SS is equal to SST-SSE = 161.67. The resulting $R^2 = 161.67/163.67 = 98.78\%$ indicates good model significance.
6. While the SS of all factors and interactions remain the same as the original L8, their F-values and P > F values are different than the original since MSE originates from CenterPoint replications.

7. CenterPoint mean should be shown in all model coefficient plots as well as interaction plots such as those in Figs. 6.1 and 6.3.

After completing CenterPoint statistical analysis before pooling, model reduction can proceed by pooling non-significant factors and interactions, one at a time, until only significant factors remain. DoE outcome prediction should be calculated based on regression equation analysis. A confirming experiment should be performed to validate the DoE outcomes.

6.1.7 Error Handling and Significance for Some Experiment Outcome Replications

This is another shorter method for handling error than replicating L8 experiments. A single or multiple L8 experiment line(s) can be replicated providing error from experiment outcome redundancy. The advantage of replicating some lines versus replicating CenterPoint is lack of additional experiment setup to provide for CenterPoint outcomes. The disadvantage is that model coefficients and ANOVA are different than the original L8 analysis results and must be recalculated. If a single L8 experiment was conducted, followed by additional replicated lines to provide error, all previous analysis will have to be performed again based on this method.

An example using this error handling method for Plastic Injection Molding DoE is given in Table 6.1. The original L8 experiment is given in the first eight lines, and experiment lines 2 and 6 are replicated two times each, for a total of four replications as shown in lines R9–R12 in Table 6.1. Replication outcomes for L8 experiment 2 are 9 and 8 and for replicating L8 experiment 6 are 9 and 8. Table 6.7 shows the reconfigured L8 with some repetition of experiment lines, and Table 6.8 shows statistical analysis summary of results before pooling. Several observations can be made of the data analysis of replicating some L8 experiment lines:

1. Model coefficients and SS of all factors and interactions are all different than the original L8, since the four additional experiments influence their calculations.

Table 6.7 Plastics DoE with replication of some experiment outcomes

Velocity	Pressure	Time	Original Outcome	Replicated Outcomes(s)	Outcome(s) Mean	Total Results	V
1	2000	0.2	6	6	6	6	0
1	2000	0.5	9	9, 9, 8	8.67	26	0.33
1	4000	0.2	6	6	6	6	0
1	4000	0.5	5	5	5	5	0
4.5	2000	0.2	10	10	10	10	0
4.5	2000	0.5	8	8, 9, 8	8.33	25	0.33
4.5	4000	0.2	16	16	16	16	0
4.5	4000	0.5	18	18	18	18	0

Table 6.8 Plastics DoE with replication of some experiment outcome analysis before pooling

Model Coefficients

Source	Effect	Coef.	SE Coef.	95% CI	t-Value	$P > t$
DoE Mean	9.750	0.186		(9.23, 10.27)	52.32	0.000
Velocity	6.667	3.333	0.186	(2.82, 3.85)	17.89	0.000
Pressure	3.000	1.500	0.186	(0.98, 2.02)	8.05	0.001
Time	0.500	0.250	0.186	(-0.27, 0.77)	1.34	0.251
Velocity*Pressure	4.833	2.417	0.186	(1.90, 2.93)	12.97	0.000
Velocity*Time	-0.333	-0.167	0.186	(-0.68, 0.35)	-0.89	0.422
Pressure* Time	0.000	0.000	0.186	(-0.52, 0.52)	0.00	1.000
Velocity*Pressure* Time	1.833	0.917	0.186	(0.40, 1.44)	4.92	0.008

ANOVA Before Pooling

Source	DOF	SS	Contribution	MS	F-Value	$P > F$
Velocity	1	56.33	33.80%	56.33	169.4	0.000
Pressure	1	22.04	13.22%	22.04	66.14	0.000
Time	1	0.625	0.37%	0.625	1.88	0.251
Velocity*Pressure	1	77.04	46.23%	77.04	231.18	0.000
Velocity*Time	1	1.23	0.73%	1.23	3.68	0.422
Pressure*Time	1	0	0%	0	0	1
Velocity*Pressure*Time	1	8.07	4.84%	8.07	24.20	0.008
Regression Model Total	7	165.33	99.20%	70.86		
Replication Error	4	1.33	0.80%	0.33		
Total	11	166.67	100%			

The experiment mean is 9.75 which is the mean of all 12 experiments for the reconfigured L8 in Table 6.7.

2. Model coefficients for all factors and interactions can be calculated using Table 6.7. The mean results of L8 lines 2 and 6 of 8.67 and 8.33 can be used to calculate all L8 effects and coefficients. They are different than the original L8 analysis shown in Table 6.2.

3. The MSE can be calculated using two alternate methods:

   (i) Pooled variance, where the SSE is calculated using DOF weighted mean of each experiment line variance shown in Eq. (6.2) and Table 6.7:

$$MSE = \frac{DOF_1 * V_1 + DOF_2 * V_2 + \ldots + DOF_n * V_n}{DOF_1 + DOF_2 + \ldots + DOF_n} = \frac{2 * 0.33 + 2 * 0.33}{2 + 2}$$

$$= 0.33$$

   (ii) Extracting the MSE from the ANOVA table, where SSE = SST − (model SS), and dividing the SSE by DOF = four from the four replications yields the MSE:

$$\sum(12\text{Exps}) = 112; \sum 12\text{Exps}^2 = 1212; CF = \frac{(12\text{Exps})^2}{n} = \frac{112^2}{12}$$

$$= 1045.33; SST = 166.67$$

$$SSE = SST - SSR = 166.67 - 165.33 = 1.33$$

$$MSE = \frac{SSE}{DOF_{\text{Error}}} = \frac{1.33}{4} = 0.33$$

4. The SST and SS of each factor and interaction can be calculated using the same methodology for treatments in Chap. 3. For example, SSA (Time) and SSB (Pressure) are shown below for results (y_1-y_{12}). The model SS is the sum of all SS of factors and interactions:

$$SSA = \frac{(6 + 26 + 6 + 5)^2}{6} + \frac{(10 + 25 + 16 + 18)^2}{6} - CF = \frac{43^2}{6} + \frac{69^2}{6} - 1045.33$$

$$= 56.33$$

$$SSB = \frac{(6 + 26 + 10 + 25)^2}{8} + \frac{(6 + 5 + 16 + 18)^2}{4} - CF = \frac{67^2}{8} + \frac{45^2}{4} - 1045.33$$

$$= 22.04$$

5. The remaining calculation of ANOVA in Table 6.8 can be performed using the calculations outlined in the previous sections.
6. Subsequent pooling if any, regression equation, prediction, and confirming experiments can be performed using the calculations outlined in the previous sections.

All error handling methods shown in this L8 example are applied to all two- and three-level orthogonal arrays. The DoE team must discuss error handling plans that best match project goals and constraints of staff needs, time, and resources.

6.2 Project Goal Setting and Design Space

In this section, sequential steps taken by DoE teams are explored at the beginning of the project to improve the prospect of successfully achieving project goals. They include defining the project and its goals, selecting DoE alternative, and setting the design space to be optimized.

6.2.1   Types of DoE Project Goals

Setting project goals should be the first step prior to launching the DoE project and selecting the team. The goals should be clearly set according to the assigned tasks. Typical goals and time frames of DoE projects are:

1. Solving a consistent or sudden design or process problem in the immediate or short term.
2. Improving an existing design or process over an intermediate period. The DoE project can be launched to meet company quality or productivity goal, whether quarterly or yearly.
3. Undertaking research into new technologies or materials that can yield better new design performance. A DoE project could be undertaken offline with a longer-range perspective. Not being part of current or new product design or manufacturing timeline allows for longer period to achieve project goals.

Design or process outcomes should be selected next and be accurately measured. Anticipated range of output measures for each experiment line should be reasonably variable. For variable measurements, small output changes in response to varying input variables should precipitate increased measurement accuracy. For attribute measurements such as reducing defects, multiple specimens have to be made for each experiment line to produce a reasonable quantity of defects for proper analysis. Some of these issues were discussed in Sect 5.3.2.

Goal setting should influence DoE project planning in the following manner:

1. A problem-solving DoE project should consider whether the problem is exhibited intermittently or is a permanent quality shift. Several issues to consider are:

   (a) An intermittent problem requires many replications of each experiment line to generate a reliable set of variable outcomes. If all OA experiment outcomes are of the same value or zero, then statistical analysis cannot be performed. The number of replications needed to generate an error could also be used as the DoE outcome.
   (b) Alternately, an intermittent problem might require designing test specimen at specification limits. In addition, running production equipment at the extreme of acceptable ranges can also cause intermittent problems to appear and be measured. The same number of specimens can be run for each experiment line to count the number of defects as DoE outcomes.
   (c) Screening rather than partial or full factorial designs might be best suited for solving intermittent design or process problems because of time constraints and need to consider as many causes (factors) as possible. The team should concentrate on factors which are easy to manipulate such as different materials, suppliers, and operating conditions. These may include power source, temperature, pressure, speed, geometry, and connectivity of parts in product assembly. Factors that are difficult to quickly change such as castings and plastic molded parts should not be considered.

(d) Using DoE to solve a permanent quality shift problem might not be appropriate. Other qualitative problem-solving tools such as cause-and-effect and Pareto diagrams might provide a faster resolution than performing a DoE and its subsequent analysis and interpretation of data. DoE techniques should be used after these qualitative problem-solving efforts are exhausted.

2. A design or process improvement project presents a different set of challenges to DoE teams. While similar to intermittent problem-solving in that there is good knowledge and experience about the problem, there might be uncertainty about the project goals. How much better quality can be expected in production? How much larger, smaller, or closer to target can a design be manipulated? Several issues to consider in this case are:

(a) Was the original design well researched and documented? Was there good experimentation, modelling, or analysis to back up current design or process levels? Are current design parameters (levels) just grandfathered in, or were there good reasons for selecting current design materials and processes?

(b) Is there a benchmark for design or process improvement project? Are there competitors or other parts of the company performing the same design and processes and achieving better quality or performance? Are they using industry standard methodologies for measuring quality or performance to make sure that comparisons are made under similar conditions?

(c) Are there new technology or material alternatives available that claim better quality or performance? In this case, several decisions must be made: whether to perform an initial study to narrow down the choice of new materials and then perform a two- or three-level DoE project including top 2 or 3 alternative candidates into a larger DoE project. Another decision is whether to include current (legacy) materials or technologies in the DoE or use them as separate baseline to compare against the new alternatives. These decisions depend on the team's willingness to choose large experiments or try to improve current designs while studying new alternatives at the same time.

(d) Three-level DoE might be appropriate for this type of design improvement project, with the middle level being the current design parameters to study possible improvements in an expanded design space. This chapter discussed use of other techniques such as using two-level designs with CenterPoint repetition to represent error. The use of CenterPoints as current design parameters would be useful with more factors for studying alternative materials.

(e) Partial factorial experiments with reduced interactions can also be appropriate since the team is familiar with the design and can take appropriate risks in allowing limited DoE confounding.

3. Research into new technologies or materials for better new design performance is different than the other previous two categories, since the DoE team does not have extensive knowledge or experience in the new design. Several issues to consider in this case are:

(a) Two-level DoE might be appropriate since the team might prefer to have as many factors as possible to investigate new designs. Two-level OAs offer more factor analysis through smaller number of experiments than three- or multi-level OA.

(b) Full factorial experiments with investigations of all interactions are recommended in research projects to explore factor interactions if any. Alternatively, a pre-selection of factors either by sample comparison or treatments or even a screening design to reduce the number of factors can be completed before undertaking a full factorial experiment.

(c) If a current technology or material is being obsoleted and must be replaced, then the current (legacy) technology or material can be separated from the OA and treated as baseline. The new technology can thus be compared to the baseline current technology using the statistical tests discussed in Chap. 2. The initial green soldering case study in Chap. 5 was designed in this manner using legacy solder technology as comparison baseline.

6.2.2 Design Space and Level Selection

Design space refers to the areas or volume under consideration for DoE projects. It is bounded by the levels of the factors selected in OA experiments. DoE analysis can only determine the relationships between input factor levels and the resulting model outcomes within that space. It does not give any indications about model behavior outside the design space. Further DoE projects might have to be conducted to determine model behavior outside of the design space but within the normal operating parameters of the design.

The use of two- or three-level designs will indicate the behavior of each factor in the model in the design space, whether linear or not. Earlier in this chapter, it was shown that the use of CenterPoint replications with two-level designs can help in determining the linearity of model outcomes. The curvature resulting from CenterPoint replications for two-level OA will indicate non-linearity if it is significant. For three-level designs, the linearity of the model coefficients can be easily determined by plotting the coefficients for factor levels or using a polynomial fit upon outcome changes of factor levels.

The slope of the two- or three-level factor responses within the design space also determines the direction of the steepest ascent, through the value of the factor coefficients. This was demonstrated in the L9 three-level screening DoE for bonding process discussed in Chap. 5. A follow-on experiment to explore the model coefficient estimates can be performed using the slope of one factor to see if it extends beyond the design space by increasing or decreasing current levels. It can also be used to explore if there is an inflection point of curvature within the original levels as well. Care must be taken not to extend the levels beyond the safe operations of the design or process.

Fig. 6.4 Design space for
L8 full factorial Hipot
experiment

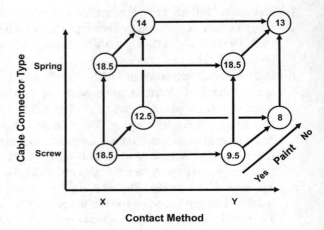

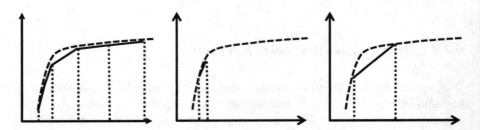

Fig. 6.5 Selection of factor levels

A design space case study from Chap. 4 was given for the two-level full factorial
L8 Hipot experiment shown in Fig. 6.4. The three factors are presented as three axes,
and their levels are shown on all three axes. The design space is presented by the
cube formed by the eight experiments, with the output response values inserted onto
the eight corners of the design space.

For two-factor experiments, the design space cube becomes an area with two
factors representing the two axes, as was shown in Fig. 1.12, multivariate charts for
two variables. Using equal outcomes for each (factor) input, a response surface could
be plotted. An optimum (maximum, minimum, or target) value can be shown in the
response surface. For four-factor design such as a screening L9 or full factorial L16
with four factors, the design space could be presented with two three-axis cubes,
whether inside or offset from each other.

The selection of factor levels determines the size of the design space. Fig. 6.5
shows alternative for two- and multi-level spacing of factors. The DoE team should
select factor levels according to the following criteria:

1. Factor levels can be non-numeric alternative methods, materials, and/or technologies. An example would be the Hipot L8 DoE discussed in Chap. 4 and plotted
   as a design space in Fig. 6.4. All the levels are not numeric variables, and the L8

could easily be analyzed based on the results of the experiments as was shown in Chap. 4.

2. A set of experiments can be conducted to reduce the number of alternative methods, materials, and/or technologies to be considered prior to conducting DoE. Alternatives can be compared using the statistical tests discussed in Chap. 2.
3. The presence or lack of a particular factor such as paint in Fig. 6.4 can also be part of level selection. This is a proper methodology to use if the design team is interested in studying whether any steps can be reduced in a process or components eliminated from a design.
4. If a response variable (outcome) of a DoE is expected to be non-linear, then multi-level factors should be used in conducting OA experiments. This allows examining of the effect of each coefficient estimate on the response while varying other factor levels in the OA.
5. Spacing of the levels is important to capture a reasonable portion of outcome responses. As shown in the two plots to the right in Fig. 6.5, a narrow level range could miss important inflection points in the response. Level spacing that is too wide would miss any curvature of the response. Using two-level OA with CenterPoint replications or three-level OAs would alleviate this issue.
6. Three-level spacing with the middle level at current process should have the other two levels around 10–20% range above and below the middle. A closer range runs the risk of the coefficient estimates being overwhelmed by the error.

A good DoE project plan is one that best matches the problem to be resolved, design or process to be optimized, or new technology or materials to be adopted. The DoE team should consider the alternatives discussed in this section to plan and execute a successful project.

6.3 Experiment Blocking (Dividing)

Another important planning task for a DoE project is sequencing and execution of OA experiments. Some experiments may take longer than few hours due to process issues such as temperature changes of large masses or specimen preparations for design conditions. In addition, some process experiments are performed on production lines that cannot be shut down to run experiments. If experiment execution must be spread out over many shifts or days, then there might be bias due to uncontrolled conditions. They might be completed by different operators, machines, or environmental levels of temperature, pressure, and humidity.

Experiments can be divided into blocks to reduce this bias effect on model coefficient estimates. This can be achieved by assigning blocks in tandem with interactions that are deemed not important. The bias shift due to blocking is thus confined to non-significant interactions and hopefully pooled into the error.

DoE practitioners suggest that three-way or higher interactions are usually not significant. For example, an L8 experiment can be divided along the levels of the

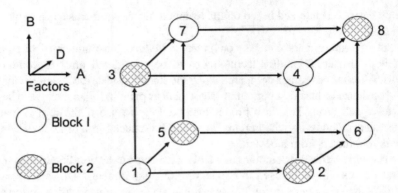

Fig. 6.6 Blocking L8 into groups of four

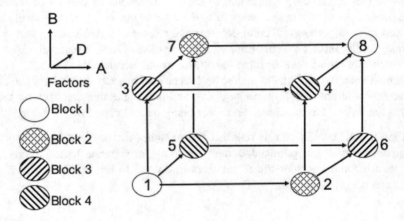

Fig. 6.7 Blocking L8 into groups of two

three-way interaction column $G = 7 = 1*2*4$. Block 1 of four experiments can be completed along lines of level 1 in column 7, and block 2 can be completed along lines of level 2 in column 7, in the following manner and shown in Fig. 6.6:

Block 1 = Experiments $1(A_1 B_1 D_1 G_1)$, $4(A_1 B_2 D_2 G_1)$, $6(A_2 B_1 D_2 G_1)$, and $7(A_2 B_2 D_1 G_1)$

Block 2 = Experiments $2(A_1 B_1 D_2 G_2)$, $3(A_1 B_2 D_1 G_2)$, $5(A_2 B_1 D_1 G_2)$, and $8(A_2 B_2 D_2 G_2)$

The two blocks and column 7 will be confounded. Hopefully, the three-way interaction is not significant, and its coefficient will not be included in the model but will be part of the error.

If the blocking needs to be smaller, then the eight experiments can be divided into four groups, with two experiments each. Each group will correspond to complementary levels of three factors and their three-way interactions, with column 7 levels alternating between levels 1 and 2, shown in Fig. 6.7:

Table 6.9 Blocking L16 into groups of four

Column #	1	2	
Factors	A	B	
Expt. #			Blocks
1	1	1	1
2	1	1	1
3	1	1	1
4	1	1	1
5	1	2	2
6	1	2	2
7	1	2	2
8	1	2	2
9	2	1	3
10	2	1	3
11	2	1	3
12	2	1	3
13	2	2	4
14	2	2	4
15	2	2	4
16	2	2	4

Block 1 = Experiments 1($A_1 B_1 D_1 G_1$) and 8($A_2 B_2 D_2 G_2$)
Block 2 = Experiments 2($A_1 B_1 D_2 G_2$) and 7($A_2 B_2 D_1 G_1$)
Block 3 = Experiments 3($A_1 B_2 D_1 G_2$) and 6($A_2 B_1 D_2 G_1$)
Block 4 = Experiments 4($A_1 B_2 D_2 G_1$) and 5($A_2 B_1 D_1 G_2$)

For an L16 OA, blocking can proceed in the same manner as L8, using two blocks of eight experiments confounding four-way interaction levels in column 15 shown in Table 4.7. Dividing L16 into four blocks of four experiments each can be made using interaction of any two factors assuming that the interaction is not significant. Confounding, which will be discussed in the next chapter, causes equivalence of some two-way versus three-way interactions in L16. For example, if using L16 columns 13 and 14 to generate the four blocks, it would be equal to using the interaction 1*2 which is formed from columns 1 and 2. Using the L16 patterns from Table 4.7, the transformation from 13*14 to 1*2 is shown below:

13*14 = ACD*BCD = A*B = 3 = 1*2

The four blocks would appear as combinations of the two levels of factor A and B levels (11, 12, 21, and 22). This L16 blocking into four groups is shown in Table 6.9. Similar blocking techniques can be used for L32. For blocking with three-level OAs, an L27 can be divided into three blocks of nine each using the pattern of any of four three-way interaction columns 9, 10, 12, and 13. An alternative method would be to use nine blocks of three each using the sequential three experiments' level pattern 123, 312, and 231 in any three-way interaction column.

6.4 Conclusions

The importance of planning for DoE project preparation in handling error, identifying design space, dividing experiments, and setting proper project goals was outlined in this chapter. These elements might not be thoroughly considered in an industrial setting where there are constraints in staffing, resources, and time. A careful examination of the use of DoE as a tool for solving problems and enhancing design performance and product quality or new materials and technologies can lead to a much better matching of the project to the underlying need.

Additional Reading Material

Berger, P., Maurer, R. and Celli G., *Experimental Design*, New York: Springer, 2018.

Box, G., Hunter, J. and Hunter, W., *Statistics for Experimenters: Design, Innovation, and Discovery*, 2nd Edition. New York: Wiley-Interscience, 2005.

Cobb, G., *Introduction to Design and Analysis of Experiments*, New York: Wiley, 2008.

Dean, A. and Voss, D., *Design and Analysis of Experiments, 2nd edition*, New York: Springer, 2017.

Hinkelmann, K. and Kempthorne O., *Design and Analysis of Experiments, Volume 2: Advanced Experimental Design*, New York: Wiley-Interscience, 2005.

Montgomery, D., *Design and Analysis of Experiments 3rd Edition*, New York: Wiley, 1991.

Narayan, S., "Optimization the Performance of Weld lines in Injection Molded Products using Robust Design Methods", Master of Mechanical Engineering Thesis, UMASS Lowell, May 1992.

Rekab, K. and Shaik, M., *Statistical Design of Experiments with Engineering Applications*, Boca Raton, FL: CRC press, 2005.

Santner, T. and Williams, B., *The Design and Analysis of Computer Experiments*, New York: Springer, 2019.

Selvamuthu, D. and Das D., *Introduction to Statistical Methods, Design of Experiments and Statistical Quality Control*, New York: Springer, 2020.

Shina, S., " An Algorithm for selecting soldering flux for cleaning and surface conductivity", NEPCON West Conference Proceedings, Los Angles, California, March 1989, pp 1064–1070.

Shina, S., "Reducing Solder Wave defects in a Printed Circuit Board Wave Soldering Process", American Supplier Institute's Sixth Symposium Proceedings, Dearborn Michigan, October 1988, pp 123–144.

Toutenburg, H., *Statistical Analysis of Designed Experiments*, 3rd Edition, New York: Springer, 2002.

Chapter 7
DoE Reduction Using Confounding and Professional Experience

Interactions and confounding are the most confusing aspects of DoE principles. They are difficult to plan and can result in DoE teams going to extremes in either using lengthy full factorial experiments or adopting screening designs. In addition, the use of non-interacting orthogonal arrays (OAs), to discussed in the next chapter, has become popular to avoid interactions and confounding.

There are two alternative approaches to DoE. The classical approach assumes that DoE teams do not have experience in the design or process to be studied or optimized and therefore should approach DoE projects with minimum assumptions. Full factorial experiments can be designed as individual factors can be studied in the context of interacting with other factors. Additional experiments can be performed to study more factors as needed or time permits.

The second approach, called interaction matrix, is to use DoE team members' professional experience to make safe assumptions about design factors and interactions, focusing on some while ignoring others. Fractional factorial DoE can be conducted with reduced experiments based on the team's assumptions. This approach might result in quicker but suboptimal results while providing reasonable success in achieving project goals.

Several DoE concepts are introduced in this chapter such as design resolutions, interaction tables, and interconnection graphs. They are excellent tools for handling interactions and confounding. They provide guidelines on which factors and interactions are to be evaluated, to the exclusion of others. They manage confounding when assigning factors to partial factorial or screening designs and selecting appropriate OA. They can help to determine whether and how to conduct follow-on experiments.

The case studies from previous chapters will be re-examined in the light of new concepts developed in this chapter. Assumptions and decisions made will be evaluated to examine their impact on DoE outcomes and conclusions. Topics to be discussed in this chapter are:

S. Shina, *Industrial Design of Experiments*, https://doi.org/10.1007/978-3-030-86267-1_7

7.1 Design Resolution and Confounding
    Resolution is a methodology to manage confounding when not using full
factorial DoE. It is used with partial factorial design when some factors are
assigned to OA interaction columns. It is also used for screening designs when
factors are assigned to all OA columns and are confounded with interactions.
Each resolution type is presented with interactions and confounding patterns and
their use in the next sections.
7.2 Interactions and Confounding for L8 for Reduced Experiments
    Using partial factorial or screening DoE has the advantage of fewer experi-
ments to perform but reduces the information obtained from DoE projects by
ignoring some interactions and confounding. The use of interaction tables and
interconnection graphs is discussed to evaluate confounding risks specific to L8
DoE. Case studies discussed in previous chapters will be analyzed anew with the
interaction decisions highlighted using these tools.
7.3 Interactions and Confounding for L16 for Reduced Experiments
    This section is like the previous section for L8, where adding factors to an
L16 DoE for half fraction, partial factorial, and screening designs are discussed
using two approaches of classical and professional experience. An additional
tool of using an interaction matrix is shown to facilitate confounding management. Interaction tables and linear interconnection graphs are shown for combinations of L16 designs with multiple factors and pre-selected interaction
scenarios.
7.4 Interaction and Confounding for Large OA
    Large OA can be used in fractional factorial DoE to provide a one-step
approach to solving design problems rather than using a sequence of screening
followed by full factorial DoE. They can study many factors and interactions
and provide an efficient reduced model to easily manage a large design or
process. The APOS case study from Chap. 4 and other examples of L32
experiments will be discussed to evaluate the merits of this approach to large
experiments.
7.5 Chapter Conclusions

7.1 Design Resolution and Confounding

Resolution is a methodology to manage confounding when using partial factorial
DoE, providing for alternative decisions based on classical DoE or interaction matrix
approaches. It is not required with full factorial OA design, since no factors are
assigned to interaction columns. It is implicit for half fraction as well as screening
designs when factors are assigned to OA columns and are confounded with
interactions.

    Handling interactions and confounding is a challenge for DoE teams. Many teams
avoid the complexity of this subject by using full factorial or non-interacting DoE.
Several plans that can be used to avoid confounding are discussed below:

1. Conduct an initial set of pairwise statistical comparison analysis discussed in Chap. 2 to reduce the number of factors in the model, and then use a full factorial OA to analyze the reduced model of factors and their interactions. This plan requires a longer time frame for DoE projects and might not uncover interesting combinations of the original factors that can achieve desired goals.
2. Select an initial set of multiple factors for the model, and use a screening design, where interactions are completely ignored. Identify the most significant three or four factors from the screening design analysis, and then assign these factors to a full factorial OA. This plan takes longer to implement since two DoE projects are required and screening design analysis might result in a large full factorial set of factors for the second DoE.
3. Select non-interacting arrays which will be discussed in the next chapter. These arrays do not provide opportunities to examine interactions or influence of one factor to another on DoE outcomes.
4. Partial factorial experiments provide an opportunity to reduce number of experiments while taking calculated risks. They can be conducted as standalone projects that can yield good outcomes. Its analysis and conclusions can steer the next DoE project if more comprehensive results are desired.

Resolution designs are denoted by 2^{k-n}_R, where k is the number of factors selected, n is the reduction (number of factors assigned to interaction columns causing confounding), and R is a resolution type in roman numerals. The first three types of resolutions are those corresponding to three-way interactions or lower. Examples of each resolution are given for half fraction designs since they are implicit with only one alternative possible. Multiple confounding alternatives are possible with partial factorial designs for higher-order experiments.

1. Resolution III: main factors confounded with two-way interactions. An example is a two-level L4 half fraction OA which is denoted as 2^{3-1}_{III}. Normally an L4 consists of two factors (A and B) assigned to columns 1 and 2 and their interaction A*B assigned to column 3 as shown in Table 4.5. An additional third factor C can be assigned to interaction column 3, confounding it with the interaction A*B. In this arrangement of screening L4 with three factors, four experiments can be used instead of the eight L8 experiments required for a full factorial design with three factors.

   Another example is screening L8 OA denoted as 2^{7-4}_{III}. All four interactions in an L8 in Table 4.6 are used for a total of seven factors including three factors assigned to primary columns (1, 2, and 4) and four additional factors assigned to interaction columns (3 = 1*2, 5 = 1*4, 6 = 2*4, and 7 = 124). In this case, confounding occurs between the additional factors and all three two-way and one three-way interactions.
2. Resolution IV: two-way interactions confound with other two-way interactions. An example is L8 OA half fraction design denoted as 2^{4-1}_{IV}. In this case, an L8 is used to study three main factors assigned to primary columns (1, 2, and 4) and one additional factor G assigned to three-way interaction column 7 (7 = 124). An L8 full factorial OA contains three main factors (A/1, B/2, and D/4) and four

Table 7.1 L8 half fraction interactions and confounding

Column Number	Column Name	Factor (f) or Interaction	Confounding Interactions	Column Confounding
1	A	f	BDG = B*D*ABD + A	247 = 2*4*124 + 1
2	B	f	ADG = A*D*ABD + B	147 = 1*4*124 + 2
3	C (not assigned)	A*B	D*G = D*ABD + A*B	3 = 1*2 + 4*7
4	D	f	ABG = A*B*ABD + D	127 = 1*2*124 + 4
5	E (not assigned)	A*D	B*G = B*ABD + A*D	5 = 1*4 + 2*7
6	F (not assigned)	B*D	A*G = A*ABD + B*D	6 = 2*4 + 1*7
7	G	f	ABD	

interactions (C/3 = A*B, E/5 = A*D, F/6 = B*D, and G/7 = ABD) as shown in Table 4.6.

For half fraction use of L8 shown in Table 4.6, the fourth factor G/7 = ABD/124 is assigned to three-way interaction column 7. If four factors are used in full factorial design, an L16 would be needed. In this manner, half of the experiments (8 instead of 16) are required. All four factors are confounded with three-way interactions (considered not important), and their two-way interactions confound with another two-way interactions as shown in Table 7.1, with two- and three-way confounding. The addition sign (+) in Table 7.1 indicates confounding between the factors and/or interactions, given in numeric column numbers or alphabetical factor assignments. When multiplying factors or interactions, individual factor multiplied by the same factor is equated to unity and is removed.

3. Resolution V: two-way interactions are confounded with three-way interactions. An example is the L16 half fraction design denoted as $2^{5-1}{}_V$. In this example, four main factors are assigned to L16 main factor columns, with an additional factor E assigned to confounding four-way interaction column (15/E). A full factorial L32 would be required for 5 factors and 32 experiments. In the L16, there are 4 main factors (A/1, B/2, C/4, and D/8), 6 two-way interactions (A*B, A*C, A*D, B*C, B*D, and C*D), 4 three-way interactions (ABC, ABD, ACD, and BCD), and 1 four-way interaction (ABCD), totaling 15 columns as shown in Table 4.7. The fifth additional factor is called E, and its two-way interactions to the other four factors (A*E, B*E, C*E, and D*E) confound four three-way interactions in columns 14, 13, 11, and 7, as shown in Table 4.8. Other confounding in the L16 half factorial include all four primary factors (A–D) confounded by four-way interactions and all two-way interactions confounded by three-way interactions including interactions of all four actors (A–D) with factor E, as shown in Table 7.2.

Resolution III is least desirable since the main factors are confounded with two-way interactions, making factor coefficients analysis venerable to confounding error. For higher-order OAs, the half fraction resolution level increases to equal the maximum number of interactions.

Table 7.2   L16 half fraction interactions and confounding

Column Number	Column Name	Factor (f) or Interaction (i)	Confounding Interactions
1	A	f	BCDE = BCD*ABCD + A
2	B	f	ACDE = ACD*ABCD + B
3	A*B	i	CDE = C*D*ABCD + A*B
4	C	f	ABDE = ABD*ABCD + C
5	A*C	i	BDE = B*D*ABCD + A*C
6	B*C	i	ADE = A*D*ABCD + B*C
7	D*E	i	D*ABCD = ABC + D*E
8	D	f	ABCE = A*B*C*ABCD + D
9	A*D	i	BCE = B*C*ABCD + A*D
10	B*D	i	ACE = A*C*ABCD + B*D
11	C*E	i	C*ABCD = ABD + C*E
12	C*D	i	ABE = A*B*ABCD + C*D
13	B*E	i	B*ABCD = ACD + B*E
14	A*E	i	A*ABCD = BCD + A*E
15	E	f	ABCD

7.1.1   Techniques for Managing Confounding

Managing DoE confounding design resolutions becomes much more complex when dealing with many factors. Three main techniques for handling confounding are in common use with specific approaches in developing further experiments and making assumption about the design model:

1. The classical approach based on the book *Statistics for Experimenters*, by Box, Hunter, and Hunter. This approach is to select confounding scenarios that lead to isolation of particular factors and their interactions. The process starts with an L8 screening design to find significant factors among seven selected factors assigned to all L8 seven columns. The most significant factor and its interactions are isolated by a process of folding the most significant factor into a second L8. If the DoE team wants to analyze other significant factors and their interactions, additional L8 experiments can be performed.

   This is a step-by-step approach, where portions of the full factorial set of experiments are performed and then more experiments are added to garner more information as needed. This technique combines the set of information collected from previous experiments with information from new additional experiments. For a seven-factor example, if three significant factors are found from the initial screening L8, three additional folded L8s would be needed to separate the three significant factors and their interactions, for a total of $4 \times 8 = 32$ experiments. This compares favorably with the L28 experiments needed for a full factorial L128 for seven factors, as shown in Fig. 4.3. This folding technique is practiced by many DoE analysis software such as Minitab© and will be further discussed in Chap. 8.

2. The Taguchi approach, based on the books published by Genichi Taguchi and practitioners of his Taguchi methods. This approach is to ignore three-way interactions and simplify confounding by selecting one of the confounding interactions to the exclusion of all others in pre-arranged scenarios as in resolution IV designs. These scenarios include combinations of factors and levels that can be selected without having to consider confounding alternatives. The DoE team can select a particular pre-arranged scenario based on their assumption of which factors and interaction are the most important. Taguchi created many significant tools for clarifying DoE design and analysis, including the following:

   1. Percent contribution (p %) as a method to quantify F-test in the ANOVA table.
   2. Interaction tables for determining interactions quickly and efficiently.
   3. Linear graphs for visualizing interactions.
   4. Signal-to-Noise ratio (S/N) for variability analysis.
   5. Quality loss function (QLF) as a methodology to optimize quality based on model coefficients' average versus variability.
      Some of these tools were already discussed in previous chapters such as p% in Chap. 3. Interaction tables and linear (interconnection) graphs are discussed in this chapter. S/N and QLF will be discussed in Chap. 9.

3. The interaction matrix approach. This a modification of the previous two methodologies where DoE teams can use their professional experience to examine all factors and their interactions and confounding and then make a reasoned selection of relevant factors and interactions while excluding others. DoE project execution is faster with this method, but the team might make wrong assumptions and select factors and interactions that are not significant, leading to suboptimal performance in achieving project goals.

   Examples of these methodologies will be given in the next sections.

7.2   Interactions and Confounding for L8 for Reduced Experiments

L8 OA is the most popular DoE used in industry. It is used as full factorial DoE for three main factors and half fraction DoE resolution IV for four factors as 2^{4-1}_{IV}. It is also used as saturated DoE for seven factors as 2^{7-4}_{III}, followed by full factorial L8 DoE for the top 3 significant factors or folded L8 for each.

7.2.1   L8 Half Fraction Interaction and Confounding

The previous section demonstrated different approaches to interactions and confounding. The full interaction and confounding list is given for half fraction L8

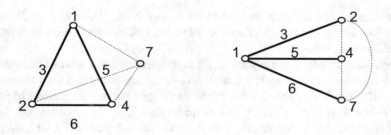

Fig. 7.1 L8 half fraction interconnection diagrams

in Table 7.1. The first three primary factors are A/1, B/2, and D/4, and the additional fourth factor (G/7) confounds the three-way interaction (124/ABD). There are three two-way interaction columns (3/C, 5/E, and 6/F) which are not assigned to factors, and each is confounded by another two-way interaction, making the half fraction L8 a resolution IV design. This results in two possible scenarios for assigning factors and interactions to a half fraction L8 and is shown in half fraction interconnection diagram in Fig. 7.1:

1. Use columns 1, 2, 4, and 7 as the main factors and the first set of two-way interactions (3 = 1*2, 5 = 1*4, and 6 = 2*4) from Table 7.1. The other confounding two-way interactions are assumed to be negligible.
2. Use columns 1, 2, 4, and 7 as the main factors and a modified set of two-way interactions (3 = 1*2, 5 = 1*4, and 6 = 1*7) from Table 7.1. The other confounding two-way interactions are assumed to be negligible.

The triangle plot on the left in Fig. 7.1 represents three main factors (1, 2, and 4) with two-way interactions (3, 5, and 6), while factor 7 interactions, shown as dotted lines, are missing and considered negligible. The star plot at the right in Fig. 7.1 shows factor 1 interacting with the other three main factors (2, 4, and 7), while they are not interacting to each other and their interactions are considered negligible, shown as dotted lines. The L8 half fraction interconnection graph in Fig. 7.1 represents only the first two of eight possible combinations of interactions:

3 = 1*2, 5 = 1*4, and 6 = 2*4 (triangle shape in left side of Fig. 7.1)
3 = 1*2, 5 = 1*4, and 6 = 1*7 (star shape in right side of Fig. 7.1)
3 = 1*2, 5 = 2*7, and 6 = 2*4
3 = 1*2, 5 = 2*7, and 6 = 1*7
3 = 4*7, 5 = 1*4, and 6 = 2*4
3 = 4*7, 5 = 1*4, and 6 = 1*7
3 = 4*7, 5 = 2*7, and 6 = 2*4
3 = 4*7, 5 = 2*7, and 6 = 1*7.

The six combinations not shown in Fig. 7.1 represent different factor assignments to columns, with no discernable difference from the two combinations shown in Fig. 7.1. The use of one of the eight interaction schemes mentioned above for L8 half fraction design indicates that the DoE team is selecting one of the eight possible combinations of interactions for four-factor design. It might be based on the team's professional experience in assigning factors to columns to select the interactions they would like to investigate.

The triangular plot to the left in Fig. 7.1 represents the condition that the first three factors assigned to columns 1, 2, and 4 are equally important and their interactions (3, 5, and 6) can be calculated, while the factor assigned to column 7 is less important with its confounding of the three-way interaction 124. In addition, the factors assigned to column 7 interactions with the other three factors (1*7, 2*7, and 4*7) are assumed to be negligible and do not confound the other three two-way interactions.

The graph to the right in Fig. 7.1 represents a star configuration where one factor (1/A) is more important than the other three. Its interactions with the other three factors (3 = 1*2, 5 = 1*4, and 6 = 1*7) can be calculated. The interactions between the remaining three factors (3 = 4*7, 5 = 2*7, and 6 = 2*4) are assumed to be negligible and do not confound the other three two-way interactions.

If the factor assigned to column 7 is shown to be significant, the second complementary half fraction L8 could be completed so that all confounding can be removed by analyzing the full factorial L16 for four factors consisting of both half fraction L8s. If not significant, the DoE analysis is concluded with no further experimentation.

7.2.2   L8 Partial Factorial Design and Confounding

For less than half fraction design (partial factorial design), the first two methodologies mentioned previously in Sect. 7.1.1 are more divergent. Their approach will be described for all L8 partial factorial designs, including screening design. For L16, partial factorial design confounding will be described up to 2^{10-6} $_{III}$. Table 7.3 shows a summary of resolution types for two-level L8 with added factors.

1. For L8 resolution 2^{5-2} $_{III}$, the original L8 three main factors A, B, and D are assigned to columns 1, 2, and 4. Two additional factors (E and F) are introduced. They could be assigned to any two of the three two-way interaction columns (3, 5, or 6). The following is the confounding table for assigning two additional factors to columns E/5 and F/6. This scheme leaves two columns (3 and 7) not assigned to factors:

Table 7.3 Resolution of two-level OA with added factors

OA	2	3	4	5	6	7	8	9	10-15
					Factors				
L4	Full	III-2^{3-1} 3rd = 3/1*2 Half Fact.							
L8		Full	IV-2^{4-1} 4th=7/123 Half Fract.	III-2^{5-2} 4th=5/1*4 5th=6/2*4	III-2^{6-3} 4th= 5/1*4 5th=6/2*4 6th=3/1*2	III-2^{7-4} 4th= 5/1*4 5th=6/2*4 6th=3/1*2 7th=7/123 Screening			
L16			Full	V-2^{5-1} 5th = 15 Half Fract. 6th = 7	IV-2^{6-2} 5th = 14 6th = 7 7th = 11	IV-2^{7-3} 5th = 14 6th = 7 7th = 11 8th = 13	IV-2^{8-4} 5th = 7 6th = 11 7th = 14 8th = 13 9th = 15	III = 2^{9-5} 5th = 14 6th = 7 7th = 11 8th = 13 9th = 15	III=2^{10-6}–2^{15-11} 5th = 14 6th = 7 7th = 11 8th = 13 9th = 15 10th = 12 11th = 10 12th = 9 13th = 6 14th = 5 15th = 3
L32				Full	VI-2^{6-1}	IV-2^{7-2}	IV-2^{8-3}	IV-2^{9-4}	III = 2^{10-5}

A + D*E = A + D*A*D	1 + 4*5
B + D*F = B + D*B*D	2 + 4*6
C + A*B + E*F = C + A*B+ A*D*B*D	3 + 1*2 + 5*6 (not assigned)
D + A*E + B*F = D + A*A*D + B*B*D	4 + 1*5 + 2*6
E + A*D	5 + 1*4
F = B*D	6 = 2*4
G = ABD + A*F + B*E = ABD + A*B*D + B*A*D	7 + 124 + 1*6 + 2*5 (not assigned)

Factor D/4 is confounded with two two-way interactions, while factors A, B, E, and F are confounded by one two-way interaction. The unassigned interaction columns (C/3 and G/7) are confounded with two two-way interactions each. The DoE team can decide that one interaction is more important than others confounding it and analyze their model accordingly. Confounding of factors with two-way interactions (resolution III) can be resolved with additional experiments.

The DoE team can alternately use column 7 (factor G = 124) as one of the additional factors and either column 5 or 6 for a second factor assignment. This will result in different confounding interactions. For example, if the main three factors are assigned to 1/A, 2/B, and 4/D and two additional factors were assigned to columns 5/E and 7/G, the resulting interactions are:

$$A + D*E = A + D*A*D \qquad\qquad\qquad 1 + 4*5$$

$$B + E*G = B + A*D*ABD \qquad\qquad 2 + 4*6$$

$$C + A*B + D*G = C + A*B + D*ABD \qquad 3 + 1*2 + 4*7 \text{ (not assigned)}$$

$$D + A*E = D + A*A*D \qquad\qquad\qquad 4 + 1*5$$

$$E + A*D + B*G = E + A*D + B*ABD \qquad 5 + 1*4 + 2*7$$

$$F = B*D + A*G = B*D + A*ABD \qquad\quad 6 + 1*7 + 2*4 \text{ (not assigned)}$$

$$G = ABD + B*E = ABD + B*A*D \qquad\quad 7 + 124 + 2*5$$

Factor E/5 is confounded with two two-way interactions, while factors A, B, D, and G are confounded by one two-way interaction. The interaction columns C/3 and F/6 have two confounded two-way interactions each. The DoE team can decide which factor is confounded with two two-way interactions in the model. This is like Fig. 7.1 interconnection graph, with one additional factor and its L8 confounding results, while ignoring three-way interactions. This approach was chosen for the underfill project discussed in Chap. 4.

The two approaches shown above for resolution designs 2^{5-2}_{III} yield similar confounding results. The difference is the three-way interaction column $(7 = 124)$ is not assigned in the first approach and is confounded by two two-way interactions in the second approach. All five factors are confounded with one two-way interaction except for factor D/4.

In the second approach, column 7 is assigned to a factor and confounded with two two-way interactions. All five factors are confounded with one two-way interaction except for factors E/5 and 7/G. The second approach has one more layer of confounding than the first approach; hence, it could be inherently less accurate.

It is advisable not to assign additional factors in an L8 to three-way interaction columns except for half fraction 2^{4-1}_{IV} or saturated design 2^{7-4}_{III} as shown in Table 7.3

2. For resolution 2^{6-3}_{III}, three additional factors (C, E, and F) are introduced into L8 with three main factors (A, B, and D). They are assigned to all three two-way interaction columns (3, 5, and 6). The resulting six factors (A–F) are confounded with two two-way interactions. Column 7/G, normally confounded by three-way interaction $7 = 124 = ABD$, which is not assigned to a factor, is confounded with three two-way interactions:

$$A + D*E + B*C = A + D*A*D + B*A*B \qquad 1 + 4*5 + 2*3$$
$$B + D*F + A*C = B + D*B*D + A*A*B \qquad 2 + 4*6 + 1*3$$

$C + A*B + E*F = C + A*B + A*D*B*D$ $3 + 1*2 + 5*6$
$D + A*E + B*F = D + A*A*D + B*B*D$ $4 + 1*5 + 2*6$
$E + A*D + C*F = E + A*D + A*B*B*D$ $5 + 1*4 + 3*6$
$F + B*D + C*E = F + B*D + A*B*A*D$ $6 + 2*4 + 3*5$
$G + A*F + B*E + C*D = A*B*D + A*B*D +$ $7 = 124 + 1*6 + 2*5 + 3*4$
$A*B*D$

(Column 7 is not assigned to a factor.)

Each of the assigned six factors to columns 1–6 is confounded by two two-way interactions. Column 7 is not assigned to a factor, but the three-way interaction ABD is confounded by three two-way interactions. The DoE team can select which of the two confounding interactions for each factor can be excluded from the model. To resolve confounding interactions, more experiments are needed, such as folding on a particular factor which will be discussed in the next chapter.

7.2.3   L8 Screening Design Confounding

For design 2^{7-4} III, L8 screening design, four additional factors (C, E, F, and G) are introduced into L8. They are assigned to all three two-way interaction columns and one three-way interaction column (3, 5, 6, and 7). The resulting seven factors (A–G) are confounded with three two-way interactions.

$A = D*E + B*C + F*G = 4*5 + 2*3 + 6*7 = $ Column 1
$B = D*F + A*C + E*G = 4*6 + 1*3 + 5*7 = $ Column 2
$C = A*B + E*F + D*G = 1*2 + 5*6 + 4*7 = $ Column 3
$D = A*E + B*F + C*G = 1*5 + 2*6 + 3*7 = $ Column 4
$E = A*D + C*F + B*G = 1*4 + 3*6 + 2*7 = $ Column 5
$F = B*D + C*F + A*G = 2*4 + 3*5 + 1*7 = $ Column 6
$G = A*F + B*E + C*D = 1*6 + 2*5 + 3*4 = $ Column 7

Significant factors from screening L8 experiment analysis and their interactions with other factors could be further investigated using one of the two options:

1. A follow-on DoE project can be initiated using the top significant 2 to 3 factors in full factorial experiments with two levels (L4 or L8) or three levels (L9 or L27). The non-significant factors should be fixed at a particular level in the follow-on DoE, depending on other issues such as cost or ease of manufacture. The outcome data collected in screening DoE cannot be used in the follow-on full factorial DoE, since it interferes with the orthogonality of the selected full factorial arrays.
2. A second L8 with the same factors can be folded using the most significant factor complementary levels to create a second L8. An L16 is thus formed from the combined two L8s and can be analyzed to determine the most significant factor coefficient and its interactions with all other factors. Additional folded L8s can be produced for each additional significant factor. This folding methodology will be discussed in the next chapter.

The use of interaction and confounding in the L8 can be managed successfully if the DoE team carefully selects the type of L8 to be used: full, half fraction, partial factorial, or screening designs. Their professional knowledge can guide the team as to which factors and interaction are considered important in order to determine how factors are assigned to columns. In addition, the team should be willing to undertake additional experiments in the case of inadequate results. They could include either as a full factorial DoE of significant factors following a screening design or continuing the initial experiment by adding more experiments to resolve confounding.

7.2.4 L8 Factor Conversion Tables for Labeling Numeric and Alphabetic Factors

The interactions shown in the L8 half factor, partial factorial, and screening designs were evaluated from multiplying the factors that are interacting by deleting factors that appear twice in the multiplication. For example, in Table 7.1, the confounding interaction for factor A was the three-way interaction BDG. This was determined by multiplying the next two main factors (B and D) with factor G which is confounded by the three-way interaction ABD. The plus (+) sign indicates confounding reduction any time a factor is counted twice in an interaction and removed as an identity value. The confounding is given in alphabetic and numeric column headings for factor A:

$$A + BDG = A + B*D*ABD = A + B*D*(A*B*D) = A + A$$
$$1 + 247 = 1 + 2*4*124 = 1 + 2*4*(124) = 1 + 1$$

For a two-way interaction confounding another two-way interaction, they can confound each other as well as another designated interaction column. For example, in Table 7.1, interaction column 3 (interaction C = A*B) is confounded by two two-way interactions:

$$C + A*B + D*G = C + A*B + D*ABD = A*B + A*B$$
$$3 + 1*2 + 4*7 = 3 + 1*2 + 4*124 = 1*2 + 1*2$$

It is easier to handle the two-way interaction using numerical rather than alphabetical column numbers. Two-way interactions for L8 OA are shown in Table 7.4. The table is triangular since interactions are symmetrical (1*2 = 2*1). The interaction table first row represents one of the factors in the interaction, and the triangular row in parenthesis (n) is the other factor in the interaction. All two-way interactions derived in this section can be immediately found from this table. Every two-way interaction in L8 Table 7.4 is repeated three times. For example:

Column 3 = 1*2 + 4*7 + 5*6
Column 5 = 1*4 + 2*7 + 3*6
Column 6 = 1*7 + 2*4 + 3*5

Table 7.4 L8 OA and interaction table

			L8 OA									Interaction table				
#	1	2	3	4	5	6	7	Col.	1	2	3	4	5	6	7
1	1	1	1	1	1	1	1	(1)	3	2	5	4	7	6
2	1	1	1	2	2	2	2	(2)		1	6	7	4	5
3	1	2	2	1	1	2	2	(3)			7	6	5	4
4	1	2	2	2	2	1	1	(4)				1	2	3
5	2	1	2	1	2	1	2	(5)					3	2
6	2	1	2	2	1	2	1	(6)						1
7	2	2	1	1	2	2	1							
8	2	2	1	2	1	1	2							

Three two-way interaction confounding of each factor are found in every column for L8 screening design. For L8 half fraction designs, every interaction column (3, 5, and 6) is confounded by two two-way interactions as shown in Table 7.1.

The classical and interaction matrix DoE approaches were discussed in Sect. 7.1.1. Both approaches use different levels (-1) and $(+1)$ as opposed to (1) and (2) as well as different columns for factor labels. This can lead to different coefficient signs in two-way interactions, based on either the multiplications of signed levels or the logical OR function equivalence of levels 1 and 2, as was shown in Table 4.4.

When using the corresponding analysis software to each approach, it is best to use the actual level names and not the signed or (1) and (2) levels to determine the coefficient sign. In addition, coefficient estimate plots should be analyzed before deciding on factor levels for outcome prediction (larger is better, smaller is better, or target value).

The L8 experiment factor and level patterns and their interactions are given in Table 7.4. The corresponding alphabetical column headings for each DoE are given in Table 7.5. If a DoE team is trained in either methodology, they can implement projects by assigning factors to columns, following the guidelines provided in this chapter. All scenarios of using L8 OA as full factorial, half fraction, partial factorial, and screening designs in classical and interaction matrix methodologies were discussed earlier. Table 7.3 provides a summary of resolution types for two-level arrays including L8, L16, and L32 with added factor assignments for each OA.

The conversion for L16 in Table 7.5 is shown for a half fraction design, based on Table 4.7, where a fifth factor is assigned to column 15 (four-way interaction column). If partial factorial L16 design is used, then different column assignment can be used to accommodate additional factors, as was shown in Table 7.3 and discussed in the next section for L16.

For the L16 Rivet DoE discussed in Chap. 4, differences between the two approaches to DoE were demonstrated. In this case study of seven factors resolution 2^{7-3}_{III}, adding three additional factors to L16 caused the model to become less accurate as a resolution III design. One factor was assigned to the four-way interaction, and two factors were assigned to three-way interactions each. Using classical

Table 7.5 DoE two-level factor conversions

Column Number	Interaction Matrix	L8 Classical Symbol	Interaction Matrix	L16 Confounding Interaction	Classical Symbol
1	A	C	A		D
2	B	B	B		C
3	C =A*B	F = B*C	A*B		P = C*D
4	D	A	C		B
5	E =A*D	E = A*C	A*C		O = B*D
6	F = B*D	D = A*B	B*C		N = B*C
7	G = ABD	G = ABC	ABC	D*E	H = BCD
8			D		A
9			A*D		M = A*D
10			B*D		L = A*C
11			ABD	C*E	G =ACD
12			C*D		K = A*B
13			ACD	B*E	F = ABD
14			BCD	A*E	E = ABC
15			ABCD = E		J = ABCD

DoE approach of assigning two to four additional factors to three-way interactions, shown in Table 7.3, can be a more accurate resolution IV design discussed in the next section.

7.3   Interactions and Confounding for L16 for Reduced Experiments

The previous sections demonstrated different approaches to interactions and confounding, using mostly classical and interaction matrix DoE methodologies. In this and following sections, the interaction matrix method will be presented with case studies used in previous chapters reflected with the tools introduced in this chapter. In addition, the alternative classical approach to these case studies will be discussed to show contrast with the matrix method.

The DoE team can use their professional experience to select proper interaction scenarios by assigning additional factors to columns. Factor assignments allow analysis of desired interactions, by presuming that other confounding two- and three-way interactions are negligible. If reduced interaction analysis shows significance of confounded interactions, additional experiments can be performed to resolve confounding. These principles are applicable to all DoE projects using the full range of orthogonal arrays.

7.3.1   L16 Half Fraction Interaction and Confounding

L16 half fraction design resolution V is shown in Table 7.2. All four main factors are confounded by four-way interactions, and all two-way interactions are confounded by three-way interactions. Table 7.2 can be simplified by assuming all three- and four-way confounding interactions are negligible and can be excluded from the model. The 15-column simplified L16 OA is reduced to 5 main factors (A–E) and their 10 two-way interactions shown under column names in Table 4.7. An interconnection graph for half fraction L16 is shown in Fig. 7.2. The left diagram shows a pentagon shape drawn from five main factors and their ten two-way interactions. All three- and four-way interactions and confounding shown in Table 7.2 are deemed negligible and reduced from the model.

7.3.2   Interactions and Confounding in L16 Rivet DoE Case Study

The diagram to the right of Fig. 7.2 is a modified pentagon plot used in the Rivet DoE discussed in Chap. 4. The rivet team decided to use seven factors and eight interactions into L16 by assigning one factor to the four-way interaction column 15 and two additional factors to three-way interaction columns 13 and 14, which are confounded by interactions B*E and A*E. This was not an optimal scenario since the resolution went from the original half fraction resolution 2^{5-1}_V to 2^{7-3}_{III}. The main factors are 1, 2, 4, 8, 13, 14, and 15, and their interactions are 3, 5, 6, 7, 9, 10, 11, and 12. Some of the two-way interactions are confounded by other three-way interactions as shown in Table 7.2.

The intent of using L16 interaction matrix method was to investigate 7 factors and 8 two-way interactions to fill all 15 columns in L16, according to Table 4.7. The DoE team selected the first four factors (A, B, C, and D) assigned to columns 1, 2, 4, and

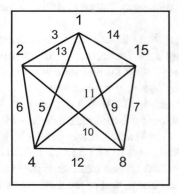
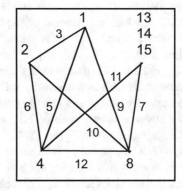

Fig. 7.2 L16 half fraction (pentagon) interconnection diagram and its use in rivet project

8 in the pentagon diagram and then added three additional factors assigned to columns 13, 14, and 15. Factor E was assigned to column 15 confounded by the four-way interaction, and the other two factors assigned to columns 13 and 14 were confounded by two three-way interactions ACD and BCD as shown in Table 4.7.

Six two-way interaction columns (3, 5, 6, 9, 10, and 12) in the original L16 were not assigned to factors. Two columns, 7 and 11, were assigned to interactions with the factor assigned to column 15. The team assumed that three- and four-way interactions are not significant. The unassigned columns provided coefficients for the following eight two-way interactions for the first five factors:

$3 = 1*2 = A*B$
$5 = 1*4 = A*C$
$6 = 2*4 = B*C$
$7 = 8*15 = D*E$
$9 = 1*8 = A*D$
$10 = 2*8 = B*D$
$11 = 4*15 = C*E$
$12 = 4*8 = C*D$

This interaction matrix design is not optimal because of the following:

1. The factors assigned to columns 13 and 14 do not have any identified interactions with the rest of the factors in the interconnection right diagram of Fig. 7.2. The factor assigned to column 15/E has only two interactions, $11 = C*E$ and $7 = D*E$, identified. The missing interactions with factors 13, 14, and 15 confound with the designated two-way interactions in the pentagon left diagram in Fig. 7.2.

   The factor assigned to column 13 is confounded by the two-way interaction $B*E$ ($13 = 2*15$), and the factor assigned to column 14 is confounded by the two-way interaction $A*E$ ($14 = 1*15$) from Table 7.2. This renders the Rivet DoE a resolution III design where main factors are confounded by two-way interactions.

2. All seven main factors are confounded by four-way interactions, and all eight two-way interactions are confounded by three-way interactions as shown by Table 7.2, which are deemed negligible.

3. A better classical DoE plan would assign three additional factors (currently assigned to 13, 14, and 15) to three-way L16 interaction columns such as 14, 7, and 11. This plan would maintain resolution IV and reduce error due to resolution III with main factors confounding two-way interactions.

4. The sequence of assigning three additional factors to L16 full factorial (2^{7-3}_{IV}) in step 3 above is shown in Table 7.3 added factors for L16. This scenario will result in analyzing all seven factors (1, 2, 4, 7, 8 11, 14) and six interactions 8*1, 8*2, 8*4, 8*7, 8*13, and 8*14 with factor 8 level sequence (12 12 12 12 12 12 12 12). The remaining two columns are $5 = 1*4$ interaction and one unassigned column of six three-way interactions. This renders the interaction of the factors like a star

Table 7.6 L16 interaction table

1	2	3	4	5	6	7	8	9	10	11	12	13	14	15
(1)	3	2	5	4	7	6	9	8	11	10	13	12	15	14
	(2)	1	6	7	4	5	10	11	8	9	14	15	12	13
		(3)	7	6	5	4	11	10	9	8	15	14	13	12
			(4)	1	2	3	12	13	14	15	8	9	10	11
				(5)	3	2	13	12	15	14	9	8	11	10
					(6)	1	14	15	12	13	10	11	8	9
						(7)	15	14	13	12	11	10	9	8
							(8)	1	2	3	4	5	6	7
								(9)	3	2	5	4	7	6
									(10)	1	6	7	4	5
										(11)	7	6	5	4
											(12)	1	2	3
												(13)	3	2
													(14)	1

diagram where the factor assigned to column 8 is the apex and it interacts with the other six factors.

5. Each of the six interactions with factor 8 and interaction 5 = 1*4 is confounded with a total of three two-way interactions. For example, two out of six confounding interactions for factor 8 from step 4 are based on interaction Table 7.6:

$$9 = 8*1 + 2*11 + 7*14$$
$$12 = 8*4 + 2*14 + 7*11$$

This 2^{7-3}_{IV} factor assignment is shown in Table 7.7. The DoE team can decide to ignore factor 8 two-way confounding interaction and only calculate the coefficients for factor 8 and its confounding with the other six factors. To resolve the confounding of three two-way interactions, four additional experiments can be conducted in the pattern of an L4 to separate the confounding interactions.

7.3.3   L16 Partial Factorial Design and Confounding

The Rivet DoE analysis in the previous section shows an important element of classical versus matrix interaction approaches. It is desirable to keep L16 additional factor assignments as resolution IV in order to minimize confounding of main factors by two-way interactions. This is accomplished by assigning additional factors to three-way interactions only as shown in Table 7.7 for six, seven, and eight factors in L16 (2^{6-2}_{IV}, 2^{7-3}_{IV}, and 2^{8-4}_{IV}). Their two-way interactions confound with other two-way interactions, and confounding with three-way interactions are ignored as follows:

Table 7.7 L16 additional factor assignments

Factors	6	7	8	9	10
Resolution	IV	IV	IV	III	III
	A/1	A/1	A/1	A + F*J	A + 2 x 2-way
	B/2	B/2	B/2	B + G*J	B + 2 x 2-way
	C/4	C/4	C/4	C + H*J	C + 2 x 2-way
	D/8	D/8	D/8	D + E*J	D + 2 x 2-way
	E/14	E/14	E /7	E /7 + D*J	E/7 + 2 x 2-way
	F /7	F /7	F /11	F /14+ A*J	F /14+ 2 x 2-way
		G /11	G /14	G /13 + B*J	G /13+ 2 x 2-way
			H/13	H /11 + C*J	H/11 + 2 x 2-way
				J/15+4*2-way	J/15 + 4 x 2-way
					K/12 + 4 x 2-way
Confounding Main (2-way)	None	None	None	8 1x2-way 1 4x2-way	8 2x2-way 2 4x2-way
Confounding Two-way	6 - 2x2-way 1 - 3x2-way	7 - 3x2-way	7 - 4x2-way	6 - 4x2-way	5 - 4x2way
Confounding Three-way	2 - 4x3-way	1 - 6x3-way			

1. Six factors with sequential classical symbols A–F assigned to columns in Table 4.7 (1/D, 2/C, 4/B, 7/F, 8/A, and 14/E) are not confounded by two-way interactions. They are each confounded with two three-way interactions which are ignored.

   – Six (2 × 2-way) interactions:
     • 4*8, 2*14 (column 12/A*B, C*E)
     • 2*8, 4*14 (column 10/A*C, B*E)
     • 1*8, 7*14 (column 9/A*D, E*F)
     • 1*2, 4*7 (column 3/C*D, B*F)
     • 1*4, 2*7 (column 5/B*D, C*F)
     • 7*8, 1*14 (column 15/A*F, D*E).

   – One (3 × 2-way) interaction: [8*14, 4*2, 1*7 (column 6/A*E, B*C, D*F)].
   – Two (4 × 3-way) interactions confounding columns 11 and 13.

2. Seven factors with sequential classical symbols A–G assigned to columns in Table 4.7 (columns 1/D, 2/C, 4/B, 7/F, 8/A, 11/G, and 14/E) are not confounded by two-way interactions. They are each confounded with three three-way interactions which are ignored.

   – Seven (3 × 2-way) interactions:
     • 4*8, 2*14, 7*11 (column 12/A*B, C*E, F*G)
     • 2*8, 4*14, 1*11 (column 10/A*C, B*E, D*G)
     • 1*8, 7*14, 2*11 (column 9/A*D, E*F, C*G)

- 8*14, 4*2, 1*7 (column 6/A*E, B*C, D*F)
- 7*8, 1*14, 4*11 (column 15/A*F, D*E, B*G)
- 8*11, 1*2, 4*7 (column 3/A*G, C*D, B*F)
- 1*4, 2*7, 11*14 (column 5/B*D, C*F, E*G)

- One (6 × 3-way) interaction column.

3. Eight factors with sequential classical symbols A–H assigned to columns in Table 4.7 (columns 1/D, 2/C, 4/B, 7/E, 8/A, 11/F, 14/G, and 13/H) are not confounded by two-way interactions. They are each confounded with seven three-way interactions which are ignored.

Seven (4 × 2-way) interactions including 8*1/A*D, 8*2/A*C, 8*4/A*B, 8*7/A*E, 8*11/A*F, 8*14/A*G, and 8*13/A*H, each confounded by three other two-way interactions. For example:

- 4*8, 2*14, 7*11, 1*13 (column 12/A*B, C*G, D*H, E*F)
- 2*8, 4*14, 1*11, 7*13 (column 10/A*C, B*G, E*H, D*F).

All L16 six, seven, and eight added factor assignments in Table 7.3 are resolution IV design. All remaining L16 added factors 9–15 are resolution III designs. For L16 partial factorial design of nine factors (2^{9-5}_{III}), the first eight factors confound with one two-way interaction, and the ninth factor confounds with four two-way interactions in column 15, as shown in Table 7.7.

For L16 partial factorial designs with tenth to the 14th factor ($2^{10-6}_{III} - 2^{14-10}_{III}$), each additional factor and interaction confound with four to six two-way interactions. The total number of factors and interactions is 15 due to the number of columns in L16, with interactions with factor 8/A getting reduced with each additional new factor. L16 with 15 factors (2^{15-11}_{III}) is a screening design with each factor confounded by seven two-way interactions.

7.3.4   L16 Partial Factorial Design and Confounding with DoE Interaction Matrix Method

For interaction matrix design, specific L16 two-way interactions are selected while ignoring other confounding interactions. This approach encourages the use of four-way interaction column 15 as one of the factor assignments, which leads to resolution III designs. The DoE team can use interconnection diagrams to select the desired interactions from the total list of available factor interactions. The team should initially decide on the number of factors, select a two- or three-level OA, and then fill in the remaining columns with the desired interactions to be analyzed. All other confounding interactions are assumed to be negligible and not included in the model. The team can decide later to perform more experiments to resolve confounding.

For five factors, an L16 half fraction (2^{5-1}_{IV}) design can be used, as shown in the left pentagon diagram in Fig. 7.2. Three- and four-way interactions shown in

Fig. 7.3 L16 six-factor
assignments and interactions

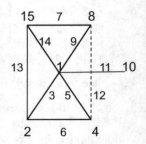

Table 7.2 can be ignored to provide a model of five equally interacting factors with ten two-way interactions. For increasing number of factors 6–8, it is advisable not to use the four-way interaction column as one of the added factor assignments, but to assign each additional factor to a three-way interaction column. This will keep the design more accurate as resolution IV, where main factors are not confounded by two-way interactions, and result in the coefficient analysis of one main factor interaction with other factors, as was shown in the preceding section.

Pre-selected interconnection diagrams of multiple factors for L16 can be used to assign factors to columns and isolate interactions of interest to the exclusion of other confounding interactions deemed not important. An interconnection diagram for six factors in L16 is shown in Fig. 7.3 with an "hourglass" diagram. The hourglass configuration was made by collapsing the pentagon half fraction factorial into a rectangle with four apexes (2, 4, 8, and 15), center (1), and a sixth factor (10). The central factor is considered to be the most important, and its interactions with all other five factors are shown. All two-way interactions of the four corners are available in the model, though missing the diagonal interactions 2*8 (column 10) and 4*15 (column 11) from L16 interaction Table 7.6. The sixth factor is assigned to columns 10 and column 11 is its interaction with factor (1). Other columns can be used as the center and corners of the hourglass diagram with corresponding interactions from Table 7.6. There are many confounding two-way interactions for some factors and interactions for Fig. 7.3 that can be found in Table 7.6, especially since each column is shown six times in the interaction table.

Figure 7.4 offers two configurations of seven factors assigned to L16 columns 1, 2, 4, 8, 10, 12, and 15 or 1, 2, 4, 8, 10, 13, and 15 with one central factor assigned to column 1 or 4 and eight two-way interactions. Six of the eight interactions are from the central factor (column 1 or 4) to the other six factors, and the remaining two interactions are 7 = 8*15 and 6 = 2*4 for the left diagram and 7 = 8*15 and 3 = 1*2 for the right diagram. All interactions are derived from L16 interaction Table 7.6 and are confounded by other replicates of the L16 columns in the table. Alternate seven factor hourglass interconnection diagrams can be used by replacing seven columns and their corresponding interactions from Table 7.6.

Both Figs. 7.3 and 7.4 are resolution III designs since they assign one of the factors to the four-way interaction column. This results in one of the factors being confounded by two-way interactions of other factors, rendering the design less

Fig. 7.4 L16 seven-factor assignments and interactions (hourglass)

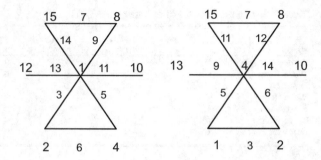

accurate than resolution IV designs, which was discussed earlier in this section and shown in Table 7.7.

7.3.5 Interaction Matrix L32 Case Study: Solder Wave DoE Design

The right hourglass diagram in Fig. 7.4 was used in the solder wave DoE which will be analyzed in Chap. 10. The goal was to minimize solder defects with a complex conveyorized solder wave machine with multiple operational adjustments. The defect history of the solder process was shown in Fig. 1.7. After the introduction of statistical process control (SPC), the soldering process defect was stabilized at 800 ppm. The DoE goal was to achieve world-class six sigma quality level of tens of ppm solder defects which was benchmarked at the company's Japanese facility.

The DoE team examined all possible machine parameters and selected seven controls that can be manipulated with varying difficulty through their operational range. In prior attempts to minimize defects, members of the team attempted to vary each control independently using one factor at a time (OFAT) with little success. The team believed that there were strong interactions between machine controls. The critical indicator for soldering quality of printed circuit boards (PCBs) was the temperature profile at each solder joint as it travelled across the machine. This profile can be manipulated with machine controls such as conveyer speed, preheat temperature, its conveyer angle, and direction of components before passing over the molten solder wave. The solder wave width and height can be adjusted by the solder drive pump rotational speed.

The team decided on a DoE project for fast resolution to this problem, since many machine controls affected the temperature profile individually or through interactions with other controls. For example, speeding the conveyer would provide less heat to the profile, while raising the preheat and solder temperature at the same time would have the opposite effect. The preheat temperature would pre-condition the PCB to the soldering operation before hitting the wave. Levels had to be carefully selected, so that the design space of seven factors does not result in heat damage to

Table 7.8 L16 interaction matrix solder wave DoE

Factor	Col #	Wave Temp	PCB Direction	Wave Height	Preheat Temp	Wave Width	Conv. Speed
Conveyer Angle	1/A	0	0	x	0	0	x
Wave Temp	13/B*E	0	0	0	0	0	x
Direction	10/B*D		0	0	0	0	x
Wave Height	2/B			0	0	0	x
Preheat temp	8/D				x	x
Wave Width	15/E					x
Conveyer Speed	4/C					

the PCBs or their mounted components. After multiple discussions, the following seven machine controls were selected as factors:

1. Conveyer Angle
2. Wave Height
3. Conveyer Speed
4. Preheat Temperature
5. Wave Width
6. PCB Direction
7. Wave Temperature

The DoE team decided to use an L16 partial factorial design for 7 factors and 8 interactions to fill all of the 15 columns. The assignments of factors to L16 columns were based on the perceived importance rank of factors and the difficulty in changing the levels. Conveyer Angle was the most difficult to change, and it was assigned to L16 column 1/A from Table 4.7. The pattern of levels in column 1/A provided the least changes, with eight experiments in level (1) followed by eight experiments in level (2). Alternately, Conveyer Angle could have been assigned to any column and the run order of experiments altered to minimize level changes.

The DoE team went through all possible 21 two-way interactions between the 7 factors in an L16, shown as an interaction matrix in Table 7.8. The matrix is triangular showing one half of the interactions, since they are symmetrical (1*2 = 2*1). The six interactions with speed were considered the most important for its effect on the temperature profile as discussed earlier. Two other geometry-based interactions were also selected since they influence each other through their dimensions: Conveyer Angle*Wave Height and Preheat Temperature*Wave Width.

The interaction matrix entries matched the hourglass interconnection and resulted in the right diagram in Fig. 7.4. In order to fit all selected factors and desired interactions, several confounding two-way interactions were ignored as well as all confounding with three-way interactions as shown in Table 7.9. Two factors, PCB Direction and Wave Temperature, were confounded with one two-way confounding, degrading the model to resolution III design. The desired interactions are confounded with one or two two-way interactions as shown in Table 7.9.

Table 7.9   L16 confounding for solder wave DoE resolution III

Factors	Col. #	Interaction	Confounding
Conveyer Angle	1		
Wave Height	2		
Conveyer Speed	4		
Preheat Temperature	8		
Wave Width	15		
PCB Direction	10	(2*8)	
Wave Temperature	13	(2*15)	
Interactions			
Conv. Angle*Wave Height	3	(1*2)	
Speed*Conveyer Angle	5	(1*4)	(8*13) + (10*15)
Speed*Wave Height	6	(2*4)	(11*13)
Preheat Temp*Wave Width	7	(8*15)	(10*13)
Speed*Wave Temperature	9	(4*13)	(1*8)
Speed*Wave Width	11	(4*15)	(1*10)
Speed*Preheat Temperature	12	(4*8)	(1*13)
Speed*Direction	14	(4*10)	(1*15)

The use of the two diagrams in Fig. 7.4 for design of seven factors and eight interactions was chosen from two opposite perspectives. The figure at the left is a top-down L16 interconnection graph of seven factors, indicating one most important factor interacts with the other six factors, while two other interactions are from other factors and can be used as a pre-selected set of seven factors and eight interactions. The solder wave interconnection diagram at the right is a bottom-up diagram selected by the DoE team after discussing the merits of all possible interactions by using the interaction matrix. Specific interactions were selected while ignoring other confounding interactions and then fitted into the diagram. The two diagrams in Fig. 7.4 can be interchangeable due to the orthogonality of L16 array patterns. Any set of factors and interactions can be used, if it can fit into an L16 and adhering to the interaction Table 7.6.

7.3.6   Resolving Confounded Interactions

A resolution IV design for the solder wave DoE 2^{7-3}_{IV} requires a different set of factor assignments to columns as was shown in Table 7.7. The original four factors would be assigned to columns 1/A, 2/B, 4/C, and 8/D. The additional three factors would be assigned to columns 14/E, 7/F, and 11/G. The resulting interactions and confounding are shown in Table 7.10, when ignoring three-way interactions. None of the assigned seven factors are confounded with two-way interactions. Each of the two-way interactions shown is confounded by two other two-way interactions.

Seven out of eight desired interactions are available for coefficient analysis in the columns shown in Table 7.10.

Two desired interactions, Preheat Temp*Wave Width ($12 = 2*14$) and Speed*Conveyer Angle ($12 = 8*4$), confound each other, as well as another interaction Direction*Wave Temperature ($12 = 7*11$). If the two desired confounded interactions in column are shown to be not significant, then the confounding interactions are negligible and can be ignored. If they are significant, the DoE team can perform additional experiments to separate the three two-way confounding as shown in Table 7.11, since L16 is 1/8 fraction design of the full factorial equivalent 2^7 or 128 experiments.

The additional experiments can be designed by picking two of three factors assigned to three-way interactions and sequencing them (with the most important factor 8) in the same manner as L4. The additional experiments should keep all other factors constant as shown in Table 7.11. The first 16 rows in Table 7.11 are rearranged L16 with 7 assigned factors to columns as indicated in Table 7.10. The three confounding interactions have the same sequence for the first 16 experiments. The last three columns in Table 7.11 indicate the three confounding interactions corresponding to column 12 $(8*4) + (2*14) + (7*11)$. Their coefficients are the same when analyzing the outcomes since they all have the same level patterns.

Additional experiment 17 is the same as the first experiment and is used as the baseline for the L4 following sequence 18–20. The first experiment and the additional three experiments (18–20) can thus be used to separate the confounding and estimate the coefficients for each interaction.

The analysis of the solder wave case study will be presented in Chap. 10, where another factor (Flux Material) is added for a total of eight factors and two types of

Table 7.10 L16 confounding for solder wave DoE resolution IV

Factors	Col. #	Interaction	Confounding
Conveyer Speed	8		
Conveyer Angle	4		
Preheat Temperature	2		
Wave Height	1		
Wave Width	14		
PCB Direction	7		
Wave Temperature	11		
Interactions			
Speed*Conveyer Angle	12	(8*4)	(2*14) + (7*11)
Speed*Wave Height	9	(8*1)	(2*11) + (7*14)
Speed*Wave Temperature	3	(8*11)	(1*2) + (4*7)
Speed*Wave Width	6	(8*14)	(2*4) + (1*7)
Speed*Preheat Temperature	10	(8*2)	(4*14) + (1*11)
Speed*Direction	15	(8*7)	(4*11) + (1*14)
Conv. Angle*Wave Height	5	(4*1)	(2*7) + (11*14)
Preheat Temp*Wave Width	12*	(2*14)	(8*4) + (7*11)

Table 7.11 Additional runs to resolve confounding in L16 seven-factor resolution IV design

Col/Exp.	4	2	1	11	8	7	14	4*8	2*14	7*11
1	1	1	1	1	1	1	1	1	1	1
2	1	1	1	2	2	1	2	2	2	2
3	2	1	1	1	1	2	2	2	2	2
4	2	1	1	2	2	2	1	1	1	1
5	1	2	1	2	1	2	2	1	1	1
6	1	2	1	1	2	2	1	2	2	2
7	2	2	1	2	1	1	1	2	2	2
8	2	2	1	1	2	1	2	1	1	1
9	1	1	2	2	1	2	1	1	1	1
10	1	1	2	1	2	2	2	2	2	2
11	2	1	2	2	1	1	2	2	2	2
12	2	1	2	1	2	1	1	1	1	1
13	1	2	2	1	1	1	2	1	1	1
14	1	2	2	2	2	1	1	2	2	2
15	2	2	2	1	1	2	1	2	2	2
16	2	2	2	2	2	2	2	1	1	1
1(17)	1	1	1	1	1	1	1	1	1	1
18	1	1	1	1	1	2	2	1	2	2
19	1	1	1	1	2	2	1	2	1	2
20	1	1	1	1	2	1	2	2	2	1

defects were analyzed concurrently through a single L32. Though it was implemented as a resolution III design with confounding of two main factors, it was successful in achieving the project goal of six sigma defect rate. The DoE team interaction selections provided guidelines for selecting non-significant factor levels using interaction plots like Fig. 4.5 for Hipot DoE and Fig. 6.3 for Plastics DoE. A resolution IV design for this project would have resulted in more accurate resolution to find the optimal level settings for six sigma quality defects.

7.3.7 L16 Partial Factorial Design and Confounding for Eight or More Factors

A summary of resolution types for two-level OA with added factors was given in Table 7.3. A more detailed L16 added factor assignment was given in Table 7.7, allowing for resolution IV to be maintained while adding up to eight total factors to the L16 half fraction design with five factors. L16 with 15 columns for 8 factors results in 7 unassigned columns to investigate up to 7 interactions. The full factorial equivalent for eight factors is $2^8 = 256$, resulting in L16 use as 1/16 fraction design.

As more factors are added to L16, confounding increases for each added factor (9–15 factors total), reducing the resolution from IV to III.

For nine factors assigned to L16, four main factors (A–D, assigned to columns 1, 2, 4, and 8) are added to five additional factors (E–J) which are assigned to the four three-way interaction columns (7, 11, 13, and 14), and fifth factor J is assigned to one four-way interaction column 15. Each factor is confounded with one two-way interaction, except for the last factor J, which is confounded with four two-way interactions (1*14, 2*13, 4*11, and 7*8). There are six columns with confounding four two-way interactions.

For a total of ten or more factors assigned to L16, additional factors are assigned sequentially by confounding them with additional two-way interactions beginning with column 12 (C*D). Each additional factor greater than 10 is confounded by another two-way interaction until all columns in the L16 are used for factors in screening design. As the number of factors increases, there is less room to estimate interaction coefficients. If such need arises, then more experiments have to be performed to resolve confounding as was shown in Table 7.11 for the solder wave project.

The matrix method approach is to have the DoE team decide on the number of factors, select an OA such as L16, and then assign factors and interaction as they see fit from their professional experience. Factors and interactions can be arranged in geometrical shapes such as triangle, pentagons, hourglass, or star, discussed earlier in the chapter. For eight factors and higher, factors can be clustered into groups that are separated or connected by interactions. For example, eight factors can be grouped into two clusters of five and three factors, connected by one interaction in a windmill shape: five factors with center and four wing tips and three factors forming the base, shown in two interconnecting diagrams in Fig. 7.5. Both are resolution III design, with the assignment of factors to columns depending on avoiding confounding of factors with two-way interactions. Another example would be ten separate factors in five pairs connected by five interactions.

A more appropriate approach to factor clusters in L16 and higher OA is to separate OA into two halves using the most important (significant) factor as the dividing line. The DoE team should be cognizant of the most important factor. The first L8 should have the most important factor at level (1) and up to six other factors at two levels. The second L8 should have the most important factor at level (2) and up to the same six other factors at level (2). The two L8s will form an L16 while preserving the interaction and confounding pattern of each L8.

Another common use of L16 for eight factors is the star shape with one dominant factor and seven other factors interacting with it. Figure 7.6 presents two configurations for eight factors and seven interactions, using column 1 for the most important factor. The seven other factors that are odd in the left diagram are for resolution IV design (2^{8-4}_{IV}), where only two-way interactions are confounded by three other two-way interactions and all factors are not confounded.

The star resolution IV for eight-factor DoE assigned to L16 is shown in Table 7.12. All eight factors are assigned to odd columns (1–15), and all three-way interactions and higher are ignored. There is no confounding of main factors

Fig. 7.5 L16 eight-factor assignments and interactions (windmill)

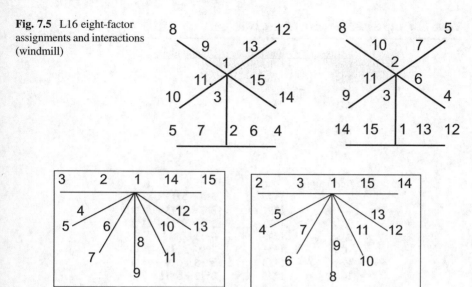

Fig. 7.6 L16 eight-factor assignments and interactions (star)

(1, 3, 5, 7, 9, 11, 13, and 15). For the remaining even columns (2, 4, 6, 8, 10, 12, and 14), they each contain one two-way interaction of factor 1 with another main factor, confounded by three two-way interactions. If one of the interaction columns is shown to be significant in ANOVA, then it can be resolved by three additional experiments in a similar method shown in Table 7.11. Three factors can be repeated using an L4 pattern, and the fourth interaction coefficient can be calculated from the coefficient of the interaction columns.

An example of resolving confounding for eight-factor resolution IV design using three additional experimental runs is given in Table 7.13. Column 2 has four confounding interactions, 1*3 + 5*7 + 9*11 + 13*15, and its coefficient can be estimated in the model analysis. However, the contribution of four possible confounding interactions cannot be resolved by the L16, and three additional runs must be performed. Experiment 17 does not need to be repeated since it is the same level pattern as experiment 1. Three out of four interaction coefficients can be estimated from the four experiments, and the fourth coefficient 9*11 can be calculated from the four confounding interactions as follows:

Coefficients (9*11 = 2 − 1*3 − 5*7 − 13*15)

The right diagram of Fig. 7.6 is a resolution III design (2^{8-4}_{III}), with main factor (1) and seven other even factors. All factors are confounded by three two-way interactions as shown in Table 7.14, and their interactions are not confounded by other two-way interactions. This approach is a resolution III design where the seven other even factors are confounded with three two-way interactions each as shown in Table 7.13. The interactions from the most important factor assigned to column 1 to all other seven factors are not confounded. This is the case when the DoE team is

Table 7.12   L16 eight-factor star resolution IV design

Factors	Interactions	Confounding
1/A		
3/B		
5/C		
7/D		
9/E		
11/F		
13/G		
15/H		
Interactions		
2	1*3	5*7 + 9*11 + 13*15
4	1*5	3*7 + 9*13 + 11*15
6	1*7	3*5 + 9*15 + 11*13
8	1*9	3*11+ 5*13 + 7*15
10	1*11	3*9 + 5*15 + 7*13
12	1*13	3*15 + 5*9 + 7*11
14	1*15	3*13 + 5*11 + 7*9

Table 7.13   Additional runs to resolve confounding in column 2 for eight-factor star resolution IV design

Expt. #								Factor Columns									Interaction Columns (confounding Col 2)			
	1	5	11	9	13	3	7	15	1*3	5*7	13*15	9*11
17/1	1	1	1	1	1	1	1	1	1	1	1	1
18	1	1	1	1	1	2	2	1	2	2	1	1
19	1	1	1	1	1	1	2	2	1	2	2	1
20	1	1	1	1	1	2	1	2	2	1	2	1

confident based on their professional experience in assuming that three confounding interactions for each factor assigned to an even column are negligible.

7.4   Interaction and Confounding for Large OA

DoE projects using large OA such as two-level L32 shown in Table 4.8 and three-level L27 shown in Table 5.2 were discussed in earlier chapters. The APOS case study in Chap. 4 illustrated the use of L32 for analyzing palletizing efficiency for 13 factors and 11 interactions. While interaction diagrams were not given for L32, it was shown that based on factor assignments to L32 columns, each selected

Table 7.14 L16 eight-factor star resolution III design

Factors	Interactions	Confounding
1		
2		4*6 + 8*10 + 12*14
4		2*6 + 8*12 + 10*14
6		2*4 + 8*14 + 10*12
8		2*10 + 4*12 + 6*14
10		2*8 + 4*14 + 6*12
12		2*14 + 4*8 + 6*10
14		2*12 + 4*10 + 6*8
3	1*2	
5	1*4	
7	1*6	
9	1*8	
11	1*10	
13	1*12	
15	1*14	

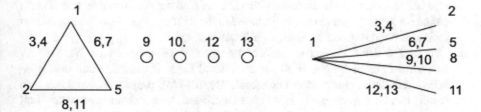

Fig. 7.7 L27 factor assignments and interactions

interaction was confounded by five two-way interactions. If a selected interaction is significant, then additional experiments will be needed to separate the confounding.

The Green Electronics Phase II case study demonstrated the use of L27 for zero-defect goal. Five factors were selected, with three factors assigned to L27 primary columns (1, 2, and 4) and two factors assigned to two out of four three-way interaction columns (9 and 10), thus confounding them. The interconnection diagram for L27 is shown in Fig. 7.7 and is similar to L8 half fraction in Fig. 7.1. The triangle plot on the left in Fig. 7.7 represents three main factors (1, 2, and 5) with three sets of two-way interactions (3,4; 6,7; and 8,11). The four three-way interaction columns (9, 10, 12, and 13) are shown as additional factors, while their interactions with each other and primary factors are missing and considered negligible. The star plot at the right in Fig. 7.7 shows factor 1 interacting with the other four factors (2, 5, 8, and 11), while they are not interacting to each other, and the missing interactions are considered negligible. It is recommended that additional factors assigned to three-way interactions are those considered by DoE teams to be least important and hopefully shown to be non-significant through DoE analysis. In the L27 case study, Profile factor assigned to column 9 was not significant, while Atmosphere factor assigned to column 10 was significant for Nitrogen level. It was later rendered non-significant through the phase II full factorial DoE discussed in Chap. 3, using the multiple range test.

Large OAs are difficult to implement especially in industrial settings where there are time and budget constraints. Many DoE teams elect to make several replicated samples (test vehicles) to ensure that every experiment line has at least one valid result for analyzing a design or create a useful range of defects for process DoE. Large OA projects might use production equipment to create samples for each experiment line, making it difficult to shut down production for a long period. Identifying, processing, and recording data for large numbers of samples increase the probability of error. Several alternatives to using large OA were discussed in this and earlier chapters, including the following:

1. Use an initial screening DoE project to reduce the number of DoE factors such as L8, followed by a full factorial DoE project on the most significant three or four factors. The equivalent approach in classical DoE is to identify the most significant factor, complement its levels, and perform a second screening DoE (second L8) to isolate the factor and its interactions with the remaining factors. This is called foldover technique to be discussed in the next chapter. Additional foldover DoE can be performed for each significant factor.

2. Use half fraction design to reduce the number of experiments by half for two-level OAs or by one third for three-level OAs. If the one additional confounded factor is analyzed as non-significant, the DoE project is completed, and results can be implemented. If it is not significant, then the complementary half fraction DoE can be performed to provide for a full factorial analysis. An example for L32 division into two half fraction L16 is given in Table 7.15, where the fifth factor in L16 is confounded by the four-way interaction. The first 1/2 fraction L16 is created by using the fifth factor confounding four-way interaction level with the same row pattern as the half fraction L16, and the second L16 is created by using the fifth factor confounding the complement (negative) four-way interaction, with the remaining 16 rows in the L32. Table 7.16 shows the division of L16 into two L8 complementary half fractions OA, and Table 7.17 shows L8 division into two L4 half fractions OA.

3. Use partial factorial DoE fitted into smaller OA. Confounding and design resolution issues will have to be addressed by the DoE team similar to the issues discussed in previous sections for L16. Two partial factorial approaches can be used:

   (i) Resolution IV of assigning additional factors to three-way interactions first and then to higher-order interactions last. This was demonstrated in L16 seven- and eight-factor partial factorial designs in previous sections. This approach allows for additional experiments to resolve significant confounding interactions as was shown in Tables 7.11 and 7.13. The

Table 7.15 L32 half fraction division into two L16s

Expt. #	L32 Full Factorial					L16 (1/2 Fraction)					L16 Comp. (½ Fraction)				
	1	2	3	4	5	1	2	3	4	5(+1234)	1	2	3	4	5(-1234)
1	1	1	1	1	1	1	1	1	1	1					
2	1	1	1	1	2						1	1	1	1	2
3	1	1	1	2	1						1	1	1	2	1
4	1	1	1	2	2	1	1	1	2	2					
5	1	1	2	1	1						1	1	2	1	1
6	1	1	2	1	2	1	1	2	1	2					
7	1	1	2	2	1	1	1	2	2	1					
8	1	1	2	2	2						1	1	2	2	2
9	1	2	1	1	1						1	2	1	1	1
10	1	2	1	1	2	1	2	1	1	2					
11	1	2	1	2	1	1	2	1	2	1					
12	1	2	1	2	2						1	2	1	2	2
13	1	2	2	1	1	1	2	2	1	1					
14	1	2	2	1	2						1	2	2	1	2
15	1	2	2	2	1						1	2	2	2	1
16	1	2	2	2	2	1	2	2	2	2					
17	2	1	1	1	1						2	1	1	1	1
18	2	1	1	1	2	2	1	1	1	2					
19	2	1	1	2	1	2	1	1	2	1					
20	2	1	1	2	2						2	1	1	2	2
21	2	1	2	1	1	2	1	2	1	1					
22	2	1	2	1	2						2	1	2	1	2
23	2	1	2	2	1						2	1	2	2	1
24	2	1	2	2	2	2	1	2	2	2					
25	2	2	1	1	1	2	2	1	1	1					
26	2	2	1	1	2						2	2	1	1	2
27	2	2	1	2	1						2	2	1	2	1
28	2	2	1	2	2	2	2	1	2	2					
29	2	2	2	1	1						2	2	2	1	1
30	2	2	2	1	2	2	2	2	1	2					
31	2	2	2	2	1	2	2	2	2	1					
32	2	2	2	2	2						2	2	2	2	2

APOS DoE discussed in Chap. 4 is an example of resolution IV design with many confounding two-way interactions, while the main factors were not confounded.

(ii) Interaction matrix approach where the DoE team selects some interactions as more important than other (confounding) interactions. It is a bottom-up approach where OA selection and interconnection graph are created according to the DoE team professional experience. The choice of assigning factors to columns should be made carefully in order to maintain resolution IV design. It was shown in many DoE that assigning additional factors to four-way interactions in L16 leads to resolution III designs. The rivet and

Table 7.16 L16 half fraction division into two L8s

Expt. #	L16 Full Factorial 1	 2	 3	 4	L8(1/2 Fraction) 1	 2	 3	4 (7 = 123)	L8 Comp. ½ Fraction 1	 2	 3	4 (7=-123)
1	1	1	1	1	1	1	1	1				
2	1	1	1	2					1	1	1	2
3	1	1	2	1					1	1	2	1
4	1	1	2	2	1	1	2	2				
5	1	2	1	1					1	2	1	1
6	1	2	1	2	1	2	1	2				
7	1	2	2	1	1	2	2	1				
8	1	2	2	2					1	2	2	2
9	2	1	1	1					2	1	1	1
10	2	1	1	2	2	1	1	2				
11	2	1	2	1	2	1`	2	1				
12	2	1	2	2					2	1	2	2
13	2	2	1	1	2	2	1	1				
14	2	2	1	2					2	2	1	2
15	2	2	2	1					2	2	2	1
16	2	2	2	2	2	2	2	2				

Table 7.17 L8 half fraction division into two L4

Expt. #	L8 Full Factorial 1	 2	 3	L4(1/2 Fraction) 1	 2	3(=12)	Complementary L4 (½ Fraction) 1	 2	3(= -12)
1	1	1	1	1	1	1			
2	1	1	2				1	1	2
3	1	2	1				1	2	1
4	1	2	2	1	2	2			
5	2	1	1				2	1	1
6	2	1	2	2	1	2			
7	2	2	1	2	2	1			
9	2	2	2				2	2	2

solder wave DoE discussed in Chap. 4 and in this chapter both demonstrated this approach.

4. A modified version of half fraction design is to have the DoE project team identify the most important factor for large OA such as an L32. These factor-two levels can divide L32 into two L16 halves accordingly. Each L16 is performed with one of two levels of the most important factor. The 32 experiments can be analyzed as 2 separate L16s and 1 combined L32. This triple analysis can help in selecting the proper levels to meet the goals of targeting the DoE outcomes.

The advantage of using this strategy is to reduce the number of confounding two-way interactions. Each L16 half fractional experiment can be independently designed and interactions selected based on the interaction matrix methodology. The two levels of the most important factor can be optimized locally where some factors that are not significant in L32 might become significant in one of two L16s analyzed.

In the solder wave project discussed in the previous section, the design team selected eight factors and seven two-way interactions. The process was using a legacy flux material, and a new formulation became available with superior quality performance. The team was interested either in optimizing the current flux or adopting the new one while optimizing it with the same set of factors. This was a unique opportunity to combine two L16s into a single L32 with analysis of the three DoE combinations. Analysis of the solder wave experiments is provided in Chap. 10.

7.5 Conclusions

It has been shown in this chapter how to reduce the number of experiments in a DoE project by using different techniques for classical and interaction matrix methodologies. The use of two-level orthogonal arrays in half fraction, partial fraction, and screening design methodologies was discussed with different confounding resolutions leading to various sequences for completing DoE projects.

As the number of factors selected in the DoE project increases, the complexity of handling interactions and confounding increases at a much higher rate. In addition, many industrial DoE projects are constrained by project schedule and available resources. The project team might have the opportunity to conduct only one or two DoE projects within the scope of product or process lifecycle. In these situations, alternative methodologies for interaction handling and resulting confounding can be simplified by pooling the team experience to either ignore some interactions to reduce confounding or use non-interacting arrays.

A tool for managing confounding called the interaction matrix was shown with an example case study. It is a combination of using professional experience in the choice of confounding interactions. It is presented as an alternative to the interaction tables and interconnection graphs discussed earlier.

Case studies from previous chapters were re-examined in light of the tools discussed in this chapter and shortcomings highlighted for better understanding of the complexity of implementing DoE projects.

Additional Reading Material

Box, G., Bisgaard, S. and Fung C., *An explanation and Critique of Taguchi's Contribution to Quality Improvement*, University of Wisconsin, 1987.

Box G., "Studies in Quality Improvement: Signal to Noise Ratio, Performance Criteria and transformation". *Report No. 26, Center for Quality and Productivity Improvement*, University of Wisconsin, 1987.

Cochran, W. and Cox, G., *Experimental Designs (2nd Edition)*, John Wiley and Sons Inc., New York, 1981.

Diamond W., *Practical Experiment Design*, Van Nostrand Reinhold, New York, 1981.

Ealy, L., "Taguchi Basics," *Quality Journal*, November 1988, pp 26–30.

Guenther W. *Concepts of Statistical Interference*, New York: McGraw Hill, 1973.

Hicks, C., *Fundamental Concepts in the Design of Experiments*, New York: McGraw Hill, 1964.

Holusha, J., "Improving Quality the Japanese Way," *The New York Times*, July 20th, 1988, page D7.

John, P., *Statistical Analysis and Design of Experiments*, New York: The MacMillan Company, 1971

Kackar R. and Shoemaker A., "Robust Design: A Cost Effective Method for Improving Manufacturing Processes," *AT&T Technical Journal*. 1985.

Lipson, C and Sheth N., *Statistical Design and Analysis of Engineering Experiments*, New York: McGraw Hill, 1973.

Ranjit, R., Design of Experiments Using The Taguchi Approach: 16 Steps to Product and Process Improvement, New York: Wiley, 2001

Ross, P., *Taguchi Techniques For Quality Engineering*, New York: McGraw Hill, 1987.

Roy, R., *A Primer on the Taguchi Method*, New York: Van Nostrand Reinhold Co. 1990.

Phadke, M., *Quality Engineering using Robust Design*, Engelwood Cliffs, NJ: Prentice-Hall. 1989.

Shina, S., "Reducing Solder Wave defects Using the Taguchi Method", *American Supplier Institute Sixth Symposium*, Dearborn Michigan, October 1988.

Shina S. and Capulli, K., "Alternatives for Cleaning Hybrid Integrated Circuits Using Taguchi Methods", NEPCON East Conference, June 1990, Boston MA.

Taguchi, G., *Introduction to Quality Engineering*, Tokyo: Asian Productivity institute, 1986

Taguchi, G., *System of Experimental Design*, UNIPUB- Kraus International Publications, 1976.

Chapter 8
Multiple-Level Factorial Design and DoE Sequencing Techniques

Previous chapters discussed DoE design and analysis including setting goals, factor and level selection, error handling, and reducing experiments by using confounding and professional experience. Most of the discussions in Chaps. 4, 5, 6 and 7 focused on these techniques for specific two- or three-level orthogonal arrays (OAs) performed as same level experiments. In this chapter, techniques for multiple-level design as well as DoE sequencing techniques for analysis of a large number of factors will be presented, using two alternative methodologies.

The concepts of interaction and confounding discussed in Chap. 7 are difficult to grasp, leading DoE design teams to use either screening or full factorial designs where interactions are either ignored or fully recognized. An alternative to interactions and confounding is the use of non-interacting OA, and their applications in DoE projects are examined illustrating benefits and shortcomings.

Industrial DoE case studies discussed in this chapter explored techniques mentioned above for quickly solving design problems and new material or process selection. The decisions made by DoE teams will be discussed and results analyzed using methodologies highlighted in this and previous chapters.

Topics to be discussed in this chapter are:

8.1 Multi-level OA Arrangements

In previous DoE discussions, two-level factors were discussed in Chap. 4 and three-level factors in Chap. 5. In some cases, the design team may decide to use multiple-level factors, when surveying alternative materials to be evaluated in new product design or process improvements. The use of multi-level OA arrangements and how to handle interactions and confounding will be explored in this section with industrial case studies as well as decisions made and results obtained.

8.2 DoE Sequencing Techniques

Industrial DoEs have multiple constraints of time and resources to resolve design problems, incorporate new materials or processes, or improve manufacturing outcomes. These constraints could encourage DoE teams to achieve project goals through multiple successive experiments to gain incremental knowledge from each, rather than implementing a single large full factorial experiment.

Several techniques have been successfully used, starting with a screening design to evaluate multiple factors, identifying the most significant ones, and then using partial or full factorial follow-on DoE to evaluate significant factors and their interactions. Another technique, called "folding," includes the use of screening DoE to identify the most significant factor, followed by another screening DoE to measure that factor and its interactions with other factors. This process is repeated as necessary for all remaining significant factors, resulting in less experiments than full factorial design. Either of these techniques can be preceded by reducing some of the levels through the completion of statistical comparison tests or treatments discussed in Chaps. 2 and 3.

8.3 Non-interacting Orthogonal Array Use in DoE

The use of non-interacting OA as screening designs is popular since interactions are distributed uniformly across all columns. These non-interacting arrays, such as L12, L18, and L36, have columns with two or three levels, with each column level orthogonal to every other column. Interactions may exist, but they are not assigned to any columns and do not fully confound any of the other columns in the original arrays. All factors can be analyzed independently to determine DoE outcomes.

Some non-interacting OAs are in multiples of four factors with two levels such as L12, L20, L24, etc. They are called Plackett-Burman (PB) designs. They are intermediate screening arrays between two-level OA such as L8, L16, and L32. They can be followed by full or partial factorial DoE to investigate significant factors and interactions.

Factor and level arrangements of non-interacting arrays will be discussed with case studies. They are used by DoE teams as screening designs to whittle down large numbers of factors and levels for sequential DoE projects. In addition, they are used when DoE teams are wary of analyzing interactions and confounding and decided to ignore them.

8.4 Conclusions

8.1   Multi-level OA Arrangements

Factors with multi-levels can be accommodated in two- or three-level OA by combining columns to create multiple levels. Any (n) columns in a two-level OA can be combined to create up to 2^n levels. Similarly, combing (n) columns in three-level arrays yields 3^n levels. For example, if a four-level factor is desired, then two two-level columns could be combined to form a new column with four levels. If an eight-level factor is desired, then three two-level factors could be combined to form a new column with eight levels. When using multi-level DoE, the design team should consider the following issues and make the appropriate decisions:

Table 8.1 Multi-level arrangements for L8 with four levels

Expt. #	1	2	3	4	5	6	7	#	A	4	5	6	7	
				L8						Modified L8 (4-Levels)				XOR Table
1	1	1	1	1	1	1	1	1	1	1	1	1	1	Columns
2	1	1	1	2	2	2	2	2	1	2	2	2	2	1 2 A
3	1	2	2	1	1	2	2	3	2	1	1	2	2	1 1 1
4	1	2	2	2	2	1	1	4	2	2	2	1	1	1 2 2
5	2	1	2	1	2	1	2	5	3	1	2	1	2	2 1 3
6	2	1	2	2	1	2	1	5	3	2	1	2	1	2 2 4
7	2	2	1	1	2	2	1	7	4	1	2	2	1	
8	2	2	1	2	1	1	2	8	4	2	1	1	2	

1. Multi-level factor degrees of freedom (DOF) should be maintained: An eight-level factor (DOF = 3) can be made up from three two-level columns of DOF = 1 each, not considering interactions.
2. When multi-level factors that are not multiple of two or three levels are desired, then the closest higher multiple of two-level factor can be selected, and some levels could be repeated. For example, if a five-level factor is desired, then an eight-level factor can be created (from three two-level factors), with five levels assigned, and the three most important levels can be repeated within the eight-level column while maintaining orthogonality.
3. Additional level(s) in a multi-level factor can be assigned to combinations of the constituent column levels. For example, the sequence of levels in a four-level column could be set to the following combinations in the two constituent columns: 1 = 11, 2 = 12, 3 = 21, and 4 = 22. For eight levels, the combinations from the three constituent columns are 1 = 111, 2 = 112, 3 = 121, 4 = 122, 5 = 211, 6 = 212, 7 = 221, and 8 = 222.

    Table 8.1 shows the arrangements for a four-level column in an L8. A modified XOR logic can be applied to the two two-level columns to create a four-level column (A) containing factors 1 and 2, as well as their interaction 3 = 1*2. The new column A has four levels repeated twice, while their interactions and orthogonality are maintained with all other factors and levels.
4. Orthogonality must be maintained in multi-level OA. If a four-level factor is selected, the minimum OA to be used including that factor is L8. In the five-level factor OA example mentioned above, L16 should be the minimum OA selected.
5. An interaction and confounding strategy must be addressed based on Chap. 7. For a four-level factor OA using two columns, the DoE team must decide whether to include the two-way interaction of the two columns as part of four-level column. For eight levels implemented in L16, the three columns (1, 2, and 4) and their four interactions (3, 5, 6, and 7) could be included in the eight-level column. Using none, part or all interactions for the eight-level column could be managed by the assignment of multi-level factors to columns.

6. Multi-level statistical analysis can be altered slightly from two- or three-level analysis shown in earlier Chaps. 4 and 5. Factor plots can show all multi-levels, and ANOVA sum of squares can be assigned the proper DOF value based on the interaction and confounding strategy used.

Multi-level OA arrangements are not easily analyzed using commercial software, and the DoE team must consider the issues above in determining proper model multi-level coefficients and ANOVA.

8.1.1   Multi-level Arrangement DoE L8 Case Study: Machining I Pin Fin Heat Sinks

This case study shows the efficacy of using a particular tooling method to produce efficient pin fin heat sinks for electronic components. It demonstrates that even if the experiment layout is faulty in the assignment of factors to columns, DoE can make significant improvements to current design.

Pin fin heat sinks, which are used for heat dissipation of electronic components, are made from extruded bars, and cut crosswise as sections to form heat dissipating pin towers. They are made from Aluminum 6063—TS extrusions and are sized for $0.070 \times 0.070''$ cross section. Prior to DoE, machine shop technology could consistently achieve production pin heights of $0.375''$, occasionally reaching a maximum of $0.5''$. Higher pin towers are desirable since more heat can be dissipated from electronic components.

The difficulty in achieving higher pins is due to the vibrations that occur when saw cutting extruded aluminum. There are several forces acting on the saw blades including the vibrations resulting from the natural frequency of cutting saws and teeth, as well as the vibration amplified by the natural frequency of the pin towers being formed. In addition, there is a bending force due to cutting. When all these forces were studied, including compression and tension of aluminum cutting, the forces begin to diverge at around a pin height of 0.5 inches and 540 lbs./sq. inch force exerted on the pins, which is the natural yield force of aluminum. This mathematical analysis of machining forces matches the results of maximum achievable pin height in production.

A special alignment fixture was developed to reduce the vibrations, and it was decided to use DoE to model a combination of speeds and feeds as well as use the fixture to achieve higher pin towers. An L8 was selected with the goal of reducing the number of bent pin outcomes. Figure 8.1 shows extreme cases of success and failure of machining pin fins for use as heat sinks.

The selection of factors and levels for the Machining I DoE was as follows:

1. The machining pin height at four levels (0.48, 0.6, 0.72, and $0.78''$) which corresponded to the height of the extruded aluminum stock available. This

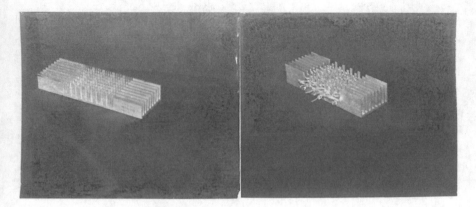

Fig. 8.1 Pin fin Machining I outcomes

Table 8.2 DoE Machining I Pins factor-level assignments and outcomes

Factor Expt. No. Column	Height A&B 1&2	Speed C 3 = A*B	Feed D 4	A*Feed E 5 =A*D	Align. F 6 = B*D	Height*Feed G 7=124	Bent Pins Outcomes					Mean
1	0.48	1	1	1	1	1	4	2	4	1	4	3
2	0.48	1	2	2	2	2	24	23	21	23	22	22.6
3	0.60	2	1	1	2	2	25	14	19	17	26	20.2
4	0.60	2	2	2	1	1	0	0	0	0	2	0.4
5	0.72	2	1	2	1	2	0	0	0	0	0	0
6	0.72	2	2	1	2	1	5	0	0	0	0	1
7	0.78	1	1	2	2	1	27	27	27	27	27	27
8	0.78	1	2	1	1	2	27	27	27	27	27	27

required a combination of two columns to produce four levels. This factor was assigned to primary columns A/1(column 1) and B/2 in L8 array.

2. The cutting saw feed was set at high (2) and low levels (1). Actual values were not recorded due to proprietary information. This was considered an important factor and assigned to column D/4 which is a primary column in the L8 array.

3. The alignment fixture at two levels (yes or no) was assigned to column F, given that all primary factors in the L8 array were already assigned. This column assignment confounds with the interaction of Height (column B) * Feed (column D).

4. The cutting saw speed with high and low levels was assigned to column C, which would confound the interaction of the two primary columns (A, B) combined to form the Height column.

5. The remaining three-way interaction column G (ABD) was assigned to interaction of Height (A&B)* Feed.

The assignment of factors to columns for the DoE Machining I is shown in Table 8.2 with bent pin outcomes. The L8 was repeated five times to obtain

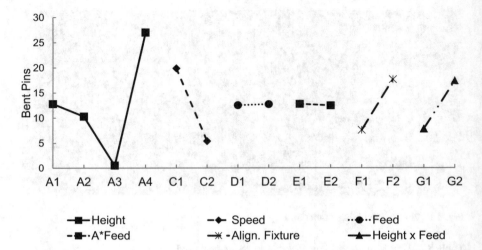

Fig. 8.2 DoE Machining I bent pins mean analysis

variability analysis (to be discussed in Chap. 9) as well as mean analysis. The bent pin mean analysis is shown in Fig. 8.2. Based on Table 8.2 as well as Fig. 8.2, zero bent pin outcomes were recorded for 0.72″ pin height, with low variability. Greater variability was shown on the pin heights of 0.48 and 0.6. The 0.78″ pin height was not feasible to use based on the high number of bent pins.

Full model coefficient analysis is not required for the pin DoE since near-zero defects were achieved in experiment lines 5 and 6 corresponding to Pin Height 0.72, while all other factor levels were assigned to levels 1 or 2 except for Speed level 2 as shown in Table 8.2. The ANOVA of multi-level OA arrangements might be difficult to perform using commercial DoE analysis software.

The L8 arrangement shown in Table 8.2 can be analyzed with columns 1 and 2 labeled Height 1 and Height 2, and their associated statistics can be summed for the four-level Height factor with DOF = 2 and four-level Height sum of the squares equal to sum of the squares for Heights 1 and 2. The DoE Machining I Pins ANOVA is shown in Table 8.3, using the mean of the five repetitions. Two factors, Feed/6 and Height A*Feed/5, were not significant and pooled into the error. Three factors were significant, either by using F-tests or $p\%$, and were ranked as Speed, Height, and Alignment. One interaction, Height* Feed/7, was significant, showing that Feed plays a role in the zero outcomes achieved. The two factors of Speed and Alignment accounted for 56% of the effect of all factors in the experiment, while the Pin Height accounted for 27%. It was a surprise to the design team that Feed, which was considered an important factor, was not significant, though its interaction with Height is significant. This combined choice of speeds and feeds with 0.72″ aluminum stock used in combination with the pin height of 0.72 and alignment fixture was consistently achieved in production.

The original purpose of DoE Machining I was to quantify the contribution of the alignment fixture to reduce vibrations and increase pin fin heat sink tower heights.

Table 8.3 DoE Machining I Pins ANOVA

Source	DOF	Sum SQ	Mean SQ	F value	SS'	p%
Height	2	297.70	148.85	1145	297.44	27
Speed	1	420	420	3231	420	38
Feed	1	0.08	0.08			
Height A*Feed	1	0.18	0.18			
Alignment	1	204	204	1569	204	18
Height* Feed	1	184	184	1415	184	17
Pooled Error	2	0.26	0.13		0.78	0.1
Total	7	1105.96	157.86		1105.96	100

The experiment showed that the alignment fixture was significant in increasing the height of the pin fin towers to 0.72″ with a unique combination of speeds and feeds that were gleaned from DoE outcomes. It can be considered that this combination was arrived at by chance.

Another critique of the experiment is the fact that column E, which is an interaction column, was not assigned in the experiment. It represents the interaction of one of two columns of Height (A)*Feed (D). This partial interaction was already specified by leaving column G not assigned to a factor to evaluate the full interaction of Height (A&B)*Feed. A different factor to column assignment, for example, assigning alignment to column D and feed to column G, might have resulted in a more comprehensive DoE.

8.1.2   Multi-level Arrangement DoE L16 Case Study: Machining II Stencil Forming

DoE is an excellent technique to evaluate new technologies in materials or processes. DoE design teams might be faced with a dilemma of incorporating alternative new technologies into new products or enhancing existing processes. Two approaches can be made: one is to whittle down the alternatives to a more manageable two to four levels by using comparative statistical analysis tools or treatments discussed in Chaps. 2 and 3. The top selected candidates can be incorporated into a screening DoE together with other design factors that are deemed important to determine best model outcomes.

The second approach is to survey available materials or technologies and incorporate them into one multi-level DoE with other design- or process-related factors. The decision to use new technology is made in one single DoE project, with related factors providing information on interactions of alternative levels.

This case study is based on forming stencils for solder deposition onto PCB surfaces in the electronics industry. A stencil is used to deposit solder "bricks" on bare PCB using a squeegee onto the surface through precisely cut apertures in the stencil. It is desirable to have a controllable surface area and height of solder bricks to facilitate subsequent melting and attachment of the components. Several different technologies are available to cut the stencil into geometric apertures to deposit solder bricks. They vary in cost, accuracy, and smoothness of walls.

A new technology was developed for greater dimensioning accuracy and aperture wall smoothness. It consisted of metal deposition like 3-D printing called electroforming. It was decided to use DoE to evaluate this new method as compared to all available machining technologies for stencil cutting.

The DoE team considered using four different types of stencil-making technologies, band etching, chemical etching, laser cutting, and electroforming, offered by multiple vendors. Stencils were provided from two vendors for Band and Laser technologies, three from vendors of Chemical Etch, and one for Electroform. Other factors were considered including those that are particular to the solder screening operations. They include the type of solder paste to be used, stencil snap-off distance, squeegee pressure, orientation angle, and stencil lift-off pressure. The factors and levels are summarized below:

1. Stencil Types with eight levels, using four technologies from alternate suppliers: Band Etch, Chemical Etch, Laser, and Electroform.
2. Solder Paste with two levels: aqueous cleaning (C) and no-clean (NC).
3. Snap-Off Distance between stencil and surface to be printed, with two levels at 0 and 5 mils.
4. Squeegee Pressure which is the downward pressure behind squeegee, with two levels at 35 and 50 lbs.
5. Orientation Angle the deposited brick makes with the squeegee, with two levels at 0 and 90 degrees.
6. Lift-Off Pressure in the hydraulic lines which lifts the stencil off the PCB, with two levels at 60 and 75 psi.
7. Two possible factors that were not selected were resolved in prior DoE including squeegee hardness and speed which were deemed not significant. All apertures were made with the same hard stainless-steel squeegee, which was determined to be most significant earlier when compared to other softer materials.

Factor selections included one factor (Stencil) with eight levels and five factors with two levels, resulting in a total factor DOF of $8 = 3 + 5*1$. This design could be accommodated in L16 array with 15 columns, leaving 7 columns to be assigned to interactions as shown in Table 8.4. The design team decided on factor ranking based on their own professional experience. The most important factors were Stencil Technology, Paste, and Lift-Off Pressure and were assigned sequentially to primary columns of 1, 2, 4, 8, and 15. The remaining three factors were considered less important than the first three factors and confounded with two-way interactions. Four confounding interactions resulted from this arrangement, including the following:

1. Snap-Off column 12 interacting with Stencil columns 4*8 and Orientation*SQG Pressure (6*10).

Table 8.4 DoE Machining II Stencil factor and interaction assignments

Expt. #	Stencil	Paste	Snap-Off	SQG. Pres	Orient-tation	Lift-Off Pressure	Stencil* Paste	Stencil* Lift-Off	Paste* Lift-Off
Col.	2 4 8	1	12	10	6	15	3 5 9	7 11 13	14
1	Band1	C	5	35	90	75	1 = 1 1 1	1 = 1 1 1	1
2	Band2	C	0	30	90	60	2 = 1 1 2	4 = 1 2 2	2
3	Chem1	C	0	35	0	60	3 = 1 2 1	6 = 2 1 2	2
4	Chem2	C	5	30	0	75	4 = 1 2 2	7 = 2 2 1	1
5	Laser 1	C	5	30	0	60	5 = 2 1 1	7 = 2 2 1	2
6	Laser 2	C	0	35	0	75	6 = 2 1 2	6 = 2 1 2	1
7	Electro	C	0	30	90	75	7 = 2 2 1	4 = 1 2 2	1
8	Chem3	C	5	35	90	60	8 = 2 2 2	1 = 1 1 1	2
9	Band1	NC	5	35	90	60	8 = 2 2 2	8 = 2 2 2	2
10	Band2	NC	0	30	90	75	7 = 2 2 1	5 = 2 1 1	1
11	Chem1	NC	0	35	0	75	6 = 2 1 2	3 = 1 2 1	1
12	Chem2	NC	5	30	0	60	5 = 2 1 1	2 = 1 1 2	2
13	Laser1	NC	5	30	0	75	4 = 1 1 2	2 = 1 1 2	1
14	Laser2	NC	0	35	0	60	3 = 1 2 1	3 = 1 2 1	2
15	Electro	NC	0	30	90	60	2 = 1 1 2	5 = 2 1 1	2
16	Chem3	NC	5	35	90	75	1 = 1 1 1	8 = 2 2 2	1

2. Squeegee Pressure column 10 interacting with Stencil columns 2*8 and Orientation*Snap-Off (6*12).
3. Orientation column 6 interacting with Stencil columns 2*4 and SQG Pressure* Snap-Off (10*12).
4. Lift-Off Pressure column 15 = 1*2*4*8 interacting with Paste (1)*Stencil (2*4*8).

The DoE team decided that interactions of important factors with Stencil should be evaluated. Two eight-level interactions were considered and assigned to three columns:

1. Stencil (2*4*8) * Paste (1) assigned to columns 3 = 1*2, 5 = 1*4, and 9 = 1*8.
2. Stencil (2*4*8) * Lift-Off (15 = 1*2*4*8) assigned to columns 7 = 8*15, 11 = 4*15, and 13 = 2*15.

One two-level interaction Paste*Lift-Off Pressure was assigned to the remaining L16 column 14 (1*15), confounding with interactions Stencil 2*Snap-Off (2*12), Stencil 4*SQG Pressure (4*10), and Orientation*Stencil 8 (6*8).

The DoE outcome was measured as the surface area of solder bricks using an automatic optical scanner. Measurements for height and volume were not accurate since the optical system uses the focus from the PCB surface to the solder brick top to determine height. Irregular tops made it inaccurate to measure height consistently, so brick area was used as the outcome. An L16 was selected to have full orthogonality for all eight stencil levels.

Ten solder bricks from each PCB were measured at random. The L16 was replicated three times for 30 brick measurements for each of 16 experiment lines for a total of 480 replications = 10 measures x 3 times x 16 experiment lines.

Table 8.5 DoE Machining II Stencil ANOVA

Source	DOF	Sum of Squares	Mean Square	F-Ratio	p%
Stencil	3	3282033	1094011	730	33.7
Paste	1	4018	4018		
Snap-Off	1	1827638	1827638	1220	17.1
SQG Pressure	1	1574956	1574956	1051	14.7
Stencil x Paste	3	26996	8999		
Orientation	1	1403937	1403937	937	13.1
Paste x Lift-Off	1	1894905	1894905	1265	17.1
Stencil x Lift-Off	3	22041	7347		
Lift Off	1	5235	5234		
Replication Error	464	648707	1398		
Pooled Error	472	706998	1498		6.8
Total	479	10690470	22318		100

Several ANOVA analyses were performed in the Stencil DoE, including those for area, height, and volume of solder bricks. ANOVA for solder brick area was deemed the most representative of the stencil evaluation study and is shown in Table 8.5. Several factors were significant including Stencil, Snap-Off Distance, Squeegee Pressure, Stencil Orientation, and interaction of Paste*Lift-Off. As expected, the electroform stencil performed the best outcome with the highest significance and contribution and was recommended for stencil making.

The DoE team formulated an expression to account for the balance of the electroform stencil higher cost versus its higher-quality performance based on the quality loss function, which will be discussed in Chap. 9. The formula is based on the solder brick area being out of specification causing a potential quality loss. The cost is quantified by the deviation from target (mean area versus aperture specifications) and the variability of the area. A second quality cost is based on the stencil clogging, requiring frequent cleaning.

$$L = \frac{A_1}{\Delta^2}\left[(\mu - t)^2 + \sigma^2\right] + \frac{A_2}{C}$$

where:

L = total quality cost due to defects of brick area out of specifications.
A_1 = $ cost due to defects.
A_2 = $ cost due to stencil clogging.
σ = standard deviation of brick area,
μ = mean of brick area,
t = target solder brick area (aperture area),
Δ = specification of aperture area.
C = frequency of stencil cleaning.

A ranking was given to all stencil technologies based on this formula to account for overall performance both in quality and cost. The electroform stencil performed the best.

The DoE Machining II Stencil study achieved its main objectives, which were to demonstrate the efficacy of the electroform stencil machining versus multiple suppliers of existing technologies. However, the quality improvements due to this technology were not clearly evaluated. The use of the modified quality loss function was confusing when the data was presented to the public. A traditional cost-benefit analysis would be more appropriate.

The L16 DoE size was necessary since two suppliers of each stencil technology were selected. If only one supplier was chosen by pre-selection using comparative tests from Chap. 2, then the project could have reduced to L8 with four levels of Stencil while maintaining orthogonality. Orthogonality is achieved when each factor level in the eight experiments is listed against an equal number of both levels of every other factor.

It was recommended that the team should continue with a second DoE to optimize the electroform stencil material and process conditions, for those customers wishing to switch to this new higher-quality technology.

8.2   DoE Sequencing Techniques

DoE use for new designs or processes, evaluating material and method technologies, as well as problem-solving in industry is constrained by time and resources. Rather than having one extensive DoE, many industrial projects build up knowledge by using sequential DoE. The first DoE is a screening design in two or three levels to select the top 2–4 significant factors. The follow-on DoE can take one of two approaches:

1. Performing a second partial or full factorial DoE, to study the significant factors from the first screening DoE and their interactions in detail. If desired to study curvature, then the second DoE must be three-level.
2. Performing a foldover design for each significant factor from the first DoE to remove confounding of this factor and model its interactions with all other factors. Foldover should continue with more significant factors as needed.

An example of the first approach discussed earlier in this book is the Green Electronics DoE phase I, discussed in Chap. 5. It was a screening L27 to demonstrate the feasibility of green electronics and its ability to meet or exceed the performance of legacy lead-based materials. It was followed by phase II, discussed in Chap. 3, which was a multiple-level full factorial DoE to evaluate cost-effective alternative materials and processes for high-quality and reliability outcomes. Subsequent phases included a multi-level survey DoE, like the Stencil DoE discussed earlier in this chapter, to rank significant factor levels (Paste and Surface Finish) and their interactions. Another sequential DoE case study, in a different industry and dealing with design rather process issues, will be discussed next in Sect. 8.2.3. In this case study, the reverse sequence was used for selected OA: a two-level screening L8 followed by a three-level L9.

8.2.1   Foldover Sequencing Techniques: Folding on One Factor

DoE sequencing techniques described above, where a screening design is analyzed for significance, followed by a full factorial design of the top 2–4 significant factors may have shortcomings. Interactions between non-significant to significant factors can be ignored, and the final model will not be fully accurate. In addition, the design space is kept constant in this approach, and the opportunity to observe outcome curvature response is lost.

The foldover approach to DoE sequencing techniques removes these shortcomings while increasing DoE sequence. The original screening design is augmented by the addition of one screening design for each significant factor, where the factor levels are complemented. The advantage is that coefficients of each significant factor and its interactions with all other factors are determined free of confounding. Their significance can be determined either by t-test for model coefficients or F-test in ANOVA.

An example of using foldover technique for one significant factor such as assigned to column 4 can be demonstrated in Table 8.6 for a first L8 screening design. In this case, it is a resolution 2^{7-3}_{III} design, where each factor assigned to a column is confounded by three two-way interactions as shown in the top part of Table 8.6. Factor 4 is assumed to be the most significant after coefficient analysis and ANOVA.

A second L8 screening DoE is shown in the bottom part of Table 8.6. It is similar to the first screening DoE except that factor 4, designated as (4′), has its levels complemented. The confounding of each factor is similar, except that the folded column 4 results in negative interactions with other factors. The confounding of factor 4 remains the same since it contains interactions from the six other factors, except that the interactions are negative in the second screening DoE.

The coefficients of each screening L8 can be determined, and the two screening L8s are combined into one L16 as shown in Table 8.7. The confounding pattern for each factor in the L16 is determined by the mean of the addition of two screening L8s. Each factor in the L16, except for the factor assigned to column 4, is confounded by two two-way interactions, while the factor assigned to column 4 is not confounded.

The coefficient value of each L16 factor is the mean of two added coefficients of that factor in two screening L8s. Similarly, the mean of two subtracted coefficients of that factor in two screening L8 results in the coefficient of interaction of column 4 with all other factors without confounding. To illustrate this methodology, the

Table 8.6   Foldover on column (4) and confounding for two L8 screening designs

Expt. #	1	2	3	4	5	6	7	Factor confounding
1	1	1	1	1	1	1	1	1 = 2*3, 4*5, 6*7
2	1	1	1	2	2	2	2	2 = 1*3, 4*6, 5*7
3	1	2	2	1	1	2	2	3 = 1*2, 4*7, 5*6
4	1	2	2	2	2	1	1	4 = 1*5, 2*6, 3*7
5	2	1	2	1	2	1	2	5 = 1*4, 2*7, 3*6
6	2	1	2	2	1	2	1	6 = 1*7, 2*4, 3*5
7	2	2	1	1	2	2	1	7 = 1*6, 2*5, 3*4
8	2	2	1	2	1	1	2	

Expt. #	1	2	3	4'	5	6	7	Factor confounding
1	1	1	1	1	1	1	1	1' = 2*3, -4*5, 6*7
2	1	1	1	2	2	2	2	2' = 1*3, -4*6, 5*7
3	1	2	2	1	1	2	2	3' = 1*2, -4*7, 5*6
4	1	2	2	2	2	1	1	4' = -1*5, -2*6, -3*7
5	2	1	2	1	2	1	2	5' = -1*4, 2*7, 3*6
6	2	1	2	2	1	2	1	6' = 1*7, -2*4, 3*5
7	2	2	1	1	2	2	1	7' = 1*6, 2*5, -3*4
8	2	2	1	2	1	1	2	

Table 8.7   Foldover column (4) and confounding for combined L16 screening design

Expt. #	1	2	3	4	5	6	7	Factor Coefficient and Confounding
1	1	1	1	1	1	1	1	Mean (1+1') 1 = 2*3, 6*7
2	1	1	2	1	2	2	2	Mean (2+2') 2 = 1*3, 5*7
3	1	2	1	2	1	2	2	Mean (3+3') 3 = 1*2, 5*6
4	1	2	2	2	2	1	1	Mean (4+4') 4 = Not Confounded
5	2	1	1	2	2	1	2	Mean (5+5') 5 = 2*7, 3*6
6	2	1	2	2	1	2	1	Mean (6+6') 6 = 1*7, 3*5
7	2	2	1	1	2	2	1	Mean (7+7') 7 = 1*6, 2*5
8	2	2	2	1	1	1	2	
9	1	1	1	1	1	2	1	Mean (1-1') 4*5
10	1	1	2	1	2	1	2	Mean (2-2') 4*6
11	1	2	1	2	1	1	2	Mean (3-3') 4*7
12	1	2	2	2	2	2	1	Mean (4-4') 1*5, 2*6, 3*7
13	2	1	1	2	2	2	2	Mean (5-5') 4*1
14	2	1	2	2	1	1	1	Mean (6-6') 4*2
15	2	2	1	1	2	1	1	Mean (7-7') 4*3
16	2	2	2	1	1	2	2	

Table 8.8 Chemical efficiency folding screening DoE example and outcomes

Factor A Expt. # 1	1 Pitch	B 2 Diameter	C = A*B 3 = 1*2 Material	D 4 Speed	E = A*D 5 = 1*4 Angle	F = B*D 6 = 2*4 Length	G = A*B*C 7 = 1*2*4 Temp.	Chemical Efficiency
1	1	1	1	1	1	1	1	60
2	1	1	1	2	2	2	2	54
3	1	2	2	1	1	2	2	55
4	1	2	2	2	2	1	1	41
5	2	1	2	1	2	1	2	52
6	2	1	2	2	1	2	1	44
7	2	2	1	1	2	2	1	60
8	2	2	1	2	1	1	2	52

Factor A Expt. # 2	1 Pitch	B 2 Diameter	C = A*B 3 = 1*2 Material	D 4' Speed	E = A*D 5 = 1*4 Angle	F = B*D 6 = 2*4 Length	G = A*B*C 7 = 1*2*4 Temp.	Chemical Efficiency
1	1	1	1	2	1	1	1	32
2	1	1	1	1	2	2	2	62
3	1	2	2	2	1	2	2	34
4	1	2	2	1	2	1	1	71
5	2	1	2	2	2	1	2	38
6	2	1	2	1	1	2	1	68
7	2	2	1	2	2	2	1	48
8	2	2	1	1	1	1	2	66

factors assigned to columns 1 and 4 and their confounding interactions are analyzed as follows:

First screening L8:	Factor 1 = 2*3, 4*5, 6*7 and factor 4 = 1*5, 2*6, 3*7
Second screening L8:	Factor 1' = 2*3, −4*5, 6*7 and factor 4' = −1*5, −2*6, −3*7
Combined L16	Factor 1 = mean (1 + 1') = (2*3, 4*5, 6*7) + (2*3, −4*5, 6*7) = 2*3, 6*7
	Factor 4 = mean (4 + 4') = (1*5, 2*6, 3*7) + (−1*5, −2*6, −3*7) = 0 = not confounded
	Mean (1 − 1') = (2*3, 4*5, 6*7) − (2*3, −4*5, 6*7) = 4*5
	Mean (4 − 4') = (1*5, 2*6, 3*7) − (−1*5, −2*6, −3*7) = (1*5, 2*6, 3*7)

A theoretical example for a folded factor 4 sequence DoE and its outcomes is given in Table 8.8 for the design of a propeller agitator in a large chemical process tank. There are seven factors including Propeller Pitch, Diameter, Material, Speed, Angle, Length, and Tank Temperature. A summary of coefficient analysis for the first screening L8, the second folded screening L8, and the combined L16 is given in Table 8.9 as well as the interactions of factor Speed assigned to column 4. The coefficient of Speed is not confounded by any two-way interaction.

Table 8.9   Chemical efficiency example of folded factor coefficients

Factor	F-Effect 1st Screening L8	F'-Effect 2nd Folded L8	Effect L16 (Mean(F+F')	(Interaction 4*F) L16 (Mean (F-F')	Confounding
1 (Pitch)	-0.500	5.25	2.375	-2.875 (4*5)	2*3, 6*7
2 (Diameter)	-0.500	4.75	2.125	-2.625 (4*6)	1*3, 5*7
3 (Material)	-8.50	0.75	-3.875	-4.625 (4*7)	1*2, 5*6
4 (Speed)	-9.00	-28.75	-18.875	9.875 (2*6,1*5,3*7)	None
5 (Angle)	-1.00	4.75	1.875	-2.875 (4*1)	2*7, 3*6
6 (Length)	2.00	1.25	1.625	-0.375 (4*2)	1*7, 3*5
7 (Temp.)	2.00	-4.75	-1.375	3.375 (4*3)	1*6, 2*5,

The factor coefficients shown in Table 8.9 are calculated using levels (1) and (2) from the first screening and second folded DoE. The coefficients for all factors assigned to columns can also be calculated from L16 by the subtraction of eight level (1) means from eight level (2) means. This method should result in the same seven factors and seven interaction coefficients as calculated in Table 8.9.

When using DoE analysis software with levels (+) and (−) as opposed to (1) and (2), the coefficient signs for the primary factors and three-way interaction assigned to columns 1, 2, 4, and 7 are the same, while the coefficient signs for two-way interaction are opposite. This is due to the discrepancy in the formula for interaction levels as follows $(2*2) \neq (-* -)$ as shown in Table 4.4.

Folding on one factor assigned to column 4 resulted in removing confounding for that factor coefficient and reducing confounding for the other six factors from three two-way interactions to two. In addition, all interaction coefficients of the factor assigned to column 4 with all other factors are not confounded.

The process of sequential folding can continue for each significant factor in the first screening L8, as determined by either coefficient t-tests or ANOVA F-tests. Table 8.9 indicates that Speed assigned to column 4 had the largest coefficient (−9) in the first screening L8, followed by Material assigned to column 3 (−8.5). Speed and Material were determined to be the only significant factors in the first screening L8. A third screening L8, folded on column 3, can be analyzed to remove confounding from Material coefficient and its interactions with all other factors.

The agitator design DoE example provided a demonstration of using folding technique as an alternative to full factorial DoE. The number of experiments for seven factors was reduced from L128 = 2^7 for full factorial design to 24 = 3*L8, given that there were only two significant factors. All coefficients of the two factors and their interactions with all other factors were determined without confounding.

8.2.2 Foldover Sequencing Techniques: Folding on All Factors

A special consideration of using folding in DoE sequencing is folding on all factors for the second screening DoE. When the two screening L8s are combined into an L16, the resolution increases to IV, and all seven factors are not confounded. Their two-way interactions are confounded by two other two-way interactions.

If an eighth factor is added to the first screening L8, with all experiments having level 1 of that factor, and all eight factors are folded in a second screening design, the resulting L16 will have eight factors with no confounding as shown in the left-hand diagram 2^{8-4}_{IV} in Fig. 7.6. Factors 1–8 are not confounded, and every two-way interaction is confounded with four two-way interactions, as shown below and in Table 7.12:

$$2 = 1*3, 5*7, 9*11, 13*15$$
$$4 = 1*5, 3*7, 9*13, 11*15$$
$$8 = 1*9, 3*11, 5*13, 7*15.$$

The confounding can be resolved with three additional experiments as was shown in Chap. 7.

8.2.3 DoE Sequencing Technique Case Study: Printer Design DoE

A thermal printer was redesigned for a new application of printing glossy letters on customer photos and cards for photo departments in a consumer store chain. During late design and early production phase, the printer design team discovered inconsistent printing. The design team was made up of many different disciplines, including mechanical, electrical, and software engineering. Their problem-solving efforts included the use of multiple one-factor-at-a-time (OFAT) experiments with no conclusive results. It was decided to attempt DoE techniques to resolve the printing consistency problems to enable shipping and delivery before the busy holiday season.

The Printer DoE team first task was to identify defective outcomes. A list of different defects and their definitions was generated. Some defects were opposites of each other: voids meant lack of printing material on the cards, while fills meant excess printing material. Each of the three major defect categories was divided into finer definitions as shown in Table 8.10. The third defect category was the adhesion of print pigment to the card substrate. It was decided to test for this defect by printing dots and then applying scotch tape on printed dots and pulling scotch tape off. The number of dots pulled off by the scotch tape would determine the number of defects.

It was decided to print 100 cards per experiment line to make sure that a reasonable range of defects would be generated, since the printing was intermittent and required large opportunities for defects. In addition, a printer test pattern was

Table 8.10 Printer DoE defect definitions

I. Voids (lack of printed materials)	
1. Transition lines (parallel to print line)	One per card
2. Non-transition lines (90′ to print line)	One per card
3. Perpendicular line missing	One per card
4. Edge voids (mostly leading edge)	One per card
5. Fine detail missing (lines and dots)	Circle, cross, lines, dots
6. Voids (others)	Frame and corner
II. Fills (excess printed materials)	
1. Bleeding fills (excess material next to lines or shapes)	
2. Bridging fills (excess material between lines or shapes)	
III. Adhesion (pigment did not stick to substrate)	
1. Number of dots removed per scotch tape pull	One per card

Fig. 8.3 Printer DoE test pattern

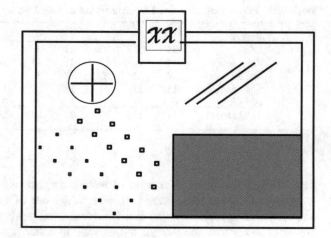

designed to generate the most defects possible for the categories outlined in Table 8.10. The pattern contained lines and graphics at the extremes of the printing specifications and was shown in Fig. 8.3.

The test pattern contained a feature on the top left side of a cross with a circle around it with minimum clearances specified for printing. Other features that tested printing limits included the solid quadrant on the lower right side and series of smallest dots and smallest filled squares with a blank hole in the middle. In the right upper corner, three oblique lines were to be printed at the minimum possible spacing and two digits on top indicating the year.

It was decided to follow a two-step DoE analysis: the first DoE is a screening design, followed by a full factorial design of the most significant factors resulting from the first DoE. The design team selected L8 for the screening design with seven different factors at two levels while keeping other factors constant. Factors not selected in screening DoE included molded paper path tooling and printing media, which were already approved by the customer and would be difficult to change given

Table 8.11 Printer DoE screening L8 DoE

Expt. #	Head Align	Print Energy	Roller Hardness	Foil Tension	Dot Compress.	Head Temp.	Head Force
Col. #	1/A	2/B	3/C	4/D	5/E	6/F	7/G
1	N	0.97N	45 cP	N	Off	35°C	N
2	N	0.97N	45 cP	O	On	45°C	N-4
3	N	1.03N	60 cP	N	Off	45°C	N-4
4	N	1.03N	60 cP	O	On	35°C	N
5	20	0.97N	60 cP	N	On	35°C	N-4
6	20	0.97N	60 cP	O	Off	45°C	N
7	20	1.03N	45 cP	N	On	45°C	N
8	20	1.03N	45 cP	O	Off	35°C	N-4

Table 8.12 Printer DoE screening L8 outcomes versus defect types

Expt. #	1	2	3	4	5	6	7	8
Lines	33	33	33	21	0	7	1	2
Fine Details	44	44	44	41	13	28	5	34
Voids	22	22	22	22	1	4	0	9
Squares Filled	11	11	11	10	3	11	8	11
Bleeding	0	0	0	3	7	4	6	0
Total Defects	110	110	110	97	24	54	20	56

tight delivery schedules. The selected factors are shown in Table 8.11 for printer screening L8. The team selected two levels where one of the levels is the current or normal design specification (N) as in head force, foil tension, and head alignment or higher or lower than the current setting such as roller hardness, print energy, and head temperature. The dot compression software was either turned on or off.

The selection of the factors to columns indicated the preference of the design team to the importance of each factor. Head Alignment, Print Energy, and Foil Tension were considered the top 3 factors and subsequently assigned to columns 1, 2, and 4. Head Force was considered the next important factor after the first three and was assigned to column 7, confounding the three-way interaction (1*2*4). The remaining factors, Roller Hardness, Dot Compression, and Head Temperature, were considered less important and were assigned to two-way interaction columns 3, 5, and 6, respectively, and are confounded by two-way interactions of the primary factors. This resolution III design confounding is like the top L8 in Table 8.6 with each factor confounded by three two-way interactions.

Several possible factors were excluded from the L8, including pre-travel, foil suppliers, guided foil peel, and print speed. These factors required tooling changes or customer approval which is too time-consuming given the project schedule.

The printer L8 screening DoE outcomes are shown in Table 8.12 for five types of defects. Experiment line numbers 5 and 7 produced the lowest defects, while

Table 8.13 Printer DoE first screening L8 ANOVA summary

Defects	Head Align	Print Energy	Roller Hardness	Foil Tension	Dot Compress	Head Temp.	Head Force
Lines	L 2	x	x	x	L2	L1	x
	20	x	x	x	On	35	x
	91%	x	x	x	3%	2%	x
Fine	L 2	x	x	L1	L2	x	x
Details	20	x	x	N	On	x	x
	66%	x	x	13%	17%	x	x
Voids	L 2	x	x	L1	L2	x	x
	20	x	x	N	On	x	x
	93%	x	x	2%	2%	x	x
Large	L 2	x	L2	L1	L2	L1	x
Squares	20	x	60	N	On	35	x
	19%	x	5%	19%	28%	5%	x
Filled	L 1	x	L1	L2	L1	x	L2
Above	0	x	45	O	Off	x	N-4
Quad	40%	x	12%	7%	30%	x	7%
Recommend	L2	100%	L1	L1	L2	L1	L2
Level	20		45	N	On	35	N-4

experiments 1–3 produced the largest defects, with a mean of 73 defects for all 8 experiment outcomes. These results showed that defects can be significantly manipulated by selected factors and levels. The printer test pattern and the selection of 100 cards per experiment line proved to be effective in producing varying defects.

It was decided to analyze each defect type independently by selecting significant factors and recommended levels for minimum defects from model coefficient plots. The screening L8 outcomes versus defect types are shown in Table 8.13. Some of the defect categories in Table 8.10 were combined for ease of outcome interpretation for analysis. For each defect type, factors are marked with an "x" if they are not significant. If a factor is significant, the recommended level (the level with least defect) and its value are shown as well as the p%. Different levels were recommended, and their contribution varied as well based on defect types.

Table 8.13 indicates that there is no universal set of factor levels that can reduce all defect types. A compromise must be made between competing level recommendations for each defect type. The DoE team analyzed the ANOVA summary in Table 8.13 and recommended universal factor levels based on the highest contribution or the preponderance of certain levels. For example, level 2 of Head Alignment was universally recommended based on its significance in four defect types, two of which had contribution greater than 90%, while level 1 was recommended for one defect type with 40% contribution.

Table 8.14   Printer DoE second screening L9 DoE

Set following factors: Roller = 45 cP, Foil tension = N (0.25 lbs) and Dot Compression = On

A = Head Alignment		6	16	26 Degrees.			
B = Print Energy		165	175	185 mw			
C = Head Temperature		35	45	55 °C			
D = Head Force		-4	-6	-8 lb.			

Exp #	A	B	C	D	A	B	C	D
1	1	1	1	1	6	165	35	N-4
2	1	2	2	2	6	175	45	N-6
3	1	3	3	3	6	185	55	N-8
4	2	1	2	3	16	165	45	N-8
5	2	2	3	1	16	175	55	N-4
6	2	3	1	2	16	185	35	N-6
7	3	1	3	2	26	165	55	N-6
8	3	2	1	3	26	175	35	N-8
9	3	3	2	1	26	185	45	N-4

Head Alignment showed significance in every defect type. Print Energy, though only varied by 3%, did not show any significance. Head Force showed significance in only one defect type. Head Temperature showed significance only at 35 °C. Dot Compression (on) showed that it was significant for four defect types. Roller Hardness and Foil Tension showed significance but with smaller percentage contribution.

Three factors were selected at certain levels and were excluded from the model: Roller Hardness at 45 cP, Foil Tension at normal value, and Dot Compression software on. The remaining four factors were further investigated into a second DoE. Rather than choosing a full factorial OA, the team decided on a screening L9 DoE with four factors at three levels to expand model design space and examine any curvature of factor outcomes. The middle level of each factor was set to the current design, with appropriate deviation for the top and bottom levels.

The second L9 DoE factor and level settings are shown in Table 8.14. Three factors, Head Alignment, Print Energy, and Head Force were selected again with expanded levels. The design team felt that Head Temperature should still be considered for the second DoE for the range of 35–55 °C, even if the first DoE indicated that it was significant at 35 °C. The second DoE confirmed that Head Temperature is recommended for 35 °C as well.

The second DoE outcomes, shown in Table 8.15, did not show consistency in the levels that produced lowest defects. The only consistent level that yielded good results for experiments 1, 5 and 9 was Head Force at N-4, which was originally selected based on the L8 screening experiment. Some factor levels would decrease defect types, as in Bleeding in experiments 1–3, while the other levels would decrease a different set of defects as in Voids in experiments 7–9, necessitating a compromise for recommended universal factor levels.

Table 8.15 Printer DoE second screening L9 outcomes

Expt. #	1	2	3	4	5	6	7	8	9
Lines	0	0	0	10	3	6	1	0	8
Fine Details	1	40	39	8	0	9	0	9	0
Voids	9	20	20	0	4	0	0	0	0
Squares Filled	3	0	0	10	2	10	10	10	9
Bleeding	0	0	0	2	1	1	7	1	8
Total Defects	13	60	59	30	10	26	18	20	25

Table 8.16 Printer DoE second screening L9 ANOVA summary

Defects	Head Align	Print Energy	Roller Hardness	Foil Tension	Dot Compress	Head Temp.	Head Force
Lines	L1	L 2				L3	x
	6	175				55	x
	47%	15%				29%	x
Fine	L3	L1				x	L1
Details	26	165				x	-4
	41%	10%				x	21%
Voids	L 2	L1				L1	x
	16	165				35	x
	75%	12%				7%	x
Large	L1	L2				L3	x
Squares	6	175				55	x
Filled	66%	7%				6%	x
Bleeding	L 1	L2				L1	L1
	6	165				35	-4
	52%	5%				5%	50%
Recommend	L2	L1	L1	L1	L2	L1	L1
Levels	16	165 mw	45 cP	N	On	35 °C	-4 lb.

The L9 ANOVA summary, shown in Table 8.16, indicates that the mid-level Head Alignment is best, with the lowest Print Energy and Print Head Temperature and a minimum reduction in Head Force (N-4). The recommended factor and levels were chosen with the same criteria as the L8 screening experiment, with preponderance of highest contribution factor levels.

The recommended factors and levels at the bottom of Table 8.16 were confirmed by printing 100 cards with the test pattern. Only seven defects were obtained in the hundred cards, which was the lowest of any experiment line performed beforehand. All seven defects were the same, which were square dots with the smallest circle voids filled inside. This is a tenfold reduction from the first screening experiment average of 72 defects. The DoE team felt that this defect type would rarely happen in typical customer applications. The recommended settings at the bottom of Table 8.16 were used as final printer specifications. Printers were distributed to photo departments in consumer store chains and successfully used in printing salutations on photos until they were obsoleted by direct photo printing technology.

The use of a two-step DoE sequence of a screening followed by partial or full factorial design in this case study is an efficient method for solving design problems. Though a perfect optimum selection of factors and levels to generate zero defects was not achieved, the DoE sequence method provided for a balanced resolution to the printer design problem. Increasing the design space in the second L9 resulted in reducing print energy which contributed to resolving this problem.

8.3   Non-interacting Orthogonal Array Use in DoE

The previous section on DoE sequencing techniques outlined two methodologies starting with a screening DoE, followed either by full factorial DoE using the top 2–4 significant factors or folding sequentially on each significant factor. A case study and an example were given for each methodology using L8 as the first screening DoE with seven factors. The next two-level OA that can be used for screening is L16 with 15 factors. For three-level OA, the initial screening DoE is L9 with four factors, and the next OA that can be used for screening is L27 with 13 factors.

Three popular non-interacting OAs such as L12, L18, and L36 offer alternatives to higher-order OA used for screening, with two-level factors for L12 and a mix of two- and three-level factors for L18 and L36. Interactions in these arrays are not assigned to any columns and do not fully confound any of the other columns in the original arrays. All main factors can be analyzed independently to determine DoE outcomes. Non-interacting arrays are also used by DoE teams that find the topics of interactions and confounding discussed in Chap. 7 difficult to implement or decided that interactions are not significant.

Three non-interacting arrays are commonly used in industry:

1. L12, shown in Table 8.17, contains 11 columns, each with two levels (2^{11}).
2. L18, shown in Table 8.18, is a mixed level array with one column of two levels and seven columns of three levels and can be labeled as $2^1 \times 3^7$.
3. L36, shown in Table 8.19, has two implementations, both with varying number of columns of two and three levels. The L36 ($2^{11} \times 3^{12}$) as shown in Table 8.19 contains 2 levels in the first 11 columns and 3 levels in the next 12 columns. The second L36 implementation is 16 columns of mixed 2 and 3 levels $2^3 \times 3^{13}$. All three non-interacting arrays are orthogonal, with each factor level matched against all other levels of adjacent columns.

Table 8.17 Non-interacting L12 (2^{11}) OA

Expt #	1	2	3	4	5	6	7	8	9	10	11
1	1	1	1	1	1	1	1	1	1	1	1
2	1	1	1	1	1	2	2	2	2	2	2
3	1	1	2	2	2	1	1	1	2	2	2
4	1	2	1	2	2	1	2	2	1	1	2
5	1	2	2	1	2	2	1	2	1	2	1
6	1	2	2	2	1	2	2	1	2	1	1
7	2	1	2	2	1	1	2	2	1	2	1
8	2	1	2	1	2	2	2	1	1	1	2
9	2	1	1	2	2	2	1	2	2	1	1
10	2	2	2	1	1	1	1	2	2	1	2
11	2	2	1	2	1	2	1	1	1	2	2
12	2	2	1	1	2	1	2	1	2	2	1

8.3.1 Non-interacting OA Case Study I: L18 Painting DoE

This case study is a DoE to improve painting quality on flexible plastic. The painting process is the last stage in the construction of animated robots for theme parks. Quality defects are caused by poor adhesion and repetitive robot motions. Urethane is the most common use of the plastic material, and its chemical characteristics cause it to be painted with silicone-based paint. The painting is performed by skilled artists, with no uniformity in the application of paint or processing parameters. It was decided to conduct a non-interacting DoE with the goal of adapting a universal set of painting mix design and processing parameters to develop a low defect process. In addition, the effect of each factor on the painting outcomes can be evaluated with the use of three levels to investigate any curvature due to factor levels.

An L18 was selected, and the outcome was identified as the number of defects on the surface of 3 × 3 centimeters square plastic panel, repeated three times for variability analysis, to be discussed in the next chapter. Defects were identified as pigment accumulation, air bubbles, and loss of paint smoothness leading to nonuniform surfaces. These defects negatively impact the artistic quality of the product. The DoE goal is to minimize these defects and reduce rework of painted robots before shipping while providing a universal set of painting process factor and level guidelines.

The painting process is divided into five major steps: plastic base cleaning, paint preparation, primer coating, painting, and drying. After cleaning the part, the silicone-based paint requires the application of a primer coat, letting it dry followed

Table 8.18 Non-interacting L18 OA ($2^1 \times 3^7$)

Expt #	1	2	3	4	5	6	7	8
1	1	1	1	1	1	1	1	1
2	1	1	2	2	2	2	2	2
3	1	1	3	3	3	3	3	3
4	1	2	1	1	2	2	3	3
5	1	2	2	2	3	3	1	1
6	1	2	3	3	1	1	2	2
7	1	3	1	2	1	3	2	3
8	1	3	2	3	2	1	3	1
9	1	3	3	1	3	2	1	2
10	2	1	1	3	3	2	2	1
11	2	1	2	1	1	3	3	2
12	2	1	3	2	2	1	1	3
13	2	2	1	2	3	1	3	2
14	2	2	2	3	1	2	1	3
15	2	2	3	1	2	3	2	1
16	2	3	1	3	2	3	1	2
17	2	3	2	1	3	1	2	3
18	2	3	3	2	1	2	3	1

by the application of the final paint and its drying method and time. Paint is made from three materials in two steps: first by mixing the silicone base and color pigment called the base paint and then by mixing the base paint with naphtha to thin the paint. Naphtha is more volatile than other paint thinners such as turpentine or mineral spirits, so there would be no residual solvents to outgas and cause defects. Painters use different means of spraying paint through an air pressure gun with two finish coats each, applied horizontally and then vertically to achieve the desired coloration 50 cm away from the surface. Factor levels were chosen as CentrePoint and functional limits of each factor and listed in Table 8.20. The painting L18 DoE and defect outcomes are shown in Table 8.21. For each experiment line, the three defect outcomes were transformed into one variability measure in the last column for further analysis which will be discussed in Chap. 9.

The model mean coefficient plots for the painting L18 DoE is shown in Fig. 8.4, and ANOVA is shown in Table 8.22. Since most of the factors were not significant, it was decided to select factor levels based on factor ranking, as shown in Table 8.23. They are:

Table 8.19 Non-interacting L36 OA ($2^{11} \times 3^{12}$)

#	1	2	3	4	5	6	7	8	9	10	11	12	13	14	15	16	17	18	19	20	21	22	23
1	1	1	1	1	1	1	1	1	1	1	1	1	1	1	1	1	1	1	1	1	1	1	1
2	1	1	1	1	1	1	1	1	1	1	1	2	2	2	2	2	2	2	2	2	2	2	2
3	1	1	1	1	1	1	1	1	1	1	1	3	3	3	3	3	3	3	3	3	3	3	3
4	1	1	1	1	1	2	2	2	2	2	2	1	1	1	1	2	2	2	2	3	3	3	3
5	1	1	1	1	1	2	2	2	2	2	2	2	2	2	2	3	3	3	3	1	1	1	1
6	1	1	1	1	1	2	2	2	2	2	2	3	3	3	3	1	1	1	1	2	2	2	2
7	1	1	2	2	2	1	1	1	2	2	2	1	1	2	3	1	2	3	3	1	2	2	3
8	1	1	2	2	2	1	1	1	2	2	2	2	2	3	1	2	3	1	1	2	3	3	1
9	1	1	2	2	2	1	1	1	2	2	2	3	3	1	2	3	1	2	2	3	1	1	2
10	1	2	1	2	2	1	2	2	1	1	2	1	1	3	2	1	3	2	3	2	1	3	2
11	1	2	1	2	2	1	2	2	1	1	2	2	3	1	3	2	1	3	1	3	2	1	3
12	1	2	1	2	2	1	2	2	1	1	2	3	1	2	1	3	2	1	2	1	3	2	1
13	1	2	2	1	2	2	1	2	1	2	1	1	2	3	1	3	2	1	3	3	2	1	2
14	1	2	2	1	2	2	1	2	1	2	1	2	3	1	2	1	3	2	1	1	3	2	3
15	1	2	2	1	2	2	1	2	1	2	1	3	1	2	3	2	1	3	2	2	1	3	1
16	1	2	2	2	1	2	2	1	2	1	1	1	2	3	3	1	1	3	2	3	3	2	1
17	1	2	2	2	1	2	2	1	2	1	1	2	3	1	1	2	2	1	3	1	1	3	2
18	1	2	2	2	1	2	2	1	2	1	1	3	1	2	2	3	3	2	1	2	2	1	3
19	2	1	1	1	1	1	1	1	1	1	1	1	1	1	1	1	1	1	1	1	1	1	1
20	2	1	1	1	1	1	1	1	1	1	1	2	2	2	2	2	2	2	2	2	2	2	2
21	2	1	1	1	1	1	1	1	1	1	1	3	3	3	3	3	3	3	3	3	3	3	3
22	2	1	1	1	1	2	2	2	2	2	2	1	1	1	1	2	2	2	2	3	3	3	3
23	2	1	1	1	1	2	2	2	2	2	2	2	2	2	2	3	3	3	3	1	1	1	1
24	2	1	1	1	1	2	2	2	2	2	2	3	3	3	3	1	1	1	1	2	2	2	2
25	2	1	2	2	2	1	1	1	2	2	2	1	1	2	3	1	2	3	3	1	2	2	3
26	2	1	2	2	2	1	1	1	2	2	2	2	2	3	1	2	3	1	1	2	3	3	1
27	2	1	2	2	2	1	1	1	2	2	2	3	3	1	2	3	1	2	2	3	1	1	2
28	2	2	1	2	2	1	2	2	1	1	2	1	1	3	2	1	3	2	3	2	1	3	2
29	2	2	1	2	2	1	2	2	1	1	2	2	3	1	3	2	1	3	1	3	2	1	3
30	2	2	1	2	2	1	2	2	1	1	2	3	1	2	1	3	2	1	2	1	3	2	1
31	2	2	2	1	2	2	1	2	1	2	1	1	2	3	1	3	2	1	3	3	2	1	2
32	2	2	2	1	2	2	1	2	1	2	1	2	3	1	2	1	3	2	1	1	3	2	3
33	2	2	2	1	2	2	1	2	1	2	1	3	1	2	3	2	1	3	2	2	1	3	1
34	2	2	2	2	1	2	2	1	2	1	1	1	2	3	3	1	1	3	2	3	3	2	1
35	2	2	2	2	1	2	2	1	2	1	1	2	3	1	1	2	2	1	3	1	1	3	2
36	2	2	2	2	1	2	2	1	2	1	1	3	1	2	2	3	3	2	1	2	2	1	3

1. Ranking based on the factor effects. This is the span from the maximum level mean to the minimum level mean. It is not recommended to use this ranking since it does not account for the outcome curvature.
2. Ranking based on the F-value of each factor using ANOVA. This is an indication of the significance of the factor.

Table 8.20 Painting L18 DoE factors and levels

Control Factors	Levels
A. Primer Coat	Mist, Light
B. Mix Base	80:1, 80:3, 80:5 (Grams)
C. Mix Naphtha	1:1, 1:1.5, 1:2
D. Cleaning Chemical	MEK, Alcohol, Toluene
E. Pressure	45, 60, 80 psi.
F. Primer Dry Time	1, 2, 3 (Hours)
G. Paint Dry Time	10, 35, 60 (Minutes)
H. Drying Method	Heat, Fan, Air

Table 8.21 Painting L18 DoE and outcomes

Expt. #	Primer Coat	Mix Base	Mix Naphtha	Clean Chemical	Spray Press.	Primer Dry	Paint Dry	Dry Method	Defects			Variability
1	Mist	80:1	1:1	MEK	45	1	10	Heat	108	137	154	-42.56
2	Mist	80:1	1:1.5	Alcohol	60	2	35	Fan	160	246	236	-46.75
3	Mist	80:1	1:2	Toluene	80	3	60	Air	139	127	115	-42.10
4	Mist	80:3	1:1	MEK	60	2	60	Air	30	45	49	-32.49
5	Mist	80:3	1:5	Alcohol	80	3	10	Heat	314	280	366	-50.16
6	Mist	80:3	1:2	Toluene	45	1	35	Fan	41	30	38	-31.28
7	Mist	80:5	1:1	Alcohol	45	3	35	Air	64	45	35	-33.89
8	Mist	80:5	1:1.5	Toluene	60	1	60	Heat	54	70	73	-36.42
9	Mist	80:5	1:2	MEK	80	2	10	Fan	90	171	113	-42.23
10	Light	80:1	1:1	Toluene	80	2	35	Heat	383	380	243	-50.67
11	Light	80:1	1:1.5	MEK	45	3	60	Fan	93	103	118	-40.44
12	Light	80:1	1:2	Alcohol	60	1	10	Air	8	14	11	-21.04
13	Light	80:3	1:1	Alcohol	80	1	60	Fan	38	55	30	-32.53
14	Light	80:3	1:1.5	Toluene	45	2	10	Air	73	70	67	-36.91
15	Light	80:3	1:2	MEK	60	3	35	Heat	42	35	23	-30.69
16	Light	80:5	1:1	Toluene	60	3	10	Fan	204	191	226	-46.34
17	Light	80:5	1:1.5	MEK	80	1	35	Air	24	40	43	-31.28
18	Light	80:5	1:2	Alcohol	45	2	60	Heat	86	116	102	-40.18

3. Ranking based on the selected coefficient that provides the best desired outcome. In this case study, this would be the coefficient of the factor level that produces the lowest defects. This is the recommended ranking when selecting factor levels for desired outcome.

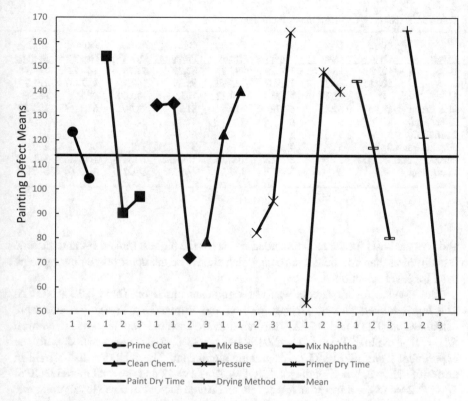

Fig. 8.4 Painting DoE L18 mean coefficient analysis

Table 8.22 Painting L18 DoE mean ANOVA

Factor	DOF	SS	Contribution	MS	F-Value	P-Value	Rank
Primer Coat	1	1618	1.01%	1618	0.29	0.642	8
Mix Base	2	14763	9.26%	7382	1.34	0.428	5
Mix Naphtha	2	15557	9.75%	7779	1.41	0.415	4
Clean Chem.	2	12008	7.53%	6004	1.09	0.479	7
Pressure	2	23103	14.49%	11552	2.10	0.323	3
Primer Dry Time	2	32662	20.48%	16331	2.96	0.252	2
Paint Dry Time	2	12426	7.79%	6213	1.13	0.470	6
Drying Method	2	36321	22.77%	18161	3.29	0.233	1
Error	2	11027	6.91%	5514			
Total	17	159486	100.00%	9381.53			

The rankings of the top 5 factors based on the ANOVA F-test on Table 8.22 include paint Drying Method, Primer Dry Time, paint gun Pressure, Naphtha Mix, and silicone Base Mix. This ranking matches the original consensus of the DoE team

Table 8.23 Painting L18 DoE mean coefficient ranking

Level	Prime Coat	Mix Base	Mix Naphtha	Clean Chem.	Gun Pressure	Primer Dry Time	Paint Dry Time	Drying Method
1	123.33	154.17	134.28	**78.78**	**82.22**	**53.78**	144.28	164.78
2	**104.37**	**90.33**	135.00	122.56	95.39	147.78	117.11	121.28
3		97.06	**72.28**	140.22	163.94	140.00	**80.17***	**55.50**
Max Effect	18.96	63.83	62.72	61.44	81.72	94.00	64.11	109.28
Effect Rank	8	5	6	7	3	2	4	1
Coef. Rank	8	7	3	4	6	1	5	2
F-Value Rank	8	5	4	7	3	2	6	1
Selected Level	2/Light	2/80:3	3/1:2	1/MEK	1/45 psi	1/1 hr.	1/10 min.	3/Air
Coefficient	-9.5	2/-23.5	3/-41.6	1/-35.1	1/-31.6	1/-60.1	3/+30.4	3/-58.4

and was the basis for the recommended levels for the highest ranked F-value factors. It is different than coefficient ranking which should be the appropriate ranking to be used for level selection.

The P-value for all factors was not significant based on Table 8.22 ANOVA. The top 2 ranked factors, Drying Method and Primer Dry Time, P-values were calculated at 0.233 and 0.252. The Painting DoE model was significant with an $R^2 = 1 - SSE/SST = 1–11,027/159486 = 93\%$. This was verified with the experiment mean of 113.85 defects being significant. The ANOVA has a built-in error of DOF = 2, based on the total DOF = 18 − 1 = 17 and the total model DOF of 1 + 7 * 2 = 15 (one factor at two levels and seven factors at three levels). Sample calculations of significant L18 mean are as follows:

$V_{error} = MSE = 5514.$

Standard error (SE) $= \sqrt{(V_{error}/n)} = \sqrt{(5514/18)} = 17.5.$

Computed t of L18 mean = mean/SE = 113.85/17.5 = 6.5, very significant for DOF = 1.

The factor levels with the least defects are shown in bold in Table 8.23. All were recommended except for Paint Dry Time (marked with an asterisk) which ranked low in F-value, but level 1 (10 min) was selected for faster processing since it was low ranked in mean (6) and variability (8). The recommended levels for five of the top F-value ranked factor coefficients are shown in Table 8.23 based on the F-test ranking, even if the factors were not significant based on mean ANOVA. The remaining factors of application of primer coat, cleaning method, and painting dry time were selected for ease of processing and operator safety, as well as variability analysis to be shown in the next chapter. They were light primer coat, MEK cleaning, and 10-min paint drying time (even if the coefficient increases defect rate). The predicted value of setting all factors to the selected levels would be negative defects, since the mean paint defect is 113.85 and the coefficient of each selected level (except for Paint Dry Time) would reduce the mean paint defect by the coefficient of each factor shown at the bottom of Table 8.23.

Experiment line 12 consisted of three repetitions (8, 14, 11) representing the lowest defect rate mean (11) and the least variability with sample standard deviation s = 3.0 as shown in Table 8.21. The DoE team decided to verify the recommended levels in the hope of achieving comparable low defect rate with lower variability. Three confirming experiments were performed using selected levels based on F-value rank with outcomes of 14, 15, and 17 defects, which are lower than mean defects of all experiment line = 113.85. Different pigment colors were used when preparing the paint for each confirming experiment to observe any deviation due to pigment colors, but none was detected.

The confirming level outcome mean was 15.33 defects which is higher than the best experiment line (12) mean of 11 defects. The confirming experiment variability as measured by sample standard deviation s = 1.53 or a variability reduction of 50%. These outcomes indicated that the operating levels of all selected factors have been adequately chosen, even without the use of the recommended coefficient ranking, with defect rate reduction of $(1 - 15.33/113.85) = 87\%$ from all experiment mean and 50% from the best experiment line variability.

The successful use of DoE in standardizing the painting process for theme park robots was effective in reducing the process variability and increasing final product quality. In addition, reducing the primer drying time to 1 h and drying painted parts in air and for 10 min were beneficial in streamlining the painting process.

8.3.2 Non-interacting OA Case Study II: L36 Air Knife DoE

This case study is typical of a multi-function machine that incorporated new technology and how to evaluate this technology within the current machine performance. The most common operation in the electronics industry is wave soldering of mixed packaging using different component technologies. The two most common component types were:

1. Surface-mount technology where component leads are attached to PCB directly by matching component features such as leads, balls, or bumps to pads or targets directly on the surface. A solder paste is deposited on the pads prior to component placement and melted through passing in a temperature-controlled conveyorized oven. This operation is completed before the wave soldering machine.
2. Through-hole (TH) technology where component leads are inserted into holes in PCBs and then passed through a wave soldering machine to fill holes with solder to attach components.

This sequence of operations causes solder defects with shadowing effects of SMT components. Equipment providers used several techniques to resolve quality problems by adding a turbulent wave prior to the soldering wave. An air knife was developed to reduce solder defects by blowing heated and pressurized air on PCB immediately after the solder wave. It was decided to use DoE to evaluate different settings of machine elements for highest quality outcome and recommendations to customers.

Two types of solder defects were measured by industry standards and collected separately by visual inspection:

1. Excess solder material more than specified including catastrophic bridges (shorts).
2. Insufficient material less than specified including catastrophic lack of solder (misses).

A total of 11 factors were selected, 8 factors with 3 levels for the current machine design and additional 3 air knife factors with 4 levels each. The air knife pressure included a zero-pressure level to simulate the impact of not installing the air knife. Machine factors and levels are as follows:

1. Solder temperature   440/470/500 °F
2. Preheat Rate   1.5/2.0/2.5 °C/sec
3. Conveyer Speed   3/6/9 ft/min
4. Fluxer Air Knife Angle   45/60/90′
5. Fluxer Air pressure   10/20/30 psi
6. Flux Type   Suppliers 1/2/3
7. Turbulent Wave   Off/Normal/High
8. Z Wave   Shallow/Normal/High
9. Air Knife Angle   0/55/70/85′
10. Air knife Pressure   0/6/10/14 psi
11. Air Knife Temperature   100/500/700/900 °F

The DoE team decided to use non-interacting orthogonal array L36 ($2^{11} \times 3^{12}$), considering the large number of factors and their levels and the difficulty in managing their interactions and confounding. Eight of 12 three-level columns were used to model the current machine design, using L36 columns 12–19. Six of 11 two-level columns were used to model three air knife factors with four levels each. The L36 was modified to accommodate the four levels by combining two columns for each of three four-level air knife factors in a multi-level arrangement for a total of six columns 1–6.

Six PCB test vehicles were used for each experiment line to generate a reasonable range of defects for model coefficient and ANOVA statistical analysis. The defects were averaged per PCB and were collected into four types of TH and SMT packaging technologies, with excess and insufficient solder defects for each. Two L36 repetitions were conducted, one set of 36 PCBs soldered directly without prior cleaning and another set chemically cleaned before soldering to remove any oxidation of PCB surface finish due to storage. This was to study the effects of PCB cleaning prior to soldering, which is not under the control of the machine supplier and can be tested by using a simple z-test comparing the two L36 outcome means. Fortunately, PCB cleaning prior to soldering was shown to be not significant. Total PCB test vehicles used for L36 Air Knife DoE = 36 × 2 × 6 = 432.

Table 8.24   Air Knife DoE defect outcome level analysis summary

	TH Excess	TH Insuf.	SMT Excess	SMT Insuf.	Recommended Levels	
Solder Temp	2	3	2	2	2	470° F
Preheat	3	1/2	3	1	2	2 °C/sec
Conveyer Speed	2	2	2	2	2	6 ft/min
Fluxer Angle	1	1	1	1	1	45'
Fluxer Pressure	3	1	3	1	3	30 psi
Flux Type	3	3	1	3	3	Supplier 3
Turbulent Wave	1	2	1	3	2	Normal
Z Wave	2	2	1	2	2	High
Air Knife Angle	2	1	4	2	3	70'
Air Knife Pressure	3	1	3	4	3	10 psi
Air Knife Temperature	1	1	1	2	2	500 °F

Table 8.25   Air Knife DoE defect prediction and validation

	TH Excess	TH Insufficient	SMT Excess	SMT Insufficient
Mean Defects (Before DoE)	1.5	2.0	14	23
Predicted Defects	-1.5	-9.0	7.5	2
Confirming Expts.	0	0	13	0.3

Statistical analysis was performed for the four types of defect outcomes and summarized in Table 8.24. The analysis was conducted like the Printer DoE case study earlier in this chapter, with the dominant level shown for each defect type based on coefficient analysis. The recommended level for each factor was selected based on the propensity of a particular level in the four defect categories or the highest percentage contribution in ANOVA. There was no universal level selection to reduce both excess and insufficient defects for each type of component packaging technology. The DoE team made universal level selection compromise based on the data analysis and their professional experience.

DoE defect model prediction showed a remarkable improvement from previous machine mean defects without the use of the air knife as shown in Table 8.25. A confirming experiment of six test vehicles was performed for validation of selected levels. All defects were reduced to near zero with the exceptions of SMT excess solder defects. These defects could be lowered by selecting dominant factor levels particular to outcome analysis of SMT excess solder column in Table 8.24. For example, air knife angle, pressure, and temperature could be adjusted to levels 4, 3, and 1, respectively, to favor reducing this defect type at the expense of other defect types. This will be dependent on the customer bias in reducing certain defect level types as opposed to others.

8.4   Conclusions

Three different enhancements to DoE projects were discussed in this chapter. Multi-level arrangements using different levels concurrently within orthogonal arrays (OAs) were illustrated using two case studies. The impact of this technique on DoE statistical analysis of model coefficients and ANOVA was demonstrated. This technique can be used for surveys of new materials, tooling, and technologies to be investigated while comparing it to legacy designs. The handling of interactions and confounding is left to the discretion of the DoE team to maintain DoE orthogonality.

DoE sequencing techniques were discussed as alternatives to large full factorial DoE. Screening designs followed by either full factorial or folded OA can be used to successfully reduce factor interactions. The risk of factor coefficient error due to confounding or ignoring interactions can be reduced while conducting less experiments than full factorial DoE.

The use of non-interacting arrays was demonstrated to handle a large number of factors and levels without the complexity of interactions and confounding. Two case studies of L18 and L36 arrays were discussed to standardize processes for high-quality custom painting and wave soldering using complex multi-function machines.

These three DoE tools can be used in industrial settings where time and resources are constrained. Their proper use is dependent on DoE teams' experience with design enhancements or problems to be resolved. A universal solution of selected levels can be pursued as a compromise for different outcome types.

Additional Reading Material

Bullard, H., "Failure Prediction Analysis in Machining Pin Fin Heat Sinks, Hoke Bullard, Master of Mechanical Engineering Thesis, UMASS Lowell, May 1997.

Calvarho, V. and Shina, S., "Evaluation of SMT Paste and Stencil Technologies", NEPCON West, Anaheim CA, February 1997.

Cardenas, R., "Application of Robust Design to Defect Reduction in the Silicon Painting Process", Master of Management Science Thesis, UMASS Lowell, May 1990.

Gagne D., Quaglia, M. and Shina, S., "Methods for Paste Selection and Process Optimization for Fine Pitch SMT", Journal of the SMART (Surface Mount and Related Technologies) Group, No. 24, October 1996, pp. 8–11.

Ross, P., *Taguchi Techniques for Quality Engineering*, New York: McGraw Hill, 1987.

Roy, R., *A Primer on the Taguchi Method*, New York: Van Nostrand Reinhold, 1990.

Phadke, M., *Quality Engineering Using Robust Design*, Englewood Cliffs, NJ: Prentice-Hall. 1989.

Wu J. and Shina, S., "Optimizing the new HOLLIS Wave Solder Machine", American Supplier Institute Seventh Symposium Proceedings, Phoenix, AZ, October 1989, pp 101–115.

Chapter 9
Variability Reduction Techniques and Combining with Mean Analysis

Most of the DoE topics discussed in previous chapters focused on creating models of designs or processes and conducting experiments with control factors and levels for analyzing model coefficients. In addition, several techniques for handling model error were discussed, including replication of some or all experiments. In this chapter, the variability inherent in factors and levels will be discussed and combined with mean analysis. Factors that are not in control by DoE teams such as materials from multiple mandated suppliers or production personnel working in processes can be investigated through replication of experiments.

Several quality initiatives in the 1980s, especially in the automobile industry, focused on reducing variability. The American Supplier Institute (ASI) was created to focus on quality tools including variability reduction through DoE for auto suppliers. QS9000 standard was developed by the big three auto suppliers, General Motors, Ford, and Chrysler, in 1994, and its implementation was mandated. It is currently superseded by ISO/TS16949. DoE and process capability measurements of mean and variability are part of the system requirements for all auto suppliers.

Methodologies for measuring variability will be introduced in this chapter. They differ based on DoE goals to have the minimum, maximum, or target desired results. Measures of variability could be used as DoE outcomes and analyzed using model coefficients and ANOVA as illustrated in previous chapters. DoE case studies showing trade-offs between mean and variability analysis for recommended levels will be presented and discussed based on decisions made and results obtained. Topics to be discussed in this chapter include the following:

9.1 Controlled and Noise Factors in DoE

Controlled factors are those selected by DoE teams with two, three, or multi-levels to investigate their effect, if any, on DoE outcomes. Noise factors are uncontrolled factors not selected for DoE but used in creating multiple repetitions. They could be materials from qualified suppliers or internal operations with varying specifications, operating conditions, process capability, manufacturing location, operator skills, or shift scheduling. Factors can be investigated by selecting them as controlled or noise factors within or outside of orthogonal arrays (OAs).

9.2 Variability Reduction Definitions and Analysis

This section presents DoE outcome mathematical transformation to variability measures and subsequent analysis. Different definitions and equations for variability are used depending on desired outcomes. Once a definition is agreed upon, DoE outcomes can be transformed into a single measure of variability, which can be used as an input to variability statistical analysis such as model coefficients and ANOVA. This intersection of mean and variability analysis will be demonstrated using case studies with desired minimum, maximum, or target outcomes combined with minimum variability.

9.3 Balancing Mean and Variability Outcomes

Balancing DoE outcome analysis for mean and variability may require expanding DoE projects to encompass operational benefits from either shifting design or process mean or reducing variability. Design components or process steps could be eliminated with reduced variability. Greater efficiency using either scheme should be included in the overall DoE project recommendations. This section will give examples of analyzing these additional benefits within DoE project goals.

9.4 Conclusions

9.4.1 Controlled and Noise Factors in DoE

Some factor levels can be manipulated to study their effect on DoE outcomes, while others are not, since they are specified in actual operating condition for designs or processes. For example, in the half-adder chip design case study in Chap. 5, four repetitions were made to L27 screening DoE. They represented the range of specified operating conditions including temperature (0–70°) and supply voltage (4.5–5.5 volts).

Factor designation as controlled or noise can be relevant in DoE analysis as shown in Fig. 9.1. Controlled factors are part of orthogonal arrays (OAs) and can be analyzed as model coefficients and rated for significance in ANOVA. If a factor is assigned as a noise factor in repetitions, then its effect, interactions with other

Fig. 9.1 Controlled and
noise factors for variability
measurements

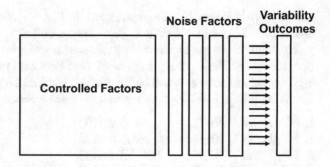

controlled factors, and significance cannot be determined. Noise factors contribute to determining variability as part of DoE error and allow design teams to model their design within advertised specifications. In addition, repetitions due to noise factors can be transformed to a single variability outcome that can be analyzed separately from the mean. DoE teams can make universal factor and level recommendations based on the two analyses.

In the chip design case study, temperature and supply voltage could have been assigned as controlled factors in the L27. Their effect on DoE outcomes can be determined if significant and measured accordingly. By assigning them as noise factor, the DoE team considered these advantages:

1. A smaller OA can be used, since they are two less controlled factors.
2. A universal set of recommended levels can be used for the range of specified operating conditions.
3. Variability analysis can be performed and combined with mean analysis for universal level selections.

The use of noise factors to simulate different operating conditions in DoE can be mitigated by other statistical tools discussed in Chaps. 1, 2 and 3. Outcome variability can be reduced by completing quality projects prior to DoE. Significant differences in process capability based on operator skills, machine capability, factory location, and work shifts can be addressed by training, proper equipment maintenance, and incoming material inspection. Environmental conditions such as temperature and humidity variability within the factory can be addressed through better equipment and controls or by conducting surveys to select less environmentally susceptible materials and processes.

Reducing Noise Factor Repetitions

DoE outcome replication can accommodate multiple noise factor levels. If there are two noise factors at three levels, the DoE for controlled factors can be replicated nine (3^2) times. For three noise factors at two levels, DoE outcomes can be replicated eight (2^3) times. For two noise factors at two levels, the DoE for controlled factors can be replicated four (2^2) times as shown in Fig. 9.1. These replications are like inverted L9, L8, and L4 at right angle to the controlled factor OA.

The replications can be reduced by using methodologies discussed in Chap. 7 for DoE reduction. The eight replications of three two-level factors can be reduced by half fraction replications based on L4. For full factorial L8 replication of three noise factors NF_1, NF_2, and NF_3 at two levels, the eight replications would be as follows:

1. NF_1(Level 1) NF_2(Level 1) NF_3(Level 1)
2. NF_1(Level 1) NF_2(Level 1) NF_3(Level 2)
3. NF_1(Level 1) NF_2(Level 2) NF_3(Level 1)
4. NF_1(Level 1) NF_2(Level 2) NF_3(Level 2)
5. NF_1(Level 2) NF_2(Level 1) NF_3(Level 1)
6. NF_1(Level 2) NF_2(Level 1) NF_3(Level 2)
7. NF_1(Level 2) NF_2(Level 2) NF_3(Level 1)
8. NF_1(Level 2) NF_2(Level 2) NF_3(Level 2)

The full factorial eight experiment replications can be reduced to four using L4 half fraction design, as follows:

1. NF_1(Level 1) NF_2(Level 1) NF_3(Level 1)
2. NF_1(Level 1) NF_2(Level 2) NF_3(Level 2)
3. NF_1(Level 2) NF_2(Level 1) NF_3(Level 2)
4. NF_1(Level 2) NF_2(Level 2) NF_3(Level 1)

Similar reductions can be made for any repetition of multiple noise factors at two or three levels. Four noise factors at two levels resulting in 16 (2^4) repetitions can be reduced to one half fraction L8 repetition. For larger two-level noise factors, repetitions could be reduced using screening design L8 for noise factors as was discussed in Chap. 7.

The DoE team should first consider reducing variability by performing quality projects prior to DoE and then consider whether to assign factors as controlled or noise depending on the conditions described above. If noise factors are selected for repetitions, DoE reduction techniques should be used for conducting project quickly and efficiently.

9.4.2 Variability Reduction Definitions and Analysis

Variability reduction is an important Six Sigma quality goal and can be achieved by DoE. Figure 9.1 demonstrates the transformation of DoE replicated outcomes into a single measure for variability. Two replications are required as a minimum to indicate variability. The single variability outcome for each experiment line can be used for DoE coefficient and ANOVA factor analysis as described in earlier chapters. The combined mean and variability analyses can be evaluated to produce a single set of recommended factor levels to achieve project goals with minimum variability. Goals can be minimum, maximum, or target values for a design or process being analyzed by DoE.

The transformation of each DoE experiment line replication resulting in a variability outcome can use different formulations, depending on DoE project goals. These formulas use an additional reduction transformation function such as Log_{10} to reduce the data range due to square terms inherent in the standard deviation (σ) Eq. 2.2. The formula for σ is made independent of the mean (μ) by the correction factors (CF), which forces the mean to be equal to zero. In addition, the transformation terms are adjusted by having higher value of variability indicating best level setting for reduced variability. The transformation formulas are different depending on relationship of mean to variability and specific project goals. They are as follows:

1. Log Variance for Smaller Is Better or Target Is Best I Goals. This transformation is equivalent to using one outcome as a source of for coefficient analysis. The sample standard deviation (s) of each experiment line is used to transform replicated outcomes to a single variability term, as follows:

Variability Outcome for Nominal Is Best I Goals

$$= -10 * \log_{10} (\text{Variance}) = -10 * \log_{10} (s^2) \qquad (9.1)$$

This formula can be used for project goals of minimum (smaller is better) or target values (target is best). The term $(-\log_{10})$ reduces the range of this variability outcome and ensures that the most positive outcome indicates reduced variability. In the case of the replications having the same outcome value in an experiment line, standard deviation (s) will be equal to zero, and Eq. 9.1 will result in infinite variability, due to the logarithm term. In that case, the replication values should be made slightly different to mitigate this result.

For DoE project smaller is better goal, the formula can be adjusted to eliminate the CF, which would be equal to zero when minimum is achieved for (n) repetitions:

Variability Outcome for Smaller Is Better Goals

$$= -10 \, \log_{10} \left(\frac{1}{n} \sum Y_i^2 \right) \qquad (9.2)$$

Inversely, a modification of Eq. 9.2 for project goals of maximum (larger is better) results can be used. In this case, the CF goal is infinity, and the inverse of the square term can be used for (n) repetitions:

Variability Outcome for Larger Is Better Goals $= -10 \, \log_{10} \left(\frac{1}{n} \sum \frac{1}{Y_i^2} \right)$ (9.3)

2. Coefficient of Variance Square for Nominal Is Best II Goals. This transformation is based on the relationship of sample average ($\overline{Y}$) versus standard deviation (s),

especially for scaling factors, which are factors where mean and variance are proportional. The transformation equation modifies the mean square by dividing it with variability square. Individual repetitions for each DoE line are not in the formula:

$$\text{Variability Outcome for Nominal Is Best II Goals} = 10 \; log_{10} \left(\frac{\overline{Y}^2}{s^2} \right) \quad (9.4)$$

3. Mean Square Error (MSE). In this transformation, the distance from all experiment line outcomes to the desired outcome or target (T) is minimized, whether it is minimum, maximum, or nominal. Selecting the level with the smallest value in the MSE variability coefficient analysis reduces MSE value and renders other factors more significant based on the F-test in ANOVA.

$$\text{Variability Outcome Using MSE} = -10Log \left[\frac{1}{n} \sum (Y_i - T)^2 \right] \quad (9.5)$$

4. Signal-to-Noise (S/N) Transformation. This transformation is favored by Taguchi method practitioners. It uses Eqs. 9.1–9.4 for all four types of desired goals (smaller is better, larger is better, and two nominals are best options: I (log variance) or II (log coefficient of variance square)).

Variability analysis DoE can be used for tolerance analysis of all factors. OA can be replicated three times for each tolerance set of each factor (nominal, USL, and LSL). If there are 4 factors, DoE results can be repeated 12 times to investigate DoE response to each factor tolerance. While this technique is lengthy, it could indicate whether some of the tolerances are significant for increasing or decreasing variability and therefore could be altered accordingly.

Analysis using the most positive variability coefficient should be used as a guide for selecting non-significant factor levels in mean analysis. This might conflict when mean significant interactions indicate a different level is more beneficial for desired outcome, as shown in Bonding II DoE of neoprene to steel case study discussed next. The DoE team can make appropriate decision based on their professional experience. ANOVA of the variability should not be considered for determining model reduction, since it is based on data transformation of DoE repetitions, though it is used for illustration in the next three case studies.

Mean and Variability Analysis L8 Case Study I, Larger Is Better: Bonding II DoE of Neoprene to Steel

This case study was conducted to improve light rail vehicle maintenance by providing a secure bonding mechanism. It was investigated an alternative procedure to using metal hardware for clamping power and signal cables that require disconnection by means of a split neoprene block attached to a steel plate. Ideally the bonded assembly would be stronger than neoprene itself. A project team was assembled to conduct a DoE to implement the new bonding process in simulated harsh environment of vehicle operation.

The team began by narrowing down the bonding material selection. Several brands of epoxies and contact cements were investigated for use in bonding steel to neoprene using grit blasted steel and degreased and abraded neoprene according to manufacturers' recommended settings. The only adhesives that met the peel force threshold were cyanoacrylate (superglue) based. Two different suppliers were selected to be used as factor levels in the Bonding II DoE.

The DoE project objective was to provide the best possible material and processing combinations that could be manufactured with the lowest cost and highest quality while performing beyond customers' expectations. The customer is the vehicle maintenance facility, and the cable clamp assembly must have zero defects with no neoprene to steel separation under any conditions. A secondary DoE goal was to determine if some of the manufacturers' recommendations could be omitted as means of reducing assembly time. This is in the case of material variability or accidental assembly shortcuts.

Four two-level factors were selected for DoE analysis:

1. Superglue-based adhesives from two suppliers. These were selected after prior experiments to reduce the number of levels.
2. RTV sealing. Superglue materials are prone to breakdown with moisture. An RTV barrier can be applied around the mating edges of the bonded materials. Two levels of seal/no seal will examine the effect of not performing abrading operations to save time or account for improper processing.
3. Neoprene cleaning. Neoprene was abraded per the manufacturer's recommendations and then cleaned with alcohol. Two levels of clean/no alcohol clean will examine the effect of not performing cleaning operations to save time or account for improper processing.
4. Steel abrading. Steel is abraded to remove any finish oxidation and then degreased to remove residual protective surface oils. Two levels of abrade/no abrade will examine the effect of not performing abrading operations to save time or account for improper processing.

The DoE team selected L8 half factorial analysis for the four factors. They were assigned to columns 1/A, 2/B, 4/D, and 7/G, respectively, as shown in Table 9.1. The neoprene test vehicle was 50–60 durometer and $1'' \times 1'' \times 2\ 1/4''$ size. The steel was SAE 1020 angle and $1'' \times 1'' \times 1/8''$ size. Both materials were of the same size and finish as intended for use in the final assembly. All neoprene test vehicles were abraded before bonding with 20 strokes of 100 grit emery cloth. Steel was abraded with hundred strokes of 100 grit emery cloth to descale the finish before degreasing in magnaflux solvent. RTV was applied 24 hours after the bonding process and allowed to cure for another 24 h. All test vehicles were fully submerged in water for 5 weeks to accelerate the moisture effect on the bonded joints.

ASTM test procedure D1876–72 "Peel Resistance for Adhesives" was used as guideline. Experiment lines were performed at random, and outcomes were

Table 9.1 Bonding II DoE L8 factors, levels and outcomes

Expt. # Col. #	Adhesive 1/A	RTV Seal 2/B	A*B 3/C	Neoprene Clean 4/D	A*D 5/E	B*D 6/F	Steel Abrade 7/G	1	2	3	4	Mean	Variability
1	Supplier 1	NO	1	NO	1	1	NO	10	12	2	5	7.25	11.15
2	Supplier 1	NO	1	YES	2	2	YES	10	12	17	15	13.50	22.07
3	Supplier 1	YES	2	NO	1	2	YES	30	33	31	27	30.25	29.55
4	Supplier 1	YES	2	YES	2	1	NO	30	35	20	25	27.50	28.22
5	Supplier 2	NO	2	NO	2	1	YES	7	10	11	8	9.00	18.67
6	Supplier 2	NO	2	YES	1	2	NO	5	9	11	9	8.50	17.39
7	Supplier 2	YES	1	NO	2	2	NO	21	15	10	12	14.50	22.28
8	Supplier 2	YES	1	YES	1	1	YES	16	11	8	11	11.50	20.45

Table 9.2 Bonding II DoE mean coefficient analysis

Model term	Effect	Coef.	SE Coef.	95% CI	t-value	P-value
DoE Mean		15.250	0.691	(13.824, 16.676)	22.08	0.000
A (Adhesive)	−8.750	−4.375	0.691	(−5.801, −2.949)	−6.33	0.000
B (RTV Seal)	11.375	5.688	0.691	(4.262, 7.113)	8.23	0.000
C = A*B	7.125	3.562	0.691	(2.137,4.988)	5.16	0.000
D (Clean)	0.000	0.000	0.691	(−1.426, 1.426)	0.00	1.000*
E = A*D	1.750	0.875	0.691	(−2.301, 0.551)	1.27	0.217*
F = B*D	2.875	1.438	0.691	(−2.863, −0.012)	2.08	0.048
G (Abrade)	1.625	0.813	0.691	(−0.613, 2.238)	1.18	0.251*

measured using 100 lbs. spring scale in one single day to reduce variability. The recorded peel force is the maximum value for up to 1/4″ length of mating surface peel. Four replicated outcomes with their mean and variability are listed in Table 9.1. Variability was calculated from Eq. 9.3, for larger is better goals.

The bonding mean coefficient analysis is shown in Table 9.2 and plotted on the top of Fig. 9.2. Both factor plots and mean coefficient analysis show that RTV seal, adhesive, and their interaction are the most significant. The other two factors, Clean and Steel Abrade, were not significant. Interaction RTV Seal and Neoprene Clean was significant. These significance results were verified by mean ANOVA in Table 9.3 with pooled factors indicated by an asterisk in the P-value column. The R^2 was calculated at 83.72%, indicating reasonable DoE model significance.

The recommendations for significant factor levels can be made by observing levels in the top plot of Fig. 9.2 for mean and bottom plot for variability. Significant factor levels Adhesive (A) supplier one and RTV Seal (Yes) were selected. The two other factors Clean (D) and Steel Abrade (G) showed non-significance mean analysis. Significant interaction F = B*D or RTV Seal (B)*Neoprene Clean (D) can be examined for its four-level selections by averaging the two experiment lines for each level combination. The second largest number (B_2D_2) corresponding to Neoprene Clean (Yes) was selected, since its variability plot in Fig. 9.2 showed that level

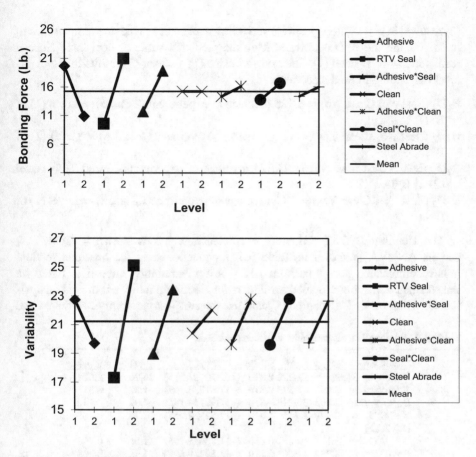

Fig. 9.2 Bonding II DoE mean and variability coefficient plots

Table 9.3 Bonding II DoE mean ANOVA

Source	DOF	SS	Contribution	MS	F-value	P-value
A (Adhesive)	1	612.50	24.19%	612.50	40.11	0.000
B (RTV Seal)	1	1035.12	40.88%	1035.12	67.78	0.000
C = A*B	1	406.12	16.04%	406.13	26.59	0.000
D (Clean)	1	0.00	0.00%	0.00	0.00	1.000*
E = A*D	1	24.50	0.97%	24.50	1.60	0.217*
F = B*D	1	66.13	2.61%	66.12	4.33	0.048
G (Abrade)	1	21.12	0.83%	21.12	1.38	0.251*
Error						
Pooling	3	45.62	1.80%			
Replication	24	366.50	14.47%	15.27		
Total error	27	412.12	16.28%	26.61		
Total	31	2532.00	100.00%			

(2) resulted in better (more positive) variability than level 1. This selection will result in subtracting the B*D coefficient from the predicted value of peel force since it results in F= B*D level (1). Selected Adhesive (1) * Clean (2) results in E =A_1D_2 level (2), adding to predicted value.

B_1D_1 (Seal No, Clean No) = 8.125 (average of experiments 1 and 5) = Level (1) of B*D

B_1D_2 (Seal No, Clean Yes) = 11 (average of experiments 2 and 6) = Level (2) of B*D

B_2D_1 (Seal Yes, Clean No) = 22.375 (average of experiments 3 and 7) = Level (2) of B*D

B_2D_2 (Seal Yes, Clean Yes) = 19.5 (average of experiments 4 and 8) = Level (1) of B*D

The Bonding II DoE variability coefficient analysis is shown in Table 9.4, and its ANOVA is shown in Table 9.5. As mentioned in the previous section, it is recommended not to perform ANOVA for variability since it is based on transformed terms, but provided for illustration. DoE variability coefficient analysis indicates that factor D (Neoprene Clean) has the smallest variability coefficient and

Table 9.4 Bonding II DoE variability coefficient analysis

Model term	Effect	Coef	SE Coef	95% CI	T-value	P-value
Variability Mean		21.222	0.810	(10.930, 31.515)	26.20	0.024
A (Adhesive)	−3.050	−1.525	0.810	(−11.817, 8.767)	−1.88	0.311*
B (RTV Seal)	7.805	3.902	0.810	(−6.390, 14.195)	4.82	0.130
C = A*B	4.470	2.235	0.810	(−12.527, 8.057)	2.76	0. 221
D (Clean)	1.62	0.813*				
E = A*D	3.175	1.588	0.810	(−11.880, 8.705)	1.96	0.300*
F = B*D	3.200	1.600	0.810	(−11.892, 8.692)	1.98	0.298*
G (Abrade)	2.925	1.463	0.810	(−8.830, 11.755)	1.81	0.322*

Table 9.5 Bonding II DoE variability ANOVA

Factor	DOF	SS	Contribution	MS	F-value	P-value
A (Adhesive)	1	20.480	8.41%	20.480	3.90	0.298*
B (RTV Seal)	1	121.836	50.06%	121.836	23.21/7.46	0.130/0.041
C = A*B	1	39.962	16.42%	39.962	7.61/2.45	0.221/0.178
E = A*D	1	20.161	8.28%	20.161	3.84	0.300*
F = B*D	1	18.605	7.64%	18.605	3.54	0.311*
G (Abrade)	1	17.111	7.03%	17.111	3.26	0.322*
Error						
D (Clean)	1	5.249	2.16%	5.249		
Pooled factors	5	81.61	33.53%	16.32		
Total	7	243.404	100.00%			

can be pooled into the initial variability ANOVA error. A total of five factors and interactions are shown to be non-significant and noted by an asterisk in Table 9.4. They were pooled to reduce the model in variability ANOVA shown in Table 9.5. Only factor B (RTV Seal) was shown to be significant after pooling five factors. The next important factor is the interaction C = A*B (Adhesive*RTV Seal) with 0.178 significance after pooling. These results show divergence of significance in mean versus variability analysis.

The recommended levels based on mean and variability coefficient analysis having more positive values are the same when observing Fig. 9.2 (except for Clean with zero mean coefficient), though their significance and coefficients might be different. This shows the benefits when using variability analysis to complement mean analysis by recommending appropriate levels when mean analysis indicates non-significance of some factors.

The recommended mean level for Steel Abrade cannot be determined since it is not significant and none of its interactions with other factors are significant. Variability plots in the bottom of Fig. 9.2 indicated that Abrade level 2 should be selected, since its value was more positive than level 1, even if it was found to be non-significant in Table 9.3, with significance at 0.251. The selection of factor levels and their coefficient contribution to the desired outcome is shown below:

Experiment Mean	= 15.25
Adhesive supplier 1	+ 4.375
RTV Seal (Yes)	+5.688
Interaction Adhesive*Seal	+3.562
Neoprene Clean (Yes)	0+0.875 −1.438 (from its interactions with Adhesive and RTV Seal)
Steel Abrade (Yes)	+0.813 (from its variability recommended level 2)

The predicted desired outcome based on the selected values of the four main factors and two interactions is equal to 29.125 pounds. Four replicated confirming experiments were conducted resulting in values of 24, 34, 24, and 30 pounds, with a mean of 28 pounds. This is within the range of the confidence interval of the predicted outcome.

In this case study, the use of significant interactions that included one non-significant factor can determine the non-significant factor level. Variability plots of larger is better desired outcomes can be used in level selection for non-significant factors or significant interactions in mean analysis.

Average and Variability Analysis L18 Case Study II, Smaller Is Better: Painting DoE

The mean analysis of the Painting DoE and its predicted outcomes and confirming experiments were performed in Chap. 8. This case study is like the Bonding II DoE discussed in the previous section, except that the desired outcome was minimizing painting defects rather than maximizing peel force.

Table 9.6 Painting L18 DoE variability ANOVA

Factor	DOF	SS	Contribution	MS	F-value	P-value	Rank
Prime Coat	1	42.94	4.16%	42.94	1.67	0.326	5
Mix Base	2	72.82	7.05%	36.41	1.41	0.414	6
Mix Naphtha	2	119.79	11.61%	59.90	2.33	0.301	3
Clean Chem.	2	53.80	5.21%	26.90	1.05	0.489	6
Pressure	2	111.37	10.79%	55.68	2.16	0.316	4
Primer Dry Time	2	295.26	28.61%	147.63	5.74	0.148	1
Paint Dry Time	2	24.62	2.39%	12.31	0.48	0.676	8
Drying Method	2	260.07	25.20%	130.03	5.05	0.165	2
Error	2	51.48	4.99%	25.74			
Total	17	1032.14	100.00%	60.71			

Table 9.7 Painting L18 DoE variability coefficient ranking

Level	Prime Coat	Mix Base	Mix Naphtha	Clean Chem.	Gun Pressure	Primer Dry Time	Paint Dry Time	Drying Method
1	−39.76	−40.59	−39.75	**−36.62**	−37.54	**−32.52**	−39.87	−41.78
2	**−36.67**	**−35.68**	−40.32	−37.42	**−35.63***	−41.54	−37.43	−39.93
3		−38.39	**−34.59**	−40.62	−41.49	−40.60	**−37.36***	**−32.95**
Total effect	3.09	4.92	5.74	4.00	5.87	9.02	2.51	8.83
Effect rank	7	5	4	6	3	1	8	2
Coef. rank	7	4	3	6	8	1	5	2
F-value rank	5	6	3	7	4	1	8	2
Selected level	2/Light	2/80:3	3/1:2	1/MEK	1/45 psi	1/1 hr.	1/10 min	3/air
Coefficient	−1.54	−2.54	−3.63	−1.60	−0.68	−5.70	+1.65	−5.27

The three outcomes shown in Table 8.21 were transformed into a single variability column using formula 9.2 for smaller is better. The 18 variability outcomes were analyzed using ANOVA as shown in Table 9.6. None of the factors' variability was significant, but their F-values were ranked to compare against mean F-values ranking for the top 5 factors as were performed for Painting DoE mean analysis.

The defect factor variability and their different rankings of total effect, F-value, and coefficients are shown in Table 9.7, with the recommended level with the most positive value shown in bold. The recommended levels matched for both mean coefficients (lowest value) and variability (most positive value): Prime Coat (A_2), Mix Base (B_2), Mix Naphtha (C_3), Cleaning Chemical (D_1), Primer Dry Time (F_1), and Drying Method (H_3). Two factors, noted with an asterisk, were recommend differently than the variability recommendations in bold in Table 9.7: Pressure, which had a lowest defect with level (1) and ranked number (3) in mean analysis in Table 8.23, and Paint Dry Time which was with low F-value rank in mean (6) and variability (8), where level 1 was selected for minimum dry time for easier processing.

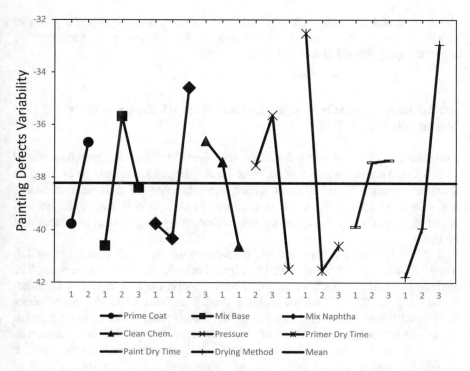

Fig. 9.3 Painting DoE L18 variability coefficient plots

The plots for Painting DoE non-interacting L18 variability coefficients are shown in Fig. 9.3, and its mean plots were shown in Fig. 8.4. The most positive level of each factor is recommended for least variability by using formula 9.2, while the smallest mean value is recommended for least defects. Recommended levels A_2, B_2, C_3, D_1, F_1, G_3, and H_3 matched for both mean and variability. Paint Gun Pressure (factor E) showed differences with level (1) for mean and level (2) for variability. The lowest pressure (level $1 = 45$ psi) was chosen since higher pressure causes paint to come out of the gun in a more pulverized state. This increases the probability it reaches target in a dry condition, facilitating the formation of small craters that cause defects by accumulating coats of paint and pigment. Paint Dry Time level (G_3) of 60 min, which was recommended for both mean and variability, was not selected, since it had a low rank in F-values, and level (1) of 10 min was selected for easier processing.

The confirming experiment for DoE painting outcomes was discussed in Chap. 8, based on recommended levels. It indicated quality improvements in lower defects of 87% from the L18 DoE mean and a reduction of 50% from the best experiment line variability using the standard deviation of three repetitions. The Painting DoE case study demonstrated the benefits of variability analysis as a complement to mean analysis, even when there were no significant factors. It used a novel method of ranking F-value of each factor and recommending factor levels based on these rankings. Variability analysis requires additional investment in repeating experiment outcomes but provides additional information for making good decisions for recommending desired outcomes of factor levels.

Average and Variability Analysis L9 Case Study III, Target Is Best: Stencil Screening II DoE

This case study is the continuation of the Machining II Stencil evaluation discussed in Chap. 8. After the selection of the stencil technology, process parameters must be specified to deposit solder bricks with a target height of 5 mils with minimum variability. An L9 screening design DoE was performed with two replications to recommend a target of 5 mils desired height outcomes through mean and variability analysis.

Four factors with three levels each were selected for the stencil screening DoE L9. They are listed in Table 9.8 with 18 height outcomes for 2 L9 replications. The variability column is the transformation of two replications to one variability outcome using Eq. 9.1 for Nominal Is Best I Goal. In the case of experiment 8, where the two replication outcomes are equal at 5.2, the standard deviation (s) is equal to zero, and Eq. 9.1 transformation of log variance (s^2) will become infinite. The variance of equal outcomes could be set to slightly greater than the most positive variability outcome or 17 versus the most positive variability of 16.99.

Mean and variability coefficient plots are shown in Fig. 9.4. The mean height is 5.8 mils, and mean variability is 13.04. Each factor level coefficient for mean and variability is shown in Table 9.9 and plotted in Fig. 9.4. Factor plots indicate the independence of mean and variability coefficients, since nominal is best formula is

Table 9.8 Stencil II DoE L9 factors, levels, and outcomes

A = Squeegee Speed	0.5	1.5	2.5	IPS
B = Squeegee Downstop	0.030	0.060	0.080	Inches
C = Snap-Off Distance	0.010	0.020	0.030	Inches
D = Squeegee Pressure	30	45	60	Lbs.

Exp. no.	A	B	C	D	A	B	C	D	Height (2 replicates)		Variability
1	1	1	1	1	0.5	0.03	0.01	30	7.0	7.4	10.97
2	1	2	2	2	0.5	0.06	0.02	45	5.5	5.7	16.99
3	1	3	3	3	0.5	0.08	0.03	60	6.2	4.2	-3.01
4	2	1	2	3	1.5	0.03	0.02	60	5.2	5.8	7.45
5	2	2	3	1	1.5	0.06	0.03	30	5.5	5.7	16.99
6	2	3	1	2	1.5	0.08	0.01	45	5.8	5.6	16.99
7	3	1	3	2	2.5	0.03	0.03	45	6.8	6.6	16.99
8	3	2	1	3	2.5	0.06	0.01	60	5.2	5.2	17.00
9	3	3	2	1	2.5	0.08	0.02	30	5.6	5.4	16.99

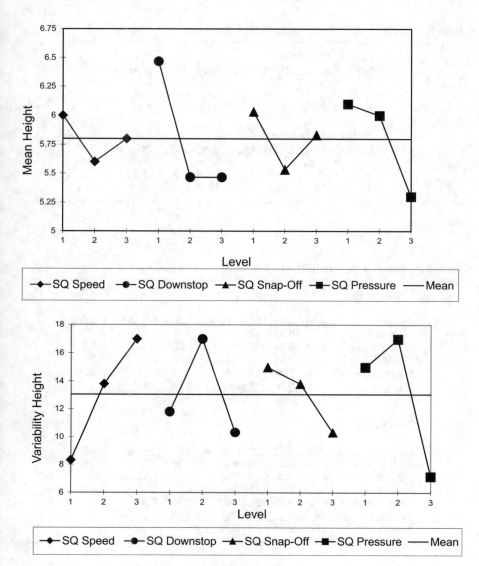

Fig. 9.4 Stencil DoE L9 mean and variability coefficient plots

concerned only with standard deviation of the replication and does not reference the mean.

Stencil L9 DoE ANOVA is shown in Table 9.10, before and after pooling. Two factors are significant, Squeegee Downstop and Squeegee Pressure, while two other factors Squeegee Speed and Snap-Off Distance were not. The recommended factor levels to achieve desired height must be contingent on observing significant factor mean levels and reduced variability indicated by the most positive variability level.

Table 9.9 Stencil II DoE L9 mean and variability coefficients

Mean level analysis

Level	Speed	Downstop	Snap-Off	Pressure
1	6.00	6.47	6.03	6.10
2	5.60	5.47	5.53	6.00
3	5.80	5.47	5.83	5.30

Variance level analysis

Level	Speed	Downstop	Snap-Off	Pressure
1	8.32	11.80	14.99	14.98
2	13.81	16.99	13.81	16.99
3	16.99	10.32	10.32	7.15

Table 9.10 Stencil II mean DoE L9 ANOVA before and after pooling

Before pooling

Factor	DOF	SS	Contribution	MS	F-value	P-value
SQ Speed	2	0.48	4.86%	0.24	0.92	0.435
SQ Downstop	2	4.00	40.49%	2.00	7.63	0.012
Snap-Off Dist.	2	0.76	7.69%	0.38	1.45	0.285
SQ Pressure	2	2.28	23.08%	1.14	4.35	0.048
Error	9	2.36	23.89%	2.36	0.26	
Total	17	9.88	100.00%			

After pooling

Factor	DOF	SS	Contribution	MS	F-value	P-value
SQ Downstop	2	4.00	40.49%	2.00	7.22	0.008
SQ Pressure	2	2.28	23.08%	1.14	4.12	0.041
Pooled error	4	1.24	12.55%	0.31	1.18	0.381
Replicate error	9	2.36	23.89%	2.36	0.26	
Total error	13	3.6	36.44%	0.28		
Total	17	9.88	100.00%			

Using Fig. 9.4, the recommended factor levels for 5 mils mean height consist of Squeegee (SG) Downstop either level 2 or 3 (coefficient − 0.33) and SG Pressure level 3 (coefficient −0.5) resulting in 4.97 outcome. Based on the variability plot in Fig. 9.4, level 2 of SG Downstop should be selected as opposed to level 3, since the value of level 2 is more positive in variability, even when the variability analysis was shown to be non-significant using ANOVA. The non-significant factors SG Speed and SG Snap-Off should be recommended at level 3 which is equal to the DoE mean outcome of 5.8, indicating near-zero contribution to the desired outcome. The recommended levels should result in 5 mils as follows:

Predicted height mean value $= 5.8 -$ contribution of $A_3 -$ contribution of $B_2 -$ contribution of $C_3 -$ contribution of D_3.

Predicted height mean value $= 5.8 - (5.8 - 5.8) - (5.8 - 5.47) - (5.8 - 5.83) - (5.8 - 5.8) = 5.00$ mils, which matched the desired outcome.

The DoE L9 Stencil II case study shows the application of recommended factor levels when desired outcomes are nominal is best. Two formulas are available, depending on whether to include mean coefficients in the variability transformation. There might be different sets of factor level combinations to achieve desired target outcome, and variability analysis can guide DoE teams to select the best combination of factor levels with the minimum variability. The high replication error contribution in Table 9.10, at 23.89% and 36.44% before or after pooling, was concerning, with R^2 values of 76% and 63%, respectively. The DoE team recommended a follow-on partial or full factorial DoE to study factor interactions with more repetitions.

The three case studies for using variability analysis in its various forms, smaller is better, larger is better, and nominal is best, were used to complement mean analysis, based on performing full replications of DoE experiments. In each case study, it was shown that variability analysis can be used to select recommended levels for non-significant or equal level factor values, making for better decisions. Variability ANOVA is not recommended since it is a reduced value transformation using ($\log_{10}$) function.

9.4.3   Balancing Mean and Variability Outcomes

The concept of mean square error $\frac{1}{n} \sum (Y_i - T)^2$ was mentioned in Sect. 9.2 as an alternative method of transforming outcome replication to variability measurements. In process capability studies, individual deviation to target can be caused by process mean (μ) not meeting specification nominal or large process standard deviation (σ). This combination of two attributes causes defects due to exceeding upper or lower specification limits. Six Sigma facilitates combining process capability, which is the responsibility of manufacturing operations, to part specifications, which is the responsibility of design functions.

Manipulating process means can be made by DoE investigating controlled factors as shown in Fig. 9.1. These factors include different design elements such as selection and preparation of materials, processing methods, and geometrical dimensions as was demonstrated in this chapter case studies. Reducing variability can be investigated using DoE noise factors, such as environmental conditions, location and timing of production, training of production operators, and proper equipment and tool maintenance for process DoE, or using specification tolerance as noise factors in design DoE.

DoE projects focus on achieving mean desired outcomes (minimum, maximum, or target) and using variability analysis to complement mean analysis. MSE can be used for improved recommendation decisions to quantify the percentage contribution of mean versus variability.

Techniques for quantifying deviation to target using population descriptors such as μ and σ can be used in two alternative methodologies:

1. Using mean square deviations (MSD) simplifies the mean versus variability calculations and provides a total MSE term:

$$MSE = MSD^2 + \sigma_n^2$$

The two elements of the MSE expression represent the percentage contribution of mean and variability in the MSE in process capability.

2. Using ANOVA where MSD represents the regression source like the correction factor and the sum of the square deviations represents the total sum of the squares. The percentage contribution can be used to quantify mean and variability.

Balancing Mean and Variability Gold Plating Example

Precious metal plating processes require significant attention to process capability and specifications due to high material costs. A theoretical example for plating gold on electronic connectors to reduce contact resistance shows the following ten samples with deviations from target value in microinches, 5, 1, 2, 3, 4, 8, 5, 3, 6, and 0, with average deviation of 3.70 and $\sigma_n = 2.28$. Using both deviations to target methodologies mentioned in the previous section results in equal outcomes:

1. $MSE = MSD^2 + \sigma_n^2 = 3.70^2 + 2.28^2 = 13.69 + 5.21 = 18.90$. MSD accounts for 13.69/18.90 or 72.43%, and variability accounts for 5.21/18.90 = 27.57%.
2. ANOVA regression $SS = MSD = (\Sigma Y)^2/n = (5+1+2+3+4+8+5+3+6+0)^2/n = 37^2/10 = 136.9$
   $SST = \Sigma Y^2 = 5^2 + 1^2 + \ldots + 0^2 = 189.0$; $SSE = SST - SSR = 189.0 - 136.9 = 52.1$

Source	DoF	SS	% contribution	MS	F-value
Regression	1	136.9	72.43%	136.9	26.28
Error	10	52.1	27.57%	5.21	
Total	11	189.0	100%		

Both methodologies give the same percentage contribution for mean and variability. The ANOVA F-value indicates significance of regression analysis at $P = F.DIST.RT (26.28, 1, 10) = 0.0004$.

The concept of quality loss function describes monetary loss if MSD and σ_n both $\neq 0$. The maximum loss occurs when products are shipped with defects and includes costs of repair and replacement as well as loss of customer satisfaction. If products are shipped within specifications, then the loss function represents inherent loss within the shipped products by not meeting the nominal specification targets.

Placing monetary options on quality alternatives is faulty as proven time and gain by mass producers such as auto manufacturers. An example is the General Motors faulty ignition switch recall of 30 million cars worldwide resulting in 124 deaths, when it was alleged that GM knew for at least a decade about the fault. Quality goals should be treated by setting low defect standards such as Six Sigma or high process capabilities, and not a trade-off between alternative mean and variability reduction.

9.4.4 Conclusions

The use of variability analysis to complement mean analysis was shown in this chapter. Techniques for quantifying variability using noise factors in DoE were illustrated, with using half factorial to reduce repetitions. Four types of variability analysis including smaller is better, larger is better, and two nominal is best transformations from repetitions to variability outcomes were discussed using case studies for each type. Variability analysis was used to enhance recommended level decisions in DoE applications in different industries and processes.

Focusing on the balance of meeting outcome targets by manipulating process mean or reducing variability was explored using two methodologies of mean square error and ANOVA of deviations with an example for precious metal plating. Caution must be exercised in not advocating a balance of the cost of quality versus harm to society at large.

Additional Reading Material

Akao, Y., *Quality Function Deployment: Integrating Customer Requirements into Product Design*, Cambridge MA: Productivity Press, 1991.

Clausing, D. and Simpson, H., "Quality by Design," *Quality Progress*, January 1990, pp. 41–44.

Crowell, R., "A Proactive Approach to Quality Through Experimental Design", Master of Management Science Thesis, UMASS Lowell, May 1993.

Hauser, J. and Clausing, D., "The House of Quality", Harvard Business Review, May-June, 1988, Vol. 3., pp. 63–73.

Phadke, M., *Quality Engineering Using Robust design*, Engelwood Cliffs, NJ: Prentice-Hall. 1989.

Ross, P., "The role of Taguchi Methods and Design of Experiments in QFD." *Quality Progress*, Vol. XXI, No. 6, pp.41–47, June 1988.

Sullivan, L.P., "Quality Function Deployment", *Quality progress*, Vol. XIX, No. 6. pp. 39-50, June 1986.

Taguchi, G., *Introduction to Quality Engineering*, Tokyo: Asian Productivity Institute, 1986

Chapter 10
Strategies for Multiple Outcome Analysis and Summary of DoE Case Studies and Techniques

This chapter will summarize the previous nine chapters with emphasis on multiple outcome analysis. Chapter 7 case study from Sect. 7.3.5 "Interaction Matrix L32 Case Study: Solder Wave DoE Design" will be analyzed using many DoE tools discussed in earlier chapters. Factors and levels are selected to satisfy mean and variability desired outcomes, as well as developing a universal set for two types of conflicting outcomes' goals.

A summary of all DoE case studies and techniques used in this book will be given as a guide to select the best strategy to conduct DoE to solve a problem or improve a design or process. Problem-solving techniques used in various case studies in this book can be applicable to similar DoE in different industries, equipment, processes, and designs with similar desired outcomes.

10.1   Summary of Previous Chapters

The first chapter in this book highlighted the origins of DoE and its connections to other aspects of quality, beginning with management tools such as total quality management (TQM) and its successor quality circles of Define, Measure, Analyze, Improve, and Control (DMAIC) for qualitative solutions to quality problems. It was followed by statistical process control tools based on discrete and continuous data distributions including normal, binomial, and Poisson distributions and associated control charts for maintaining quality in manufacturing. The advent of quality standards such as Six Sigma, which related design function specifications directly to process capability such as mean and standard deviations of production parts, was instrumental in highlighting the effective need of DoE for improved product design and more capable manufacturing operations.

Chapters 2 and 3 discussed various statistical analysis tools and significance tests for one factor at two levels for mean and variability, from normal or non-normal populations. Regression analysis was introduced for linear regression with model fit

© The Author(s), under exclusive license to Springer Nature Switzerland AG 2022   333
S. Shina, *Industrial Design of Experiments*, https://doi.org/10.1007/978-3-030-86267-1_10

of least square or normal approximation. The next level of complexity was discussion of treatments for testing one factor with multiple levels, which introduced the concept of analysis of variance (ANOVA) and significance testing. Design models were theorized and then tested using full factorial DoE with multiple factors and multiple levels. Analysis tools consisted of coefficient analysis and plots as well as ANOVA with model reduction by pooling factors from the model to the error, expressed as confidence interval (CI) of regression equation and predicted outcomes.

Chapters 4 and 5 introduced DoE simplification by using either two- or three-level factors. Orthogonal arrays (OAs) are used as a predetermined set of factors and levels to facilitate analysis and coefficient determination. Each OA can be used as full, partial, or half factorial, as well as screening designs. They can be applied individually or in sequence to gain more in-depth design or process understanding. Each additional DoE can provide more information about factors and their interactions as well as exploring expanded design space or process map. Case studies were used to illustrate each OA type and its analysis: how DoE team decisions were made, what tools were used, and conclusion drawn. Critiques for each case study were offered within the same chapter or in following chapters as more DoE tools were introduced.

Chapter 6 focused on DoE processes for goal setting, error handling, and experiment reduction. Techniques for error generation by single or multiple replications of experiments, repeating of all or some DoE experiment lines, or CenterPoint of factor levels were shown with case studies. Proper experiment dividing (blocking) was shown to reduce confounding.

Chapter 7 dealt with the issues of interaction and confounding, and their use in reducing experiments, given assumptions made by DoE teams based on their professional experience. Tools to facilitate understanding of these issues were presented such as OA interaction tables and Interconnecting diagrams. Two approaches were given: a top-down approach of selected factors and interactions grouped in certain interconnections to be used as fitted OA models and using a bottom-up approach to select certain factors and some of their interactions to the exclusion of others to build a unique model and fitting it into OA using interaction tables. The concept of resolution types was introduced to show confounding when analyzing factor and interaction coefficients. Case studies introduced in earlier chapters were revisited to show the impact of decision made on resolution types and how it can be improved for better analysis.

Chapters 8 and 9 presented additional DoE variants such as using multi-level arrangements in two- or three-level OA and use of non-interacting OA and variability analysis as an additional source of complementary information for mean analysis. Case studies were provided for four and eight factor levels, two types of non-interacting OA (L18 and L36), and three types of variability analysis for smaller is better, larger is better, and target is best outcome. In addition, sequencing techniques for substituting large full factorial OA were shown including a sequence of screening design followed by a full factorial of the top 2–4 significant factors. A classical alternate methodology of using initial screening OA to identify significant factors and then folding the saturated OA for each significant factor was demonstrated. In most case studies, handling of interactions was shown using interaction plots to find the appropriate level for the two constituent (source) factors in the interaction, even if one of them is not significant.

10.2 Combining Multiple Desired Outcomes with Mean and Variability Analysis

DoE analysis of many industrial DoE does not yield sufficient information to set factor levels, especially with multiple desired outcomes. Not all factors are significant, and some factor levels are recommended for one desired outcome, while other levels are best for other outcomes. The Printer DoE case study in Chap. 8 demonstrated the use of techniques for conflicting factor level selection for multiple outcomes.

Variability analysis techniques and combining with mean analysis were demonstrated in several DoE case studies in Chap. 9, including the L9 Bonding II, L18 Painting, and L9 Stencil II screening designs. Variability transformations using smaller is better, larger is better, or target is best presented additional information to the DoE team to select the best levels for non-significant factors.

Significant interactions, especially when one of the source factors is not significant, can also provide additional information to select the best levels. Interaction plots indicate best combination among four possible levels of two-way factor interactions or nine possible levels of three-level factor interactions. This was demonstrated with the use of significant interactions in choosing the best levels for desired outcome in the regression equations. Case studies included L8 Hipot DoE in Chap. 4 and L27 Green Electronics Phase I DoE in Chap. 5. Interaction levels cannot be set by DoE teams; they can only select source factor levels.

When factors are not significant, the most economic or environmentally beneficial levels can be recommended, as was demonstrated in the L9 Bonding I DoE in Chap. 5. The special case of ranking non-significant factors to set high-rank factors for best levels while using economically beneficial levels for low-rank factors was demonstrated in the L18 Painting DoE discussed in Chaps. 8 and 9.

A summary of topics discussed above and resulting impact is summarized below for managing multiple desired outcome DoE:

1. The model for each outcome must be universally relevant to all factors and interaction coefficients. All factors, significant or not, should be ranked and used in each outcome model, since their level coefficients will be part of predicted universal outcome. Non-significant factors should be ranked up to 0.20 significance and discarded from the model if greater than 0.20 significances.
2. Significant interactions should be part of each outcome model. They provide additional information to select both significant and high ranked non-significant factor levels for best desired outcomes. Interaction plots are additional tools to

consider best factor levels, especially when one of the source factors is not significant.

3. Variability analysis should be part of DoE projects, if resources and time allow. If a factor mean coefficient is not significant, then the least variability level (most positive coefficient) could be selected based on variability plots or analysis. The quality loss function can play a role in determining the most appropriate level selection between shifting mean to target and reducing variability, but quality concerns should be of utmost issues in not trying to balance cost of quality versus economic returns.

4. Separate DoE outcomes can be analyzed using the same OA experiment lines producing different results for each outcome. An alternative might be blocking OA along the most important factor by dividing experiments according to each level. Each half DoE can be designed locally for minimum interaction or resolution IV design and analyzed as both half DoE and full DoE to gain maximum information.

5. In all DoE cases, selection of recommended levels can be altered for low-ranking factors for economic as well as process and environmental benefits. It is up to DoE team discretion in selecting a universal set of levels to satisfy all desired outcomes.

10.2.1   Interaction Matrix L32 Case Study: Solder Wave DoE Analysis

This case study design was outlined in Chap. 7 (7.3.5) with suggestion to improve the design from resolution III to resolution IV. This analysis is based on a resolution III design that is slightly different than the one outlined in Chap. 7.

1. Summary of DoE Design. Through-hole (TH) technology was being used for making printed circuit boards (PCB) in an electronics company. Management visited competing companies and realized that their quality level was lacking. Poor quality results in high test and repair costs to remove defects. Concurrently, a soldering technology was available which was environmentally superior as well as has lower process cost. A DoE was suggested to bring soldering to Six Sigma levels, and it was a perfect opportunity to incorporate new technology as well as improve quality. The quality history in three phases is shown in Fig. 1.7.

This quality project culminating in DoE took almost 2 years to complete. It was performed during increases in sales and production of the company's products. As production volume increased, defect rates increased substantially. It was decided to approach this problem in a stepwise manner, using tools of quality in an evolutionary process. First, a TQM effort was launched to make sure all stakeholders are trained, and all documentations up to date, resulting in gradual quality improvements in 6 months. Next, statistical process control (SPC) was applied in the form of control charts to manage the process and prevent further spurious quality problems. Though the defect rate was greatly reduced, from a peak of 6000 ppm to 800 ppm during the TQM phase, it was still deemed too high compared to competition, which was boasting minimal defect rates of less than 100 ppm or four sigma. The decision was made to approve the use of DoE as a tool to obtain world-class quality levels.

The sequential quality improvement scenario of TQM, followed by SPC, followed by DoE is recommended for existing design and manufacturing processes. For new processes, it is best to start with DoE to select optimum combinations of materials and processes resulting in high-quality design and manufacturing. This high quality should be maintained with the proper application of TQM and SPC.

A DoE design team was selected and decided to survey their suppliers to investigate any alternative flux materials using new green technology. It was also decided to include the legacy flux in the DoE as a control factor.

Based on the time pressure and DoE scope, it was decided to select an L32 OA design as a single as opposed to sequential DoE. Since the flux was the most important factor, it was assigned to column 1 of the L32, and the DoE was divided into two L16s with four repetitions ensuring a range of defects and variability measurements.

2. DoE Design. The team went through the following steps for the design phase of the project:

   (i) The team selected seven additional factors in addition to Flux. They included factors dealing with temperature (Solder Wave and Preheat Temperature) and time (Conveyor Speed) and factors dealing with the geometry of the process (Conveyor Angle, Component Direction, Wave Height and Width). Most of these factors can be controlled through easy manipulation of the soldering machine operations and controls.

   (ii) The levels chosen for these factors were set by the physical limitations of the machine and the process. Levels were kept within the operating conditions of PCB and their components. For example, too high a solder wave or preheat temperatures could risk thermal damage to components. In addition, a conveyer speed that is too slow could further exasperate thermal shock. All these factors and their interactions had to be calculated to maintain DoE design space within specified operating conditions.

   (iii) The assignment of factors to columns in L16 OA was also tempered by some factor levels that are difficult to change in the overall experiment. Factors that were difficult or took a long time to change such as Conveyor Angle or Solder Wave Temperature were assigned to primary columns which required a smaller number of level changes.

3. Quality Outcomes and Test Vehicles. It was decided to focus on two types of quality defects: lead shorts and touch shorts:

(i) Lead shorts consist of defects caused by adjacent leads having a short solder bridge between them. These must be removed by visual inspection or by performing in-circuit test (ICT). These inspections and tests are very expensive since they are usually performed either by expensive technicians or automatic inspection machines. It was theorized that by changing the solder or preheat temperature as well as the conveyor speed, these defects could be reduced.

(ii) Touch shorts consist of defects caused by having a short solder bridge between leads and PCB surface. These defects are influenced by previous manufacturing cycles (PCB component automatic insertion). Defects were mitigated by expensive manual clipping of long leads. It was theorized that these defects could be reduced by changing the conveyor angle of the machine.

(iii) To reliably produce defects, a special test vehicle PCB was designed. It contained geometries of components which are located very close to each other to stress the soldering operation and produce as many defects as possible. It was decided to repeat each experiment four times to ensure a reasonable defect range and provide for variability analysis.

4. Interactions and Confounding. The interaction matrix method was used to identify all interactions of interest and shown in Table 7.8. L16 was used for each half of the total L32 DoE, and the seven factors (not including the Flux) were assigned to an hourglass interconnection in the right diagram in Fig. 7.4, and its interactions were shown in Table 7.9 as a resolution III design. A better arrangement is shown in Table 7.10 as a resolution IV design, by not assigning a factor to the four-way interaction L16 column 15.

5. DoE Analysis. The L32 DoE matrix and lead and touch defects are shown in Table 10.1. The table is divided into separate sets of columns corresponding to two L16s. The first 16 results correspond to Flux level (1), and experiments 17–32 correspond to Flux level (2). The data was divided into lead and touch mean defects per PCB as well as a log variance for both lead in touch defects. Flux level (1) is of higher quality than level (2) and resulted in minimum 0.25 lead defects in experiment 6 and zero touch defects in experiment 10, indicating that different factor levels will affect both types of defects differently.

The L32 coefficients and ANOVA are shown in Table 10.2 for lead defect analysis and Table 10.3 for touch defect analysis before pooling, and their coefficient plots are shown in Fig. 10.1. The model % contributions are almost significant with 86.19% and 83.93%, respectively, for R^2 before pooling. These analyses were performed for the two L16 OAs individually, but results are not shown. Significant factors for lead defects included Flux, Direction, and their interaction Flux*Direction. Speed was near significant, and further pooling of the non-significant interaction produced Speed significance. Significant factors for touch defects (and after pooling non-significant interactions) included Flux, Angle, and their interaction Flux*Angle, Direction, and Height. All significant

Table 10.1 L32 Solder Wave DoE project DoE matrix and results

Expt. no.	Flux	Angle	Temp	Height	Preheat	Direction	Width	Speed	Lead defects	Touch
1	1	1	1	1	1	1	1	1	6	13
2	1	1	1	1	2	1	2	2	10	26
3	1	1	1	2	1	1	1	2	10	12
4	1	1	1	2	2	1	2	1	8.5	14
5	1	1	2	1	1	2	1	2	1.5	18.75
6	1	1	2	1	2	2	2	1	0.25	16.25
7	1	1	2	2	1	2	1	1	1.75	25.75
8	1	1	2	2	2	2	2	2	4.25	18.5
9	1	2	1	1	1	2	2	2	6.5	6.5
10	1	2	1	1	2	2	1	1	0.75	0
11	1	2	1	2	1	2	2	1	3.5	1
12	1	2	1	2	2	2	1	2	3.25	6.5
13	1	2	2	1	1	1	2	1	6	7.25
14	1	2	2	1	2	1	1	2	9.5	11.25
15	1	2	2	2	1	1	2	2	6.25	10
16	1	2	2	2	2	1	1	1	6.75	12.5
17	2	1	1	1	1	1	1	1	20	29.25
18	2	1	1	1	2	1	2	2	16.5	31.25
19	2	1	1	2	1	1	1	2	17.25	28.75
20	2	1	1	2	2	1	2	1	19.5	41.25
21	2	1	2	1	1	2	1	2	9.67	21.33
22	2	1	2	1	2	2	2	1	2	10.75
23	2	1	2	2	1	2	1	1	5.67	28.67
24	2	1	2	2	2	2	2	2	3.75	35.75
25	2	2	1	1	1	2	2	2	6	22.7
26	2	2	1	1	2	2	1	1	7.3	25.7
27	2	2	1	2	1	2	2	1	8.7	30
28	2	2	1	2	2	2	1	2	9	29.7
29	2	2	2	1	1	1	2	1	19.3	32.7
30	2	2	2	1	2	1	1	2	26.7	25.7
31	2	2	2	2	1	1	2	2	17.7	45.3
32	2	2	2	2	2	1	1	1	10.3	37

factors are marked with an asterisk in the P-value column. Temperature (wave) was significant in the L16 lead analysis (with level 2 for lower defects). Two factors, Preheat and Width, were not significant, and their significance was determined using their interaction with Speed as shown in Fig. 10.2.

The regression equations for both lead and touch are shown below based on factor coefficients and pooling. All significant factors and interactions are highlighted and shown in bold. All factors were kept in the model and regression equations to consider their coefficients in the universal predicted outcomes.

Table 10.2 L32 Solder Wave DoE lead defect coefficient and ANOVA

Source	Coefficient	DOF	SS	Contribution	MS	F–value	P–value
Model	8.878	12	1146.80	86.19%	95.567	9.88	0.000*
Flux	3.581	1	410.34	30.84%	410.34	42.42	0.000*
Angle	0.341	1	3.72	0.28%	3.720	0.38	0.543
Temperature	−0.669	1	14.32	1.08%	14.325	1.48	0.239
Height	−0.370	1	4.39	0.33%	4.388	0.45	0.509
Preheat	−0.234	1	1.75	0.13%	1.753	0.18	0.675
Direction	−4.236	1	581.49	43.70%	581.49	60.11	0.000*
Width	−0.209	1	1.40	0.11%	1.399	0.14	0.708
Speed	0.986	1	31.11	2.34%	31.106	3.22	0.089*
Flux*Angle	0.325	1	3.39	0.25%	3.387	0.35	0.561
Flux*Temp	0.097	1	0.30	0.02%	0.298	0.03	0.862
Flux*Preheat	−0.343	1	3.77	0.28%	3.774	0.39	0.540
Flux*Direction	−1.685	1	90.82	6.83%	90.822	9.39	0.006*
Error		19	183.81	13.81%	9.674		
Total		31	1330.61	100.00%			

Table 10.3 L32 Solder Wave DoE touch defect coefficient and ANOVA

Source	Coefficient	DOF	SS	Contribution	MS	F–value	P–value
Model	21.10	12	3546.75	83.93%	295.56	8.27	0.000*
Flux	8.64	1	2390.00	56.55%	2390.00	66.86	0.000*
Angle	−2.11	1	142.17	3.36%	142.17	3.98	0.061*
Temperature	1.25	1	49.63	1.17%	49.63	1.39	0.253
Height	2.45	1	191.54	4.53%	191.54	5.36	0.032*
Preheat	0.29	1	2.62	0.06%	2.62	0.07	0.790
Direction	−2.48	1	196.76	4.66%	196.76	5.50	0.030*
Width	0.73	1	17.04	0.40%	17.04	0.48	0.498
Speed	0.78	1	19.39	0.46%	19.39	0.54	0.470
Flux*Angle	3.47	1	385.38	9.12%	385.38	10.78	0.004*
Flux*Temp	−1.33	1	56.84	1.35%	56.84	1.59	0.223
Flux*Preheat	−0.39	1	4.77	0.11%	4.77	0.13	0.719
Flux*Direction	−1.68	1	90.62	2.14%	90.62	2.53	0.128
Error		19	679.22	16.07%	35.75		
Total		31	4225.97	100.00%			

Lead defects = **8.88 − 4.26 Direction + 3.58 Flux − 1.69 Flux * Direction + 0.99 Speed** − 0.67 Temperature − 0.37 Height + 0.34 Angle −0.23 Preheat −0.21 Width

Touch defects = **21.2 + 8.64 Flux + 3.47 Flux * Angle − 2.48 Direction + 2.45 Height − 2.11 Angle** + 1.25 Temperature + 0.78 Speed +0.73 Width + 0.29 Preheat

Table 10.4 shows the logic of selecting a universal set of factor levels to minimize defects for both desired outcomes. Three top factor levels (Flux, Direction, and Speed) coincided with recommended levels based on coefficient plots in Fig. 10.1, while all other factors had the opposite recommendations.

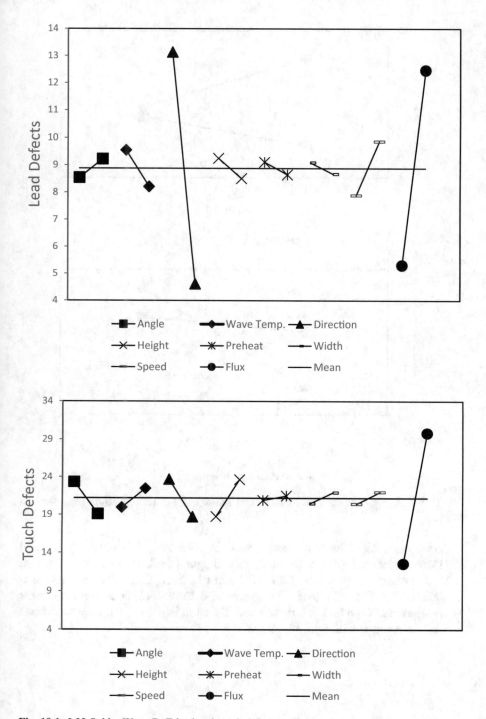

Fig. 10.1 L32 Solder Wave DoE lead and touch defect coefficient plots

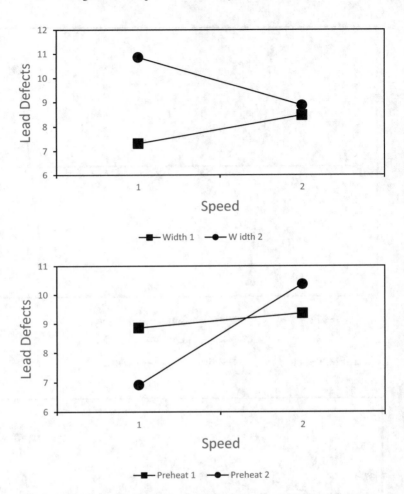

Fig. 10.2 L32 Solder Wave DoE Speed*Width and Speed*Preheat interaction plots for lead defects

Speed was significant in lead mean as well as variability, with level (1) recommended. Angle (level 2) and Height (level 1) were significant in touch analysis as shown in Table 10.3 and Fig. 10.1. Solder Temperature was significant in L16 with level (2) recommended. Preheat (2) and Width (1) levels were selected from their interaction with Speed in Fig. 10.2, with lowest defects when interacting with Speed level (1).

Table 10.4 L32 Solder Wave DoE factor level assignments

	Recommended		Significant		Variability	Recommended	
	Lead	Touch	Lead	Touch	Mean	Universal	
Flux	1	1	1	1		1	
Direction	2	2	2	2		2	
Angle	1	2		2		2	
Speed	1	1	1		1	1	
Height	2	1		1		1	
Temperature	2	1				2	Significant L16
Preheat	2	1				2	Interact.*Speed
Width	2	1				1	Interact.*Speed

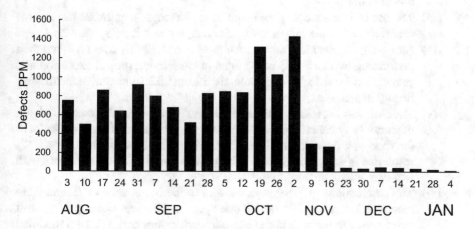

Fig. 10.3 L32 Solder Wave DoE quality performance after DoE

6. DoE Prediction and Confirmation. The defects' predicted values were calculated based on universal factor level selection in Table 10.4. Each factor coefficient was added or subtracted based on their coefficient sign in the regression equation or by observing their slope in the factor plots in Fig. 10.1. The interaction coefficient was added to the lead prediction and subtracted for the touch prediction based on the levels of its source factors.

The lead defects were predicted to be 1.68, and the touch defects were predicted to be 2.05 when used in actual production with the universally recommended levels. A confirming experiment was run with four test vehicles, and it resulted in zero lead and two touch defects.

The effect of the DoE quality transformation is shown in Fig. 10.3. The transition to the new process as recommended by DoE occurred on November 1, the start of the new fiscal year, since management was concerned about the radical change in the solder process. After the leftover stock was cleared in the first 2 weeks, the mean total defect rate was 31 ppm (four sigma) for the first

7 weeks, while the process was scrutinized heavily and then averaged 100 ppm afterward. The process was monitored for 6 months before and after DoE, as shown in Fig. 1.15. The weekly defect mean dropped from 808.5 to 98.5 ppm, or 88% reduction. The standard deviation was reduced from 213.8 to 55.3, a 74% reduction.

7. Solder DoE Postmortem. While the solder DoE was successful and had a major impact on the use of DoE methods in electronics manufacturing, there were several innovations as well as failures that should be noted:

   (i) The use of the interaction map was a useful technique to visualize all possible interactions to select from.

   (ii) The use of variability analysis provided additional information for selecting non-significant factor levels such as Speed, which ultimately became significant after pooling.

   (iii) The use of interactions of non-significant factors to significant factors was useful in setting the proper level of non-significant factors.

   (iv) Assigning an important factor such as Flux to L32 OA column 1 and then performing two L16 projects resulted in simplifying project execution and provided information to select non-significant L32 Temperature factor level from L16 analysis.

   (v) The DoE was a resolution III design with two-way interactions confounding other two-way interactions. A better assignment was not to assign a factor to the four-way interaction (1248) to column 15 in L16 to allow for a 2^{8-4}_V resolution V L16 array design with no confounding of main factors with two-way interactions.

   (vi) The assumptions of the design team as to the importance of factors were incorrect. The analysis indicated that Speed and Preheat were not important as originally thought. A better interconnection diagram could be a modified L16 pentagon shape of equal importance factors and interactions between them.

   (vii) The team did not attempt a follow-on experiment due to the success of DoE in reducing defects to Six Sigma levels.

10.3   Summary of DoE Case Studies and Techniques

This section is intended as a guide for designing and analyzing industrial DoE projects using DoE tools and techniques discussed in this book through case studies. They are described in general terms to provide insight in identifying DoE project goals within appropriate design space or process map, selecting factors and levels, developing interaction and confounding strategies, deciding on which OA to use, error handling with repetitions and pooling, performing DoE analysis by selecting the best levels for the desired outcomes even if they were not individually significant, and developing regression equations and predicted outcomes. Two lists are provided: the first for listing case studies using their OA and number of level size and the second for listing DoE techniques discussed in this book and how they were applied in some of the case studies.

10.3.1   DoE Case Studies List by OA Size for Two and Three Levels

1. Full Factorial Phase II of Green Electronics DoE (Chap. 3).
   This is a second sequential DoE following L27 Partial Factorial DoE Phase I Case Study (#21) to investigate multiple factors and levels for green technology. The factors identified in the first sequence DoE were narrowed down to three factors investigated in this full factorial DoE with five, three, and two levels, repeated two times. Coefficient analysis and ANOVA were performed and significant factors and interactions identified. This DoE was successful in achieving zero-defect outcomes. Given that desired outcomes were achieved with the most expensive levels, unique combinations of less expensive alternative levels were recommended using multiple range (REGW) tests for equivalent significant grouping of factors and interactions.
2. L8 Full Factorial Case Study, Hipot DoE: Selecting Best Alternative Among Equally Performing Designs (Chap. 4).
   This is a full factorial L8 DoE with three factors at two levels and one replication. All interactions were available for analysis with no confounding. Pooling of the least significant factor or interaction provided an initial source of error, with subsequent pooling of non-significant factors until only significant factors remain. The use of a significant interaction was highlighted in choosing the best levels for desired outcome in the regression equation.
3. L8 Partial Factorial, Underfill Voids DoE: Selecting Process Parameters for Zero Defects (Chap. 4).
   This is L8 partial factorial DoE with five factors at two levels and two replications. Two interactions were available for analysis with confounding of main factors with one- or two-way interactions, resulting in resolution III design. DoE was successful in achieving zero defects down from mean of two defects for DoE outcomes.
4. L8 Plastics Injection Molding DoE with One and Multiple OA Replications, CenterPoint, and Some Experiment Line Replications (Chap. 6).
   This is a full factorial L8 DoE with three factors at two levels. All interactions were available for analysis with no confounding. Error handling techniques were investigated with one and three replications as well as repeating CenterPoint and some experiment lines. Benefits and drawbacks of each error handling method were discussed.
5. L8 Multi-level Arrangement DoE: Machining I Pin Fin Heat Sinks (Chap. 8).

This is L8 partial factorial DoE with one factor at four levels and three factors at two levels and four replications. Two interactions were available for analysis with confounding of main factors with one- or two-way interactions, resulting in resolution III design. While factor assignments to columns were not ideal, DoE was successful in achieving zero defects in designs that were not possible to manufacture prior to DoE.

6. L8 Foldover Sequencing Techniques: Folding on One Factor Example (Chap. 8).

   In this example, a screening L8 DoE with seven factors at two levels and one replication is performed as resolution III design with every factor confounded with three two-way interactions. A second screening L8 DoE is performed with folding (complementing) the top significant factor levels in an L16 combination. The top significant factor is isolated with no confounding, and all its interaction coefficients with the other six factors can be determined. This process continues with folding every significant factor from the original screening L8.

7. L8 Foldover Sequencing Techniques: Folding on All Factor Example (Chap. 8).

   In this example, a screening L8 DoE with seven factors at two levels and one replication is performed as resolution III design with every factor confounded with three two-way interactions. A second screening L8 DoE is performed with folding (complementing) all factor levels in an L16 combination. The resulting L16 is a resolution IV design with no confounding of any factor, and each two-way interaction is confounded by three other two-way interactions. Its interconnection diagram is like the one described in Chap. 7 in Fig. 7.6 left diagram.

8. L8 Followed by L9: DoE Sequencing Techniques, Printer Design DoE (Chap. 8).

   In this example, a screening L8 DoE with 7 factors at 2 levels and 1 replication of 100 measurements for multiple outcomes was performed as resolution III design with every factor confounded with 3 two-way interactions. Separate ANOVA significance analyses were performed on each type of defect for selecting best levels for zero defects for each factor. After analysis, three factor levels were selected for final recommendations. A second screening L9 DoE is performed with three wider levels of remaining factors and analyzed similar to the L8. DoE was successful in reducing defects by 90%.

9. L8 Mean and Variability Analysis, Larger is Better: Bonding II DoE of Neoprene to Steel (Chap. 9).

   This is a half fraction L8 with four factors and three interactions repeated four times. Mean and variability coefficient analyses were performed for larger is better desired outcome, though it is recommended not to perform variability ANOVA. Two factors indicated coefficient significance, and the higher levels from coefficient plots were selected, while the other two factor levels were selected from variability analysis and significant interactions. Mean and variability plots indicated recommended levels were the same for higher coefficient value and most positive variability values.

10. L16 Partial Factorial, Rivet Design DoE: Selecting Part Dimension Design for Best Product Performance (Chap. 4).

This is L16 partial factorial DoE with seven factors at two levels and four replications. Eight interactions were available for analysis with confounding of main factors with one- or two-way interactions, resulting in resolution III design using a modified seven-factor interconnection right diagram in Fig. 7.2. DoE was successful in achieving 11% increase in desired outcome greater than mean of 16 experiments.

11. L16 Multi-level Arrangement DoE: Machining II Stencil Forming (Chap. 8).

This is L16 partial factorial DoE with 1 factor at 8 levels and 5 factors at 2 levels and 30 replications (3 L16 replications with 10 measurements each). Two interactions of the factor with eight levels are assigned to three columns each with one more two-way interaction. The recommended methodology for stencil cutting predicted and verified best quality outcomes but was the most expensive. A modified quality loss formula was used to demonstrate overall improvement in quality as well as reduced cost due to streamlined processing.

12. L18 Non-interacting OA: Painting DoE (Chap. 8).

This is L18 non-interacting DoE with eight factors, one factor at two levels and seven factors at three levels with three repetitions. Most of the factors were not significant based on ANOVA. The DoE team used F-value rank as the method for selecting factor levels. The top 5 ranked factors had their levels selected based on factor levels for maximum peel force. The analysis was then combined with variability analysis in case study #13 for selecting low ranked factor levels.

13. L18 Mean and Variability Analysis, Smaller Is Better: Painting DoE (Chap. 9).

These L18 three repetitions were transformed to a variability term based on smaller is better conversion. Variability analysis indicated that minimum variability was achieved by observing the most positive level in the variability coefficient plot. The recommended levels in mean analysis were the same as recommended variability analysis for seven out of eight factors, and two low ranked factors were recommended based on variability coefficient plots. DoE was successful in achieving 88% lower defects from the L18 DoE mean and with 50% reduced variability from the best L18 experiment line variability.

14. L32 Partial Factorial, APOS for Robotics DoE: Selecting Process Parameters for Multiple Adjustment Production Machines (Chap. 4).

This is L32 partial factorial DoE with 13 factors at 2 levels and 1 replication for 2 different outcomes. The DoE team used the interaction matrix method to identify 11 interactions of interest out of 25 available in L32. The design avoided assigning a factor to the highest five-way interaction level, resulting in resolution IV design. No factor was confounded with two-way interactions, but every two-way interaction was confounded with five other two-way interactions. To simply analysis, only the top 6 significant factors were analyzed and included in the regression equation for each outcome type. DoE was successful in achieving greater efficiency >90% of parts filled in bins greater than minimum

80% required. Results suggested that each individual part in the robotic assembly should have a unique DoE to achieve desired outcomes.

15. L16/L32 Interaction Matrix Solder Wave DoE Project Design (Chap. 7).

   This is L16 partial factorial DoE with seven factors at two levels and four replications, two each for two outcomes. One factor was assigned to the four-way interactions, and eight interactions were available for analysis with confounding of main factors with one interaction, resulting in resolution III design. The factors and interaction were selected based on interaction matrix design resulting in a unique interconnection graph shown by right diagram in Fig. 7.4. An alternate resolution IV design was presented so that main factors are not confounded, while two-way interactions are confounded by two other two-way interactions.

16. L16/L32 Interaction Matrix Solder Wave DoE Analysis (Chap. 10).

   Two L16s corresponding to two levels of the most important factor case study #15 were analyzed separately and then combined into L32 for further analysis of two desired outcomes. Factor levels were chosen from significant factors in either desired outcome from each L16 as well as L32. Variability analyses were also performed as additional information to mean coefficient analysis. Non-significant factor levels were selected by observing their interaction plots with a significant factor. DoE was successful in achieving 88% defect reduction and average 30 ppm quality in the first 7 week of full production.

17. L36 Non-interacting OA: Air Knife DoE (Chap. 8) for Multi-function Machines.

   This is L36 non-interacting DoE with 11 factors, 8 factors with 3 levels for the current machine design and additional 3 new feature factors with 4 levels each for 432 replications. Repetitions were based on six test units for each line repeated twice for one processing noise factor. This case study is like #13 with L18 Painting DoE. Four different defect types were analyzed individually, like the #8 DoE Sequencing Techniques, Printer Design DoE. All four defect analyses were combined for a universal set of recommended levels. A regression equation was developed and predicted outcomes were made. The DoE was a success since recommendations were validated with confirming experiments that produced zero defects for three out of four defect categories.

18. L9 Screening Design Case Study: Bonding I DoE (Chap. 5).

   This is L9 screening DoE with four factors at three levels and one replication. Two main factors (A and B) were confounded with two three-way interactions (AB and AB_2). One interaction was pooled into the error. DoE was successful in achieving maximum peel force greater than the largest experiment line result.

19. L9 Screening Design Case Study: Zero Defect Mixed Soldering DoE (Chap. 5).

   This is L9 screening DoE with four factors at three levels and one replication. Two main factors (A and B) were confounded with two three-way interactions (AB and AB_2). One interaction was pooled into the error. DoE was successful in achieving zero defects from the previous mean of 13 defects per unit.

20. L9 Mean and Variability Analysis, Target Is Best: Stencil II Screening DoE (Chap. 9).

This is a screening L9 with four factors repeated two times. Mean and variability coefficient analyses were performed for target is best desired outcome, though it is recommended not to perform variability ANOVA. Two factors indicated coefficient significance, and higher levels from coefficient plots were selected, while the other two factor levels were selected from variability analysis. Target is best could be achieved through multiple combination of factor levels, but the variability coefficient analysis is beneficial for recommending level combination achieving target value with lowest variability.

21. L27 Partial Factorial Case Study: Green Electronics Phase I DoE (Chap. 5).

This is L27 partial factorial DoE with five factors at three levels and two replications that was used as screening in a DoE sequence to select the top significant factors for a second full factorial DoE case study #1: Full Factorial Phase II of Green Electronics DoE (Chap. 3). Three factors were assigned to primary columns, and two less important factors were assigned to confounding three-way interaction columns. Significant interaction plot indicated the best combination among nine possible levels of two three-level factors.

22. L27 Screening Design Software DoE Case Study: Minimizing Half-Adder Chip Delay Time DoE (Chap. 5).

This is L27 DoE with 11 factors at 3 levels and 4 replications used as a screening DoE using software simulation to generate results based on two noise factors representing the extremes in operating conditions. DoE was successful in achieving 19.4% outcome desired reduction in legacy products. Additional DoE steps in increasing the design space toward the steepest level ascent response were used to investigate if more reductions can be obtained with no success.

10.3.2 DoE Techniques Used and Demonstrated in Chapters and Case Studies

1. Multiple range (REGW) tests were used to identify equivalent combinations of less expensive factors and levels as alternatives for zero-defect process design. Case study #1 Full Factorial Phase II of Green Electronics DoE (Chap. 3).
2. Use of a significant interaction in choosing the best levels for desired outcome in the regression equation. Case studies included #2 L8 Full Factorial Case Study, Hipot DoE: Selecting Best Alternative Among Equally Performing Designs (Chap. 4) and #9 L8 Mean and Variability Analysis, Larger Is Better: Bonding II DoE of Neoprene to Steel (Chap. 9).
3. Significant interaction plots indicating the best combination among four possible levels of two-way factor interactions or nine possible levels of three-level factor interactions. Case studies included #4 L8 Plastics Injection Molding DoE with One and Multiple OA Replication, CenterPoint, and Some Experiment Line Replications (Chap. 6) and #21 L27 Partial Factorial Case Study: Green Electronics Phase I DoE (Chap. 5).

4. Ranked Pareto diagram for coefficient *t*-values indicating significance or using F-value to indicate ranking when all factors are not significant. Case studies included #4 L8 Plastics Injection Molding DoE with One and Multiple Replication, CenterPoint, and Some Experiment Line Replications (Chap. 6) using *t*-value ranking and #12 L18 Non-interacting OA: Painting DoE (Chap. 8) using F-value ranking.

5. Resolution III design showing confounding of main factors with one- or two-level interactions. Case studies included #3 L8 Partial Factorial, Underfill Voids DoE: Selecting Process Parameters for Zero Defects (Chap. 4); #10 L16 Partial Factorial, Rivet Design DoE: Selecting Part Dimension Design for Best Product Performance (Chap. 4); and #15 L16/32 Interaction Matrix Solder Wave DoE Design (Chap. 7).

6. Resolution IV design with no factor confounded with two-way interactions, but every two-way interaction was confounded with other two-way interactions by avoiding assigning a factor to the highest (n)-way interaction level. Case study #14 L32 Partial Factorial, APOS for Robotics DoE: Selecting Process Parameters for Multiple Adjustment Production Machines (Chap. 4).

7. Converting resolution III design to resolution IV by avoiding assigning a factor to the highest (n)-way interaction level. Case study #15 L16 Interaction Matrix Solder Wave DoE Design (Chap. 7).

8. Multi-level arrangement DoE design and analysis was demonstrated for four and eight levels using two case studies. This technique provides evaluation surveys investigating performance of newly developed materials, designs, or processes. Case study #5 L8 Multi-level Arrangement DoE: Machining I Pin Fin Heat Sinks (Chap. 8) and case study #11 L16 Multi-level Arrangement DoE: Machining II Stencil Forming (Chap. 8).

9. Foldover sequencing technique DoE allows for using a screening L8 with seven factors to be analyzed and then folded over for each significant factor, either singly or on all factors. Folding on one factor at a time allows for isolating each factor and its interactions in L16, repeated for each significant factor. Folding on all factors allows factor coefficients to be analyzed individually in L16 resolution IV design without confounding, while each two-way interaction is confounded by three other two-way interactions. Case study #6 L8 Foldover Sequencing Techniques: Folding on One Factor Example (Chap. 8) and case study #7 L8 Foldover Sequencing Techniques: Folding on All Factor Example (Chap. 8).

10. Screening followed by full or partial factorial DoE sequencing. This is an alternate technique to the classical DoE foldover approach. The first screening DoE is performed, and a second full or partial factorial DoE is performed next to analyze the top significant factors from the screening DoE. Case study #8 L8 Followed by L9: DoE Sequencing Techniques, Printer Design DoE (Chap. 8).

11. Non-significant factors in DoE can be ranked based on F-test in ANOVA. The top ranked factor levels can be selected from the coefficient plot analysis toward the desired outcome (smallest, largest, or target), and the low ranked factor levels can be selected based on cost or processing ease. Case studies #12 L18

Non-interacting OA: Painting DoE (Chap. 8) and #13 L18 Mean and Variability Analysis, Smaller Is Better: Painting DoE (Chap. 9).

12. Non-interacting OA can be used when interactions and confounding are not considered by the design team. They can be used as screening design with two levels in L12 and combinations of two and three levels in L18 and L36. Case studies #12 L18 Non-interacting OA: Painting DoE (Chap. 8) and #17 L36 Non-interacting OA: Air Knife DoE (Chap. 8) for Multi-function Machines.

13. Variability analysis used as a complement to mean coefficient analysis to provide additional information for non-significant factors to select appropriate factor levels for reduced variability. There are three types of variability analysis (smaller is better, larger is better, and two formulas for target is best), and reduced variability is indicated by the most positive value in variability coefficient plots for all types. It is recommended not to perform ANOVA variability analysis, since it is transformed by the $(\log_{10})$ functions. Case studies #9 L8 Mean and Variability Analysis, Larger Is Better: Bonding II DoE of Neoprene to Steel (Chap. 9); #13 L18 Mean and Variability Analysis, Smaller Is Better: Painting DoE (Chap. 9); and #20 L9 Mean and Variability Analysis, Target Is Best: Stencil II Screening DoE (Chap. 9).

14. Using software simulation to generate results based on two noise factors representing the extremes in operating conditions. Case study #22 L27 Screening Design Software DoE Case Study: Minimizing Half-Adder Chip Delay Time DoE (Chap. 5).

15. Error handling techniques were investigated with one or multiple replications as well as repeating CenterPoint and some experiment lines. Benefits and drawbacks of each error handling method were discussed. Case study #4 L8 Plastics Injection Molding DoE with One and Multiple OA Replications, CenterPoint, and Some Experiment Line Replications (Chap. 6).

16. DoE experiment blocking (dividing) to reduce confounding with noise factors such as environmental conditions or operator and shift variability was discussed in Chap. 6.

17. Interconnecting diagrams showing arrangement of factors and interactions and their confounding using either resolution III or IV were discussed in Chap. 7.

18. Interaction tables showing two-way interaction patterns for L8 and L16 to quantify confounding were shown in Chap. 7.

19. Additional runs to resolve confounding in L16 factor resolution IV design were shown in Chap. 7.

20. Using half fraction division to resolve confounding for L8, L16, and L32 was shown in Chap. 7.

10.4 Conclusions

This chapter summarized the previous book chapters and their topics. It is intended as a review of relevant DoE tools and techniques outlined in each chapter with case studies. A complex soldering case study that was designed in Chap. 7 was analyzed in this chapter using most of DoE techniques discussed in the book. There were two desired outcomes, and they were analyzed individually for the most significant factors, both in half L16 and full L32 DoE, in mean as well as variability coefficients and ANOVA. Non-significant factor levels affected each desired outcome differently, and they were selected based on interactions with significant factors and variability analysis. All factors were included in the regression equation for each outcome, and a predicted value was calculated and verified with a confirming experiment. The DoE resulted in world-class quality of 31 ppm for the first 7 weeks of production after DoE or four sigma design. An 88% reduction in mean and 74% reduction in variability were achieved averaged over a 6-month production period after adopting the DoE recommendations.

The L32 soldering case study design was critiqued in Chap. 7 to show how it could be improved to a resolution IV design, with no confounding of main factors. The DoE team assumptions using matrix method of assigning factors to identify significant interactions to the exclusion of all confounding interactions, based on prevailing wisdom, were mostly incorrect. The DoE achieved four sigma quality level (31 mm) but was not followed by another DoE to achieve Six Sigma levels.

The chapter concluded with the listing of all 22 case studies in various chapters with techniques used and results obtained. It was followed by a cross listing of 20 techniques used and where they would be found either in chapters or case studies.

Index

A

Acceptance quality levels (AQL), 20
Adjusted R^2, 110
Agitator design DoE, 295
Aluminum 6063—TS extrusions, 284
American Supplier Institute (ASI), 313
Analysis of variance (ANOVA), 100, 147, 161,
    162, 334
  FF model, 125
  F-tests, 100, 116, 137
  for full factorial DoE, 126
  Green Electronics Phase II DoE, 132
  interaction, 126
  linear regression analysis, 108–111
  Minitab®, 132
  percent contribution $p\%$, 127
  significance testing, 122
  treatment, 118–120
ANOVA analysis, 286, 292, 305, 308
ANOVA analysis F-tests, 295
ANOVA-based study, 68
ANOVA significance analyses, 346
ANOVA variability analysis, 351
APOS DoE, 276
  ANOVA analysis of two parts, 182
  ANOVA pooling, 184
  assignment and experiment results, 181, 183
  benefits, 185
  brainstorming and cause-and-effect
     sessions, 183
  coefficient plots for two parts, 183, 184
  electrotechnical product, 179
  factors and interactions, 180, 183
  L32 configuration, 183
  outcome validation for two parts, 184, 185
  parameters, 179
  regression equation, 184
  SMART® robotic line concept, 179, 180
  spring and walking beam, 182, 184
  statistical analysis, 184
  structured methodology, 179
ASTM test procedure, 319
Attribute control charts, 42, 55–57
Attribute data, 55
Attribute/pass-fail test data, 55

B

Bad cholesterol, 86
Balanced orthogonal arrays, 159
Balancing DoE outcome analysis, 314
Balancing mean and variability outcomes
  alternative methodologies, 329, 330
  controlled factors, 329
  gold plating example, 330, 331
  mean $vs.$ variability, 330
  specification tolerance, 329
Band etching, 288
Base paint, 304
Binary tests, 62
Binomial distribution, 13–17, 55, 56
Binomial probability, 14
Bonding I DoE
  ANOVA analysis, 196–198
  CI, 199
  coefficient estimates, 197, 198
  design and outcomes, 196, 197
  design space/process map, 200

Bonding I DoE (*cont.*)
  goal, 195
  materials/manufacturing process, 196
  mathematical techniques, 200
  mathematical transformation, 200
  mean, 199
  nine transducers, 196
  p% contribution value, 197
  peel strength, 196
  pooling, 197
  predicted maximum peel force, 199
  process parameters, 196
  regression equation, 199
  RTV adhesive, 195
  screening factors, 196
  second-order equation, 200
  temperature and ultrasonic
      cleaning time, 201
  three-level coefficients plots, 198, 199
  three-level L9 screening design, 196
  t-values, 199
  ultrasonic bath soak chemical, 196
  ultrasonic cleaning, 200
Bonding II DoE
  ANOVA analysis, 320–322
  bonding mean coefficient analysis, 320, 321
  experiment lines, 319
  factor levels selection, 323
  interactions, 323
  L8 half factorial analysis, 319, 320
  manufacturers' recommended settings, 319
  mean and variability, 320
  mean *vs.* variability analysis, 323
  objectives, 319
  predicted outcomes, 323
  rail vehicle maintenance, 318
  recommended levels, 323
  significant factor levels, 320
  simulated harsh environment, 319
  two-level factors selection, 319
  variability coefficient analysis, 322
  variability plot, 320
Bonding mean coefficient analysis, 320
Brainstorming and cause-and-effect diagram
      techniques, 145

C
Cause-and-effect diagram, 7, 103
CenterPoint replications
  advantage, 233
  ANOVA SSE, 235
  curvature source, 235
  DoE analysis, 234
  error handling, 234
  L8 factors and interactions, 235
  model coefficients, 234, 236
  original OA experiments, 234
  SE, 235
  SS factors and interactions, 235
  statistical analysis, 234
  statistical analysis before pooling, 236
Central limit theorem, 36
CentrePoint, 304
Chemical efficiency folding screening
      DoE, 294
Chemical etching, 288
Chi-square (X^2) distribution, 112
Classical approach, 251
Coefficient, 106
Coefficient analysis, 334
Coefficient of Variance Square for Nominal Is
      Best II Goals, 317
Coefficient standard error (SD), 231
Commercial DoE analysis software, 286
Common statistical tools, 61
Complementary information, 334
Computed probability, 74
Confidence, 77
Confidence intervals (CI), 30, 62, 72, 104, 219
  baseline population, 87
  calculations, 87
  confidence interval, 84
  life sciences industry, 86
  normal distribution, 84
  pooled variability, 85
  vs. population mean, 85
  population standard deviation, 85
  purpose, 83
  regression equation, 334
  sample confidence interval, 84, 85
  spans and limits, 84
  zero point, 87
Confounding, 152, 160, 161
  alternatives, 249
  avoidance plans, 248, 249
  and factor interactions, 140
  L16, 153
  multiple layers, 160
  resolutions, 152
  rivet DoE factor, 175
  three-way interactions, 153
  two-way interactions, 153, 158, 172, 183
  underfill voids DoE, 172
Constituent columns, 283
Continuous distributions, 13, 17–19

Control charts, DoE, 5, 20, 37
  attribute, 42, 55–57
  control limits *vs.* specification limits,
      46, 47, 49
  equivalent probabilities, out-of-control
      conditions, 52
  factories and high process capability, 57, 58
  manufacturing processes, 41
  measurements, 48
  methodologies, 58
  monitoring and maintaining system, 58
  process capability, low-volume production,
      53, 54
  scenarios, 58
  selection, 42, 43
  statistical synchronization, 42
  surface cleanliness measurement, 51
  types, 2, 42–44
  variable, 42, 49, 51–53, 43–ENF
Controlled factors, 313–315
Control limits, 51, 53
Control limits *vs.* specification limits, 46, 47, 49
Conveyor Angle, 337
Correction factor (CF), 63, 108
Cramer's rule, simultaneous equations, 159

D
Data analysis, 147
Data collection, 65
Defect, 15, 17, 29, 35, 303
Defect-based Six Sigma, 12
Defect data distribution, 11
Defect rate, 35, 43, 54
Defects per million opportunities
      (DPMO), 36, 38, 58
Define, Measure, Analyze, Improve and
      Control (DMAIC), 333
Degrees of freedom (DOF), 63, 70, 79, 94, 105,
      108, 126, 127, 219, 283
Design for Manufacture (DFM), 31
Design of Experiments (DoEs)
  competitive global products, 59
  concepts, 139
  control charts (*see* Control charts, DoE)
  definitions, 141
  design team, 139
  elements, 142
  experiment, 142
  graphical presentation of data, 1
      (*see also* Graphical presentation of
      data, DoE)
  information, 143

  lifecycle process, 143–148
  mean and variability, 61
  methodology, 139, 140
  OAs (*see* Orthogonal arrays (OAs))
  objectives, 142
  outcomes and factors, 40, 142
  probability distributions (*see* Probability
      distributions, DoE)
  process quality production, 11
  project timing and error source, 148–150
  QLF, 142
  quality characteristics, 142
  quality loss, 141
  quantitative and charting analyses, 59
  reduction techniques, 61
  Six Sigma (6σ), 141 (*see also* Six
      Sigma (6σ))
  Taguchi method, 140
  two-level factorial (*see* Two-level factorial
      DoE design)
Design space, 123, 124, 144–146, 162, 175
  bonding process, 241
  definition, 241
  DoE analysis, 241
  DoE project plan, 243
  factor level selection, 242, 243
  full factorial L8 Hipot experiment, 242
  model coefficients linearity, 241
  two-factor experiments, 242
  two-/three-level, 241
Discrete distributions, 13, 16
DoE analysis error, 152
DoE analysis software, 251
DoE and process capability measurements, 313
DoE ANOVA analysis, 108
DoE case studies, 313
  APOS for Robotics DoE, 347
  designing and analyzing industrial, 344
  foldover sequencing techniques, 346
  full factorial phase II, 345
  L8 Hipot, 345
  L8 partial factorial, 345
  L9 screening DoE, 348
  L16 partial factorial DoE, 347
  L18 mean and variability analysis, 347, 349
  L18 non-interacting OA, 347, 348
  L27 partial factorial, 349
  levels selection, 344
  Machining II Stencil Forming, 347
  matrix solder wave DoE, 348
  mean and variability analysis, 346
  multi-level arrangement DoE, 346
  plastics DoE, 345

DoE case studies (*cont.*)
  printer design DoE, 346
  screening design software, 349
  variability analyses, 348
DoE defect model prediction, 311
DoE design, 61
DoE error handling techniques
  CenterPoint (*see* CenterPoint replications)
  experiment outcomes replication, 236–238
  model factors and levels, 218
  multiple replication, 230–233
  project goals dependent, 218
  regression equation, 221
  significance technique L8 case
       study, 223, 224
  single repetition (*see* Single repetition
       DoE analysis)
  statistical analysis, 219, 220
  types, 222, 223
DoE Machining II pin fin heat sinks study
  ANOVA analysis, 286, 287
  bent pin mean analysis, 285, 286
  coefficient analysis, 286
  column assignment, 287
  electronic components, 284
  extreme success and failure cases, 284, 285
  factors and levels selection, 284, 285
  heat dissipation, 284
  interaction column, 287
  mathematical analysis, 284
  purpose, 286
  special alignment fixture, 284
  speed and alignment, 286
  statistics, 286
  vibrations, 284
DoE Machining II Stencil forming study
  accuracy and aperture wall smoothness, 288
  ANOVA analysis, 290
  brick measurements, 289
  comparative statistical analysis tools, 287
  DoE design teams, 287
  DoE outcome, 289
  factors and levels, 288
  factors selections, 288, 289
  higher-quality performance, 290
  interactions, 288, 289
  materials/technologies, 287, 288
  objectives, 291
  quality cost, 290
  ranking, 291
  solder deposition, 287
DoE matrix, 144
DoE orthogonality, 312

DoE project planning preparation, 246
DoE project recommendations, 314
DoE projects enhancements, 312
DoE quality transformation, 343
DoE's ANOVA analysis, 122
DoE sequence method, 302
DoE sequencing techniques
  foldover (*see* Foldover sequencing
       techniques)
  Green Electronics DoE phase I, 291
  printer design DoE, 296–302
  problem-solving, 291
  reverse sequence, 292
  screening design, 291
DoE statistical analysis, 62
DoE Taguchi method, 109
DoE team, 313
DoE team advantages, 315
DoE team decisions, 334
DoE techniques
  additional runs, 351
  blocking, 351
  DoE design and analysis, 350
  error handling techniques, 351
  foldover sequencing technique, 350
  full/partial factorial DoE sequencing, 350
  half fraction division, 351
  interaction tables, 351
  interconnection graphs, 351
  non-significant factors, 350
  Ranked Pareto diagram, 350
  regression equation, 349
  REGW tests, 349
  resolution III design, 350
  resolution IV design, 350
  significant interaction plot, 349
  software simulation, 351
  variability analysis, 351
DoE tools, 312
DoE two-level factor conversions, 260
DoE variants, 334
Dot Compression software, 300
Dot diagram, 4, 5
Drying method, 307
d-test, 62
  critical *t* values, 82
  differences, 81
  DOF, 82
  one- or two-tailed probability, 83
  paired/blocked, 81
  randomness, 83
  significance tests, 83
  significant difference, 82

standard aging, 82
standard deviations, 82
vs. t-tests, 83

E
Electroforming, 288
Electroless Nickel/Immersion Gold
          (ENIG), 206
Electronic amplifier, 105
ENIG surface finish and nitrogen, 208
Environmental conditions, 315
Equipment variation (EV), 68
Error-based sample size determination, 87, 88
Error sum of squares (SSE), 219
Excel®-based DoE analysis, 164
Exclusive OR (XOR) logical relationship, 152
Experiment blocking (dividing)
     bias effect, 243
     four-way interaction levels, 245
     groups, 244
     L8, 243, 244
     L16 patterns, 245
     reduced confounding, 334
     three-way interaction, 244, 246

F
Factor assignments, 260
Factor coefficient (Pack), 226
Factor interaction, 150
Factor levels selection, 242
Factor velocity, 226
F distribution values, 94
FF model ANOVA, 125
FF model coefficients, 125
Finite element analysis (FEA), 174
Fishbone diagram, 7
Flux level, 338
Folded all factors, 296
Folded one factor
     analysis software, 295
     ANOVA analysis, 292
     coefficient analysis, 294, 295
     coefficient value, 292
     confounding interactions, 294
     confounding pattern, 292
     factor coefficients calculation, 295
     factor levels, 292
     full factorial, 295
     L8 screening design, 292
     L16 screening design, 292, 293
     material coefficient, 295

outcomes, 294
     two-way interactions, 295
Folding on all factor, 346
Folding on one factor, 346
Foldover, 146, 156
Foldover sequencing techniques, 276
     all factors, 296
     one factor, 292–295
Fractional factorial DoEs, 152, 153
F-tests, 99, 100, 109, 110, 116, 122,
          137, 196, 220
     critical values, 162
     data analysis function, 95
     DOF, 94, 95
     factor significance, 96
     hand calculations, 95
     normal population, 95
     ranking, 308
     sample variances, 95, 96
     two-sided values, 94
Full factorial
     array, 155
     DoEs, 152, 189
     L8, 163–170
     vs. screening, 159
Full factorial DoE design and analysis
     advantages, 100
     design space, 123, 124
     DMAIC, 123
     DoE project, 137
     DOF, 127
     factors, 100
     factors and levels, 123, 124
     Green electronics DoE case study
          (*see* Green electronics DoE
          case study)
     green lead-free soldering
          technology, 101
     green soldering technology factors, 101
     green technology, 101
     industrial case study, 101
     interactions, 124–128
     multiple factors, 100
     OFAT experiments, 123
     process map, 123, 124
     regression analysis, 123
     TQM, 123
     traditional approach, 123
Full factorial experiments, 241
Full factorial L8 Hipot experiment, 242
Full factorial L128, 251
Full interaction and confounding, 252
Full model coefficient analysis, 286

Function F.INV.RT(0.05,3,2), 95
F-value, 220

G

Gage repeatability and reproducibility
        (GR&R), 67, 149
Graphical analysis, 147
Graphical presentation of data, DoE
    bar chart, 3
    box and whisker plots, 7
    categories, 11
    cause-and-effect diagram, 7
    control chart phase, 5
    decision-making process, 6
    defect data distribution, 11
    defect-based Six Sigma, 12
    dot diagram, 4, 5
    factor level, 9
    half adder IC delay DoE outcomes, 10
    histograms, 4
    LCL, 6
    multifactor plots, 9
    multivariate charts, 8, 9
    OFAT, 8
    Pareto diagrams, 2–4
    Pie chart, 3
    SPC techniques, 5
    spider diagrams, 4
    squeegee color, 10
    steepest ascent, 10
    stem-and-leaf technique, 4, 5
    temperature *vs.* yield, 8
    TQM, 5, 12, 13
    TQM/DMIAC qualitative tools, 12
    types, 2
    typical multi-level variable control
            chart, 5, 6
    UCL, 5, 6
    uses, 11
Green electronics DoE case study
    DoE methodologies, 128
    electronic products, 128
    phase I DoE
        factors, 129
        PCB surface finishes, 128
        quality and reliability
                outcomes, 129
        results, 128
        visual defect-free results, 129
    phase II DoE
        ANOVA analysis, 132
        design matrix, 130–132

    factor multiple range test,
            total defects, 135
    factors and levels, 130, 132
    interaction plots, 136
    legacy Sn-Pb, 137
    multiple comparison tests, 132
    PCBs, 130, 132
    REGW multiple range test, 134
    reliability tests, 136
    SAC alloy technology, 129, 130
    selection process, 130
    statistical analysis, 134
    test vehicle, 130
    total defects' outcome *vs.* independent
            factor coefficients, 134
Green Electronics Phase I DoE
    ANOVA analysis, 208, 209
    coefficient analysis, 209, 210
    coefficient plots, 208, 209
    design and results, 207
    experiment lines, 207, 208
    factors, 205
    interaction A*B, 211
    interaction levels, 211
    interaction points, 211
    L27 prediction, 211
    levels selection, 206
    lowest defects, 212
    materials and process parameters, 205
    means, 209
    model factor coefficients, 211
    partial factorial L27, 205
    reduced model, 209
    Reflow Profile levels, 212
    reliability, 212
    replications, 207
    TAL, 211
    test vehicle PCB, 205, 206
    three-way interactions, 207
Green lead-free technology, 128
Green technology, 101

H

Half-adder chip design, 213
Half fraction design, 278
Half fraction factorial, 153
Half fraction interconnection diagram, 253
Handling interactions and confounding, 248
Head alignment, 299, 300
Head force, 300
Head temperature, 300
Hipot DoE

ANOVA after first pooling, 164, 165
ANOVA with no pooling, 164
ANOVA with second pooling, 165, 166
coefficient estimates, 166, 167
design and results, 163, 164
factor coefficients, 163
housing and recording, 163
Kvolt, 169, 170
L8 full factorial, 164
model coefficient, 166
predicted results based on two
    models, 167–169
regression equation, 168
shielded/grounded cables, 163
standard error (SE), 168
two-way Contact*Cable interaction
    plot, 168, 169
Histograms, 4
Hopper Vibrations, 184
Hourglass diagram, 266
Humidity, 43

I
In-circuit test (ICT), 338
Industrial DoEs, 281
Interaction, 160, 161
Interaction and confounding strategy, 283
Interaction coefficient, 343
Interaction handling, 279
Interaction level, 221
Interaction matrix, 252, 260, 277, 279
Interaction matrix L32 case study, *see* Solder
    wave DoE design
Interaction plots, 335
Interactions, 124–128, 292
Interactions handling, 335
Interconnecting graphs, 160, 161, 256, 334
Interval analysis summary, 93
ISO/TS16949, 313

K
Kvolt, 168–170

L
L4, 153, 154
L8, 141, 153–161, 171, 175, 185
    division, L4 half fractions OA, 276, 278
    Hipot DoE, 163–170
    OAs, 154
    underfill voids DoE, 170–174

L8 factor conversion tables
    analysis software, 259
    factor and level patterns, 259
    four-way interaction, 259
    half fraction design, 259
    matrix DoE approaches, 259
    resolution types, 259
    screening designs, 258, 259
    symmetrical interaction, 258
    three-way interaction, 258
    two-way interaction, 258
L8 Full Factorial Case Study, Hipot DoE, 345
L8 half fraction interaction
        and confounding, 250
    combinations, 253
    complementary, 254
    conditions, 254
    graph, 253
    interconnection diagrams, 253
    primary factors, 253
    star configuration, 254
    triangle plot, 253
L8 Hipot DoE, 335
L8 multi-level arrangement DoE, 346
L8 OA and interaction table, 259
L8 partial factorial design and confounding
    additional factors, 256, 257
    main factors, 254–256
    resolution types, 254
L8 partial factorial, underfill voids, 345
L8 plastics injection molding DoE, 345
L8 screening design confounding
    additional factors, 257
    DoE team, 258
    full factorial DoE, 258
    significant factors, 257
L9 screening DoE, 348
L16, 141, 151–153, 158–160, 171, 183, 185
    division, half fraction L8, 276, 278
    OAs, 155
    rivet design DoE, 174–179
L16 half fractional experiment, 279
L16 half fraction interactions
        and confounding, 251
    design resolution V, 261
    interconnection graph, 261
    three- and four-way interactions, 261
L16 interaction matrix method, 261
L16 interaction table, 263
L16 partial factorial design and confounding
    additional factors, 263, 264, 272
    DoE team, 272
    eight factors, 271–273

L16 partial factorial design and confounding
  (*cont.*)
    four-way interaction, 266
    hourglass configuration, 266
    interaction coefficients, 273
    interaction matrix design, 265
    interconnection diagrams, 265, 266,
        272, 273
    interconnection graph, 266
    L4 pattern, 273
    matrix method approach, 272
    nine factors, 272
    pentagon diagram, 265
    resolution IV design, 273
    sequential classical symbols, 264, 265
    seven factor assignment, 266, 267
    six factors assignment, 266
    three-way interactions, 263, 272
    two-level OA, 271
    two-way interactions, 263, 265, 266,
        272, 273, 275
L16 rivet DoE case study
    interaction matrix, 262, 263
    pentagon diagram, 261
    three-way interaction, 261
    two-way interaction, 261, 262
L18 non-interacting OA, 347, 348
L18 painting DoE, 335
    ANOVA analysis, 304, 307, 308
    factor levels, 304, 306, 308
    F-value rank, 309
    mean coefficient ranking, 304, 308
    mix design and processing parameters, 303
    model mean coefficient plots, 307
    outcomes, 303
    painting process, 303
    predicted value, 308
    quality defects, 303
    sample calculations, 308
    sample standard deviation, 306, 309
    theme park robots, 309
    urethane, 303
L27 ANOVA analysis, 208
L27 Green Electronics Phase I, 335
L27 half-adder chip design DoE
    chip performance, 213
    design team, 213
    experiment outcome summary, 214
    factor and level combinations, 214
    higher-order experiments, 212
    implementation, 213
    operating range limits, 214
    physical parts, 212

    pilot project, 213
    recommended (optimal) levels, 214
    software design and manufacturing process
        tools, 213
    software tools development, 212
    transistors, 213
L27 Partial Factorial DoE Phase I Case
        Study (#21), 345
L32, 141, 153, 154, 171, 176
    APOS, 179–185
    OAs, 156
L32 coefficients, 338, 341
L32 division, half fraction L16, 276, 277
L32 OA design, 337
L32 Solder Wave DoE
    factor level assignments, 343
    interaction plots, 342
    lead defect coefficient, 340
    matrix and results, 339
    quality performance, 343
    touch defect coefficient and ANOVA
        Analysis, 341
L32 soldering case study design, 352
L36 Air Knife DoE
    ANOVA analysis, 311
    common component types, 309
    defect outcomes, 311
    DoE defect model prediction, 311
    factor selection, 310
    multi-level arrangement, 310
    non-interacting OAs, 310
    PCB test vehicles, 310
    SMT components, 309
    SMT excess solder, 311
    solder defects, 310
    TH technology, 309
Large OA interaction and confounding
    additional factors, 275
    alternatives, 276, 277
    Green Electronics Phase II case study, 275
    implementation industrial settings, 276
    interconnection diagrams, 275
    two-level L32, 274
    two-way interactions, 275
Laser cutting, 288
Lead defects, 343
Lead shorts, 338
Lean Six Sigma, 12
Least squares regression, 102, 103
Legacy lead-based materials, 291
Legacy tin-lead (Sn-Pb) soldering
        technology, 128
Lifecycle process, DoE

characterization, 143
and collect outcomes based on selected
    OAs, 147
confirmation of predicted outcomes, 148
design the experiment, 146, 147
identify model factors and levels, 145, 146
prediction with multiple outcomes, 147, 148
project definition, 144
project process, 143, 144
project team selection, 145
select and quantify outcomes, 145
TQM/DMAIC, 143
Lift-off pressure, 288, 289
Life testing, 18, 19
Linear regression, 100–102, 105
    ANOVA, 108–110
    data transformation, 112
    histogram of observed *vs.* ranked data, 114
    model coefficient estimates, 103–107
    model fitted *vs.* residual data, 106
    normal score, 115
    normality checking, 111–113, 116
    transformed data, 113, 117
Lipid-lowering effects, 86
Log variance, 317, 318, 326
Lognormal distribution, 17
Lower control limits (LCL), 5

M
Machining II Stencil evaluation, 326
Management tools, 333
Managing DoE confounding design resolutions
    classical approach, 251
    complex, 251
    interaction matrix approach, 252
    Taguchi approach, 252
Mean square (MS), 109, 220
Mean square deviations (MSD), 330
Mean square error (MSE), 220, 318
Mean tests, 62
    CI, 72 (*see also* Confidence intervals (CI))
    d-test, 81
    t-test, 71, 77
    Wilcoxon test, 75, 76
    z-test (*see z*-test)
Mean time between failures (MTBF), 18
Measurement system error
    AV, 68
    commercially available software statistical
        packages, 68
    EV, 67, 68
    GR&R, 69

operator/equipment variation, 67
OV, 67
part variation standard deviation, 67
parts and trials, 68
PV, 69
six sigma, 67
test equipment and training, 67
theoretical study, 68
TV, 69
Median lifetime, 18
Methylene (MET), 196
Methyl ethyl ketone (MEK), 196
Micrometers, 43
Minitab$^©$, 152
Mixed soldering DoE
    ANOVA analysis, 202, 203
    cause-and-effect analysis, 201
    coefficient estimates, 203, 204
    coefficient plots, 203
    coefficient t-value, 205
    confidence limits, 205
    factor mean, 204
    factors selection, 202
    F-value, 203
    mathematical analysis, 204
    mixed soldering machine, 205
    operational settings, 201
    outcomes, 201, 202
    PCB, 201
    practitioners experiments, 202
    quality problems, 201
    regression equation, 204
    SMT, 201
    solder pot temperature, 202
    three main factors and interactions, 190
    two main factors and interactions, 189
Model coefficient analysis, 109
Model coefficient estimates, 103, 105–107
Model coefficient/F-test, 292
Model coefficient plots, 226
Modelling Tools for Data Analysis, 61
Model reductions and two-level
        OA analysis, 161–163
Modified XOR logic, 283
Moving range (*MR*) charts, 45
Moving range (MR) method, 66
MR chart control limits, 53
MSE variability coefficient analysis, 318
Multi-function machine, 309
Multi-level OA arrangements
    arrays yields, 282
    coefficients and ANOVA analysis, 284
    DoE L8 case study, 284–287

Multi-level OA arrangements (*cont.*)
   DoE L16 case study, 287–291
   issues, 282, 283
   L8 with four levels, 283
   usage, 281
Multi-level statistical analysis, 284
Multiple desired outcomes, 335
Multiple pairwise *t*-tests, 128
Multiple range (REGW) tests, 349
Multiple replication DoE analysis
   ANOVA analysis, 231
   defects range, 230
   defects/solve intermittent problems, 230
   factors/interactions, 232
   initial pooling, 232
   model coefficients, 231
   model reduction, 233
   pooling, 230
   project constraints, 230, 231
   single experiment line, 230
   statistical analysis, 231
   three-repetition Plastics DoE, 231
Multiple replications, experiments, 334
Multiple-level factors, DoE, 146
Multivariate charts, 8, 9

N
Naphtha, 304
Naphtha Mix, 307
National Institute of Standards
          and Technology (NIST), 149
Near-zero defects, 37
Neoprene cleaning, 319
New England LeadFree Consortium, 205
Noise factors, 313, 315
Non-interacting arrays, 157, 282
Non-interacting factors, 155, 192
Non-interacting L12 (2^{11}) OA, 302, 303
Non-interacting L18 OA (2^1 x 3^7), 302, 304
Non-interacting L36 OA (2^{11} x 3^{12}), 302, 305
Non-interacting OAs
   alternatives, 302
   case study I, 303–309
   case study II, 309–311
   DoE teams, 302
   levels, 282
   screening designs, 282
   types, 302
   usage, 312
Non-significant factors, 127, 257, 335, 352
Non-significant L32 Temperature
          factor level, 344
Normal distributions (ND), 30
Normality, 61

O
OA experiment lines, 336
OA interaction tables, 334
One-factor-at-a-time (OFAT), 8, 100, 123, 267
   deficiency, 150
   DoE, 150
   experiments, 150, 151
Operator variation (AV), 68, 69
Organic Solderability Preservative (OSP), 206
Orthogonal arrays (OAs), 124, 302, 312, 314
   DoE experiments, 146
   OFAT experiments, 150
   predetermined set, 334
   screening, partial/full factorial modes, 142
   two-level (*see* Two-level OA analysis)
Orthogonality, 283

P
Paint Dry Time, 308
Paint gun Pressure, 307, 325
Painting II DoE variability coefficient analysis
   advantages, 325
   mean coefficients, 324
   outcomes, 323
   recommended levels, 325
   variability coefficient rankings, 324
   variability outcomes, 324
   variability plots, 325
   variability recommendations, 324
P and *U* charts, 57
Pareto diagrams, 2–4, 7
Partial factorial, 158
   approaches, 276
   designs, 215
   DoEs, 189, 276
   experiments, 145, 153
   L8, 170–174
   L16, 174–179
   L32, 179–185
   and saturated (screening) design modes, 141
   and screening modes, 157
   two-level OA, 161
Partial/full factorial follow-on DoE, 282
σ part variation (PV), 69
Paste, 288
PCB surface finishes, 206
Peel Resistance for Adhesives, 319
Percentage contribution (p%), 121, 122
Phase II full factorial DoE, 212
Plackett-Burman (PB) designs, 161, 282
Plastics DoE, 219, 223, 224, 236, 237, 271
Plastics Single DoE, 229
Plastics three-repetition DoE analysis before
          pooling, 231, 232

Poisson distribution, 13, 15–17, 55, 56
Poisson distribution-based control charts, 15
Pooling, 141, 147, 162, 164–166, 178, 184, 185
Population statistics, 44
Precious metal plating, 330, 331
Pre-control charts, 44
Prediction, 147, 148, 185
Pre-pooled DoE ANOVA, 232
Pre-selected interconnection diagrams, 266
Primer Dry Time, 307
Printed circuit boards (PCBs), 18, 35, 128–130,
    132, 170, 201, 267, 336
Print Energy, 300
Printer design DoE
    analysis, 297
    ANOVA summary, 299
    defect outcomes, 296, 298, 299
    definitions, 296, 297
    DoE outcomes, 300, 301
    Dot Compression software, 298, 300
    early production phase, 296
    factor and level settings, 300
    factors selection, 298
    Head Alignment, 299, 300
    L9 ANOVA summary, 301
    model coefficient plots, 299
    OFAT experiments, 296
    printer specifications, 302
    recommended factors and levels, 301
    test pattern, 296, 297
    tooling changes/customer approval, 298
    two-step DoE sequence, 302
    two-way interactions, 298
    universal set of factor levels, 299
Proactive quality control, 148–150
Probability distributions, 2, 14
    binomial distribution, 13–15
    continuous, 13, 17–19
    discrete, 13
    normal distribution, 13
    Poisson distribution, 15–17
    quality methods, SND, 20–31
    reliability, 17–19
    SND, 13
    types, 13
Process capability, 38, 42, 53, 56, 59
Process map, 123, 124, 144, 146, 162, 174
Professional experience, 334
Project goal setting
    attribute measurements, 239
    design/process improvement, 240
    design/process outcomes, 239
    DoE projects, 239

    planning, 239
    problem-solving, 239
    technologies/materials research, 240, 241
Project timing and error source, DoE
    proactive quality control, 148–150
    Six Sigma quality, 148
    traditional quality control, 148
$p\%$ value, 122

Q
QS9000 standard, 313
Qualitative (attribute), 63
Quality characteristic(s), 145
Quality Function Deployment (QFD), 39
Quality initiatives, 313
Quality loss, 141, 143
Quality loss function (QLF), 32, 142, 252,
    290, 291, 336
Quality project, 336
Quality standards, 333
Quantitative (variable), 63

R
Ranked Pareto diagram, 219, 220, 227,
    232, 240
R control chart, solder paste
    thickness, 49, 50
Reduced interaction analysis, 260
Reducing noise factor repetitions
    DoE, 315
    multiple factor levels, 315, 316
    quality projects, 316
    replications, 316
Reflow atmospheres, 206
Reflow process, 206
Regression analysis, 61, 108, 333
    data variation types *vs.* model
        prediction, 102
    least squares, 102, 103
    linear regression, 100–102 (*see also* Linear
        regression analysis)
    polynomial equation, 101
    R^2, 102, 110
    scatter diagram, 111
    scatterplot, 101, 102
    standard error (SE), 100
    treatments, 99
    variables, 101
Regression equations, 221, 339
Reliability, 17–19
Repeating some lines, 236

Resolution, 142, 149, 152, 157, 158, 161,
        172, 176, 183
    definition, 248
    designs, 249
    types, 249
Resolution III, 249, 250
Resolution IV, 153, 249, 276, 336
Resolution V, 153, 250
Resolving confounded interactions
    additional experiments, 270, 271
    coefficients, 270
    Direction*Wave Temperature, 270
    DoE team interaction selections, 271
    factor assignments, 269
    outcomes, 269, 270
    solder wave case study, 270
    three-way interactions, 270
Rivet design DoE
    ANOVA analysis after
            pooling, 177, 178
    coefficient analysis, 177–179
    coefficient plots, 178
    confounding, 175, 176
    dimensions, 175
    factor and levels, 175, 176, 178
    FEA, 174
    partial factorial L16, 176
    screening L8 design, 175
    sequence and results, 176
    sleeve and mandrel, 175
    types of loads, 174
Robotics DoE, 179–185
Robust design, 139
R-squared (R^2), 102, 110, 122
RTV sealing, 319
Ryan-Einot-Gabriel-Welsch (REGW), 128

S
Sample and population statistics, 61
    attributes, 63
    calculations, 62
    control chart factors, 63
    DoE, 64
    formulas and definition, 63
    n shafts, 63
    properties, 62
    sample size (n), 64
    standard deviation estimation
            methodologies, 64–66
    statistical capabilities, 64
    system error (see Measurement system
            error (GR&R))

Sample and population tests, 62
    cholesterol-lowering test, 70
    computed value, 70
    control charts, 70
    DOF, 70
    mean and variability, 69
    mean difference, 70
    normal distributions, 70
    out-of-control condition, 70
    probability of occurrence, 69, 70
    significance α, 70
    two-sided significance, 71
Sample size (n), 62, 63
Saturated designs, 153
Scatterplot, 101, 102
Screening design, 141, 146, 155, 157–160,
        171, 186
Screening design DoEs, 153, 189, 302
Select DoE error source, 147
Sequential classical symbols, 264
SG Snap-Off, 328
Signal-to-noise (S/N) Transformation, 318
Significance tests, 76, 78, 333
Significant interactions, 335
Silicone Base Mix, 307
Simulation Program with Integrated Circuit
        Emphasis (SPICE), 213
Simvastatin, 86
Single large full factorial experiment, 282
Single repetition DOE analysis
    alternative method, 229
    ANOVA before pooling, 226
    before pooling, 224, 225
    calculations, 226, 227
    DoE team, 228
    factor plots, 228
    factors and interactions, 227
    L8 experiments, 224
    model coefficient estimates, 226
    Pareto diagram, 227
    after pooling, 225
    recommended factor levels, 230
    regression equation, 229, 230
    three-way interaction, 229
    Time, 226
    velocity and pressure interaction
            plots, 228
    velocity and pressure levels, 228
Six Sigma (6σ), 141
    Cp, 34, 36, 37
    Cpk, 32–35
    Cpk goals, 41
    customer satisfaction, 32

defect opportunities, 35
defect rates
    C*p*, 35
    C*p*/C*pk*, 35
design and mean/process capability, 34
design engineering goal, 32
design quality means, 31
design specification, 41
design team, 36
design team goal, 32
DFM, 31
electronics industry, assemblies, 35
factors, 41
formulas, 32
global competition, technology product, 31
management tool, 36
manufacturing costs, 31
manufacturing goal, 32
manufacturing process mean, 33, 41
manufacturing quality, 2, 31
multiple sigma quality level, 32
PCB assembly, 35
PCB defect rates, 36
process capability, 32, 33
    corrective action, 40
    early production parts, 39, 40
    prototype parts, 39, 40
    quality enhancement, 38, 39
product design, 31, 33
product quality, 31
QLF, 32
quality and cost, 31
specification limits, 32
testing costs, 32
traditional qualitative tools, 41
US companies, 31
Six Sigma (6σ) defect rate, 58
Six Sigma (6σ) design, 37
Six Sigma (6σ) product, 37
Six Sigma (6σ) quality, 148
SMART® robotic line concept, 179, 180
Sn/Ag/Cu (SAC) solder alloys, 128, 129
Snap-Off distance, 290
Solder DoE Postmortem, 344
Solder Paste with ENIG (B2)
    Surface Finish, 211
Solder process defect rate, 6, 51, 58
Solder wave DoE analysis
    design phase, 337
    design team, 337
    interactions and confounding, 338
    L32 DoE matrix, 338, 339, 342
    Postmortem, 344

prediction and confirmation, 343, 344
quality outcomes and test
        vehicles, 337, 338
resolution II design, 336
sequential quality improvement
        scenario, 337
summary of DoE design, 336, 337
Solder wave DoE design
    conveyer angle, 268
    defect history, 267
    geometry-based interactions, 268
    goal, 267
    interaction matrix, 268, 269
    interconnection graph, 269
    L16 array patterns, 269
    L16 partial factorial
            design, 268
    machine controls, 267, 268
    machine parameters, 267
    OFAT, 267
    PCBs, 267, 268
    preheat temperature, 267
    wave temperature, 268
Solder wave temperature, 337
Specification limits, 57
Spider diagrams, 4
Squeegee pressure, 289, 290
Standard deviation, 344
Standard deviation estimation methodologies
    between-group variation, 65
    data points, 65
    mean variations, 66
    MR, 66
    σ estimate, 66
    total overall variation, 65
    within-group variation, 65
Standard error (SE), 100, 168, 199,
        204, 209, 235
Standard normal distribution (SND), 13, 21,
        52, 62, 77, 78, 87
    acceptance rate, multiple sigma
            specifications, 31
    advantage, *z* distribution, 20
    defect rates, 20–30
    modern companies, 20
    multiple sigma, 30
    normal distributions, 30
    one- or two-sided distribution, quality
            defects, 29, 30
    quality standards, 20
    Six Sigma design, 30
    supply chain, 20
    z transformation, 20, 29

Star resolution IV, eight-factor
        DoE, 272, 274
Statistical analysis, 134, 147
Statistical analysis tools, 333
Statistical process control (SPC), 5, 58,
        267, 336
Statistical techniques, 39
Statistical tests, 61, 71
Statistical tools, 96
Steel Abrade, 323
Steel abrading, 319
Steepest ascent, 10
Stem-and-leaf technique, 4, 5
Stencil, 290
Stencil II DoE
        ANOVA analysis, 327, 329
        factor levels, 326, 328
        L9 screening design, 326
        mean and variability coefficient
                plots, 326, 327
        near-zero contribution, 328
        recommended factor levels, 328, 329
        replication error contribution, 328, 329
        standard deviation, 326
        variability analysis, 329
Stencil II screening designs, 335
Stencil Orientation, 290
Stencil Technology, 288
Student's t distribution, 78
Sum of squares (SS), 108, 196
Sum of squares of the regression
        (SSR), 108–110
Sum of the square of the error
        (SSE), 104, 108, 109, 122
Sum of the squares (SS), 219
Superglue-based adhesives, 319
Superior quality performance, 279
Surface-mount technology
        (SMT), 49, 201

T
Taguchi approach, 252
Taguchi method, 140, 152
Target Is Best I Goals, 317
t distribution, 77, 78
Temperature, 43
Thermal printer, 296
Three-level DoE analysis
        ANOVA, 194, 195
        experiments, 194, 215
        factor and level coefficients, 194
        level coefficient estimates, 194

model coefficient estimates, 194
predicted outcomes, 194, 195
screening/partial factorial, 215
significant factors, 195
statistical, 194
$vs.$ two-level designs, 215
Three-level DoE case studies
        design teams' project plans, 195
        partial factorial DoE L27 (see Green
                electronics I DoE)
        screening design L9 (see Bonding I DoE)
        screening design software L27, 212–214
        teams' knowledge, 195
        zero-defect mixed soldering
                DoE, 201–205
Three-level factorial design, 187, 188
Three-level OAs
        column assignments, 188, 190
        commonly used, 191
        DoE goals, 193
        DOF, 188
        full factorial, 192
        full fraction DoE, 192
        interaction attributes, 188
        interactions, 192
        interactions measuring methods, 189, 191
        L9, 192
        L9 and L27 compositions, 188
        partial fraction, 192
        screening design, 192
        three- $vs.$ two-level DoE, 193, 194
Three-repetition plastics DoE, 232, 233
Three-way interactions, 229, 250
Through-hole (TH), 309, 336
Time, 18
Time Above Liquidus (TAL), 206
Total overall variation, 65
Total quality management (TQM), 5, 12,
        13, 123, 333
Total sum of squares (SST), 108–110, 219
Total variation (TV), 69
Total Zocor cholesterol group, 87
Touch analysis, 342
Touch defects, 338
Touch shorts, 338
TQM/DMAIC, 143, 145, 148
TQM effort, 336
TQM tools, 43
Treatment design and analysis
        ANOVA analysis, 100, 120
        ANOVA test, 118
        between groups of, 118
        experiments, 100, 118

fitted *vs.* residual modelling, 119
F-tests, 100
F-value, 121
groups, 118
medication level, 118
outcome, 121
pairwise *t*-tests, 121
p% contribution, 121, 122
regression analysis, 118
repetition numbers, 118
significance analysis, 118
significance determination
    techniques, 121, 122
sum of the residuals, 119
variability, 118
within groups, 118
Treatments, 61
Triglyceride (TG), 86
t-test, 62, 71, 77, 99, 116, 130
actual probability value, 79
comparison, 84
complement of significance, 77
data points, 76
DOF, 79
normally distributed population, 79
sample mean, 79
sample standard deviation, 77
standard deviation, 76
t distribution, 77, 78
t-test for comparing two sample means
necessitates, 80
pooled variance, 79
population standard
    deviation, 79, 80
population variances, 80
sample standard deviation, 79
two-sided significance test, 80, 81
t-tests types, 71
t-value, calculation, 219
Two-level DoEs, 146, 193, 241
Two-level factorial DoE design
confounding, 160, 161
designs and processes, 163
full factorial, 163–170
interaction, 160, 161
interconnecting graphs, 160, 161
L16, 151
OAs, 150, 151 (*see also* Orthogonal
    arrays (OAs))
partial factorial (*see* Partial factorial)
Two-level OA analysis, 279
arrangements, 150, 151
configurations, 154
fractional factorial DoEs, 152, 153

full factorial DoEs, 152
half fraction factorial, 153
L4, 153, 154
L8, 153, 154
L16, 153–155, 159
L32, 153, 154, 156
and model reductions, 161–163
screening design DoEs, 153
two-way interactions for two-level
    assignments, 152
types, 155, 157–160
Two-way interactions, 250, 255
Types of control charts, 44

U
Ultrasonic cleaning, 196
Underfill voids DoE
cause-and-effect diagrams, 170, 171
coefficient plots, 172, 173
components, 170
confounding, 172
factors, 170, 171
L8 partial factorial, 171, 172
material selection, 170
PCB, 170
statistical analysis, 173, 174
zero-defect process, 170
Upper control limit (UCL), 5, 6

V
Variability analysis, 331, 334
DoE projects, 318, 336
mean analysis, 335
Variability measurement, 313
Variability Outcome for Larger Is Better
    Goals, 317
Variability Outcome for Smaller Is Better
    Goals, 317
Variability reduction, 139
ANOVA analysis, 316, 318
average and variability analysis L9 case
    study III, 326–329
average and variability analysis L18 case
    Study II, 323–326
definitions, 314
DoE transformation, 316, 317
mean and variability analysis L8
    case study I, 318–323
Six Sigma quality goal, 316
statistical analysis, 314
transformation formulas, 317
variability coefficient, 318

Variability tests
    F-test, 89, 94
    goodness of fit test, 91
    normally distributed
            populations, 88
    sample variances, 89
    X^2 test, 89
Variability transformations, 335
Variable control charts, 36, 42,
            46, 56–57
    +3s limits, 46
    control limits, 46
    daily samples, 43
    factors, 45, 46
    high-volume manufacturing, 44
    manufacturing process over time, 45
    monitor sample means, 44
    population statistics, 44
    pre-control charts, 44
    production quality, 44
    R chart, 45
    sample means distribution, 44
    specification limits, 44, 46
    standard variation estimators, 46
    usage guidelines, 49, 51–53
Voice of the customer (VOC), 39

W
Weibull function, 19
Weibull probability
            distribution, 17–19
Wilcoxon test, 75, 76
Within-group variation, 65

X
X^2 distribution, 89, 90
X^2 goodness of fit test
    definition, 91
    intervals, 91
    linear distribution, 92
    normal distribution, 92, 93
    normality, 92
    particular distribution, 91
X^2 (Chi-Square) significance test, 62
    computed value, 91
    distribution, 89
    probability, 90
    sample and population means, 89
    sample standard deviation, 91
    selected values, 90

Z
z distribution, 39, 112
Zero bent pin outcomes, 286
Zero-defect experiment lines, 208
Zero-defect process, 170
z-test, 62, 71
    computed probability, 74
    conditions and goals, 73
    function menu, 74
    population standard deviation, 72
    probability, 73
    sample means, 74
    shafts, 73
    significance difference, 73
    standard deviations, 72, 74
    z transformation, 72